Jinhuan Yang, Xiao Yuan, Liang Ji
Solar Photovoltaic Power Generation

Also of interest

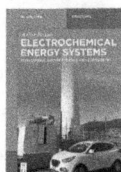

Electrochemical Energy Systems.
Foundations, Energy Storage and Conversion
Artur Braun, 2019
ISBN 978-3-11-056182-1, e-ISBN (PDF) 978-3-11-056183-8,
e-ISBN (EPUB) 978-3-11-056195-1

Polymer Solar Cells
Ram P. Singh, Omkar S. Kushwaha, 2020
ISBN 978-3-11-065268-0, e-ISBN (PDF) 978-3-11-065270-3,
e-ISBN (EPUB) 978-3-11-065285-7

Solar Cells and Energy Materials
Takeo Oku, 2017
ISBN 978-3-11-029848-2, e-ISBN (PDF) 978-3-11-029850-5
e-ISBN (EPUB) 978-3-11-038106-1

Organic and Hybrid Solar Cells
Lukas Schmidt-Mende, Jonas Weickert, 2016
ISBN 978-3-11-028318-1, e-ISBN (PDF) 978-3-11-028320-4
e-ISBN (EPUB) 978-3-11-038851-0

Photovoltaic Modules.
Technology and Reliability
Harry Wirth, Karl-Anders Weiß, Cornelia Wiesmeier, 2016
ISBN 978-3-11-034827-9, e-ISBN (PDF) 978-3-11-034828-6
e-ISBN (EPUB) 978-3-11-038398-0

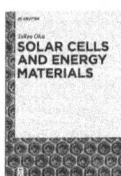

Jinhuan Yang, Xiao Yuan, Liang Ji

Solar Photovoltaic Power Generation

—

DE GRUYTER

電子工業出版社·
Publishing House of Electronics Industry
http://www.phei.com.cn

Authors
Jinhuan Yang
Institute of Solar Energy
Shanghai University of Electric Power
2588 Changyang Road, Shanghai, China
13371896215@163.com

Xiao Yuan
School of Materials Science and Engineering
East China University of Science and
Technology
130 Meilong Road, Shanghai, China
yuanxiao@ecust.edu.cn

Liang Ji
Underwriters Laboratories
33814 Farmhouse Street, Fremont
CA 94555, USA
Liang. Ji@ul.com

Co-Authors
Yongbang Xu
Wiscom System Co., Ltd
100 Jiangjun Road, Jiangning, Nanjing, China
xuyb@wiscom.com.cn

Wei Zhao
Sungrow Power Supply Ltd.
1699 Xiyou Road, Hefei, Anhui Province, China
Zhaow@sungrowpower.com

Huacong Yu
Jingergy Technology (Shanghai) Co. Ltd.
Rm M111, 5th Fl, No. 277 Huqingping Road
Minhang District, Shanghai, China
13901694698@139.com

Shitao Wang
Department of Control Science and Engineering
Harbin Institute Technology
92 Xidazhi Street, Nangang District, Harbin
Heilongjiang Province, China
brucewang@hitwh.edu.cn

Yan Xu
College of Mathematics and Physics
Shanghai University Science and Engineering
2588 Changyang Road, Shanghai, China
xuyan@shiep.edu.cn

ISBN 978-3-11-053138-1
e-ISBN (PDF) 978-3-11-052483-3
e-ISBN (EPUB) 978-3-11-052542-7

Library of Congress Control Number: 2020940178

Bibliographic information published by the Deutsche Nationalbibliothek
The Deutsche Nationalbibliothek lists this publication in the Deutsche Nationalbibliografie;
detailed bibliographic data are available on the Internet at http://dnb.dnb.de.

© 2020 Publishing House of Electronics Industry
and Walter de Gruyter GmbH, Berlin/Boston
Cover image: LL28/E+/Getty Images
Typesetting: Integra Software Services Pvt. Ltd.
Printing and binding: CPI books GmbH, Leck

www.degruyter.com

Preface

Faced with the global environmental pollution and the crisis of fossil fuel depletion, reducing greenhouse gas emissions, vigorously developing renewable energy sources, and taking a sustainable development path in the energy consumption have become a global consensus. In December 2015, the Paris Agreement was agreed at the Paris Climate Change Conference by nearly 200 Parties of the United Nations Framework Convention on Climate Change. It came into force on November 4, 2016. This reflects the common determination of almost all countries in the world.

As an important part of renewable energy, solar photovoltaic (PV) power generation has been rapidly developed in recent years, and its sunlight to electricity conversion efficiency has continued to set new records. New types of PV modules have emerged one after another, the quality and performance of balance of systems have continued to increase, the output power of PV modules has reached new heights, prices have dropped dramatically, and the application of PV continues to expand. The cumulative installed capacity of the global PV systems has reached 228 GW in 2015. PV power generation has accounted for 1.2% of the total global power generation in the same year. In many areas, the price of PV power generation is close to that of conventional power generation. It can be expected that the share of PV power generation in the energy supply structure will continue to increase in the future. According to the predictions of relevant experts, the PV power generation will play a major rule at the end of this century.

In recent years, the scale of China's PV industry has expanded rapidly, and the market share has ranked number one in the world for many major links in the industry chain. It is an important PV player in the world. However, there are still many gaps when compared with advanced countries in terms of manufacturing and application technologies. With the rapid development of science and technology, both PV industry and PV technology had great changes since the second edition of this book published 4 years ago. Some of the contents of the original book have been obsolete. In order to keep up with the trend, we are now making amendments and supplements, publishing the third edition.

This book is based on a comprehensive introduction to the basic knowledge of PV power generation, focusing on the application of PV systems and striving to reflect the latest technical achievements. For example, new trends in PV applications and the content of the solar tracking system in recent years have been added, the functions and principles of new controllers and inverters have been described, and perovskite solar cells have been added to thin-film cells. The whole process of design, manufacture, installation, and maintenance of PV system has been described comprehensively. The commonly used PV system design software and their characteristics and scopes of adaptation are described. Some new developments in the application of PV systems were also introduced.

https://doi.org/10.1515/9783110524833-202

Chapters 2, 10, and 11 were revised by Jinhuan Yang, who was also responsible for the entire book; Chapters 1, 3, and 6 were revised by Xiao Yuan, who also assisted in the publication of the book; Chapters 8 and 9 were revised by Yongbang Xu; Chapter 7 was revised by Wei Zhao; Chapter 4 was revised by Huacong Yu; Chapter 5 was revised by Shitao Wang; Yan Xu revised all exercises and part of illustration; and Liang Ji reviewed the entire book. English version was translated by Yu Wang, Jiqi Liu, and Menghong Wang, and reviewed by Liang Ji.

In the process of writing this book, we received lots of help from William A. Beckman, Sicheng Wang, Shujuan Wang, Xiaoqing Dong, Cui Liu, Huamin Gu, Xiang Chen, Chunqiu Wu, Qiang Liu, Guoliang Chen, Zhi Zhang, and many others. There were many people who have made various contributions to this book and cannot list them all. We would like to express our gratitude here.

Due to our limited academic level and writing ability, there are unavoidable mistakes and omissions. The criticisms and corrections are welcomed.

Authors

Brief Introduction

The purpose of this book is to focus on the application technology of photovoltaic power generation systems based on a comprehensive introduction to the basic knowledge of photovoltaic power generation. Since the publication of the second edition for more than 4 years, both photovoltaic technology and photovoltaic industry and applications have made great progress. Now, photovoltaic power generation costs can compete with conventional power generation in some regions. In order to reflect the current progress of photovoltaics, this book is based on the second edition, and it has a description to the whole process of design, manufacture, installation, and maintenance of photovoltaic power generation systems. It introduces some new fields of application of photovoltaic power generation systems. For example, solar tracking system, functions and principles of the new controllers and inverters, and Perovskite solar cells have been added. Each chapter is followed by references and exercises.

This book can be used as a reference book for teachers and students of research institutions and colleges. It can also be used as reference for management and engineering technicians and technology enthusiasts of solar photovoltaic enterprises.

https://doi.org/10.1515/9783110524833-203

Contents

Chapter 3
Basic principles of crystalline silicon PV cells —— 83

Chapter 10
Applications of photovoltaic systems —— 373

Chapter 1
Introduction

1.1 The significance of developing and utilizing solar energy

1.1.1 Fossil fuels are facing a crisis of depletion

With the continuous growth of the world's population and economic development, the demand for energy supply is increasing. In the current energy consumption structure, fossil fuels such as coal, petroleum, and natural gas are main sources.

The International Energy Outlook 2016 (IEO2016) issued by the US Energy Information Administration (EIA) in May 2016 had a prediction of the international energy market before 2040 [1]. According to this report, total world consumption of marketed energy expands from 549×10^{15} Btu in 2012 to 629×10^{15} Btu in 2020, and to 815×10^{15} Btu in 2040, a 48% increase from 2012 to 2040 with an average annual growth rate of 1.4%. By 2040, the share of global fossil energy in the energy consumption structure will be still more than three-fourth. According to the statistics and forecast in this report, the primary energy consumption of some of the countries and regions in the world is shown in Tab. 1.1.

IEO2016 report pointed out that in the consumption of fossil fuels, petroleum occupies a major share, and the share of natural gas grows fastest that will surpass coal around 2030; hence, the coal consumption tends to be stable. Renewable energy is the fastest growing energy source in the world, with an annual growth rate of 2.6%. By 2040, the supply of coal, natural gas, and renewable energy will achieve a balance, and the shares of world fuel requirements for electricity generation are 28–29% each.

As of December 2015, the proved world petroleum reserves, as reported by the *Oil & Gas Journal*, were estimated at 1,656 billion barrels: 2 billion barrels higher than the estimate at the end of 2014, and more than 80% of the world's proved reserves are concentrated in eight countries. In 2012, the production is 90 million barrels per day. It is expected that the production will be 100 million barrels per day in 2020 and reach 121 million barrels per day by 2040. Even calculated at 100 million barrels per day, it is estimated that petroleum can only be supplied for 45.4 years. Besides, the world's petroleum consumption is still growing at a rate of more than 1%/year. The world petroleum consumption by region in 1990–2040 is shown in Tab. 1.2.

In the past 20 years, the new proved natural gas reserves have increased by 40%, reaching 6,950 trillion cubic feet in 2016. In 2012, the world natural gas consumption was 120 trillion cubic feet, and it is expected to reach 203 trillion cubic feet in 2040. But all this reserved natural gas is only enough for more than 30 years.

https://doi.org/10.1515/9783110524833-001

Tab. 1.1: World total primary energy consumption by region, reference case, 2011–40 (quadrillion Btu).

Region	History		Projections					Average annual change (%) 2012–2040
	2011	2012	2020	2025	2030	2035	2040	
USA	96.8	94.4	100.8	102.0	102.9	103.8	105.7	0.4
Canada	14.5	14.5	15.1	15.6	16.3	17.1	18.1	0.8
Mexico/Chile	9.3	9.2	9.8	10.5	11.6	12.8	14.3	1.6
Japan	21.2	20.8	21.9	22.3	22.3	22.2	21.5	0.1
South Korea	11.3	11.4	13.9	14.7	15.4	16.1	16.9	1.4
Australia/New Zealand	6.9	6.8	7.6	8.1	8.5	9.2	10.1	1.4
Russia	30.9	32.1	33.2	34.7	35.1	35.5	34.5	0.3
China	109.4	115.0	147.3	159.4	170.4	180.7	190.1	1.8
India	25.0	26.2	32.8	38.4	44.9	52.8	62.3	3.2
Middle East	29.9	31.7	40.8	45.4	50.7	56.6	61.8	2.4
Africa	20.1	21.5	26.1	30.0	33.8	38.4	44.0	2.6
Brazil	14.8	15.2	16.3	18.1	20.0	22.0	24.3	1.7
Total world	540.5	549.3	628.9	673.9	717.7	765.6	815.0	1.4

Source: EIA International Energy Outlook 2016.

Coal reserves in the energy structure are slightly more optimistic, but serious pollution is created due to massive use of coal for energy. As of early 2012, the world's recoverable coal reserves stood at 977.9 billion short tons (1 short ton = 0.9072 tons), and the production was 8.898 billion short tons in 2012. So the coal can be supplied for 110 years, if the production per year is maintained.

Coal will remain the second largest energy source worldwide – behind petroleum and natural gas – until 2030. From 2030 through 2040, it will be the third largest energy source, behind both petroleum and natural gas. World's coal consumption increases from 2012 to 2040 at an average rate of 0.6%/year, from 153×10^{15} Btu in 2012 to 169×10^{15} Btu in 2020, and to 180×10^{15} Btu in 2040. The top three coal consuming countries are China, the USA, and India, which together account for more than 70% of world coal use. China accounted for 50% of world coal consumption in 2012, and will fall to 46% in 2040.

China's coal consumption growth averages 6%/year from 2012 to 2040. Coal consumption in China will reach a peak of nearly 90×10^{15} Btu around 2025 before gradually declining to 83×10^{15} Btu in 2040. The recoverable coal reserves in China

Tab. 1.2: World petroleum and other liquid fuels consumption by region; IEO2016 reference case, 1990–2040 (million barrels per day).

Region	1990	2000	2012	2020	2030	2040	Average annual change (%)	
							1990–2012	2012–2040
OECD	42.2	48.7	45.5	45.8	45.5	46.1	0.3	0.0
Americas	20.6	24.3	23.2	24.4	24.3	24.6	0.5	0.2
Europe	14.0	15.6	14.1	13.7	13.7	14.0	0.0	0.0
Asia	7.6	8.8	8.2	7.7	7.5	7.5	0.4	−0.3
Non-OECD	25.0	29	44.8	54.5	63.6	74.8	2.7	1.9
Europe and Eurasia	9.3	4.4	5.3	5.8	6.2	6.1	2.5	0.5
Asia	6.6	12.5	21.5	26.7	32.2	38.9	5.5	2.1
Middle East	3.3	4.5	7.7	10.0	11.3	13.2	3.9	2.0
Africa	2.1	2.5	3.6	4.5	5.5	6.9	2.6	2.4
Americas	3.8	5.0	6.7	7.5	8.5	9.6	2.7	1.3
Total world	67.2	77.7	90.3	100.3	109.1	120.9	1.4	1.0

Source: EIA International Energy Outlook 2016.

were 126.2 billion short tons in 2012, and the production was 4.256 billion short tons in that year. It was expected to supply for 30 years.

India's coal use surpasses the USA around 2030, and its share of world coal consumption grows from 8% in 2012 to 14% by 2040. Figure 1.1 shows changes in energy consumption in China, the USA, and India.

A report of *BP Statistical Review of World Energy 2016* released in June 2016 pointed out that global primary energy consumption increased by 1.0% in 2015, and well below the 10-year average of 1.9%, the lowest since 1998. Consumption growth was below the 10-year average for all regions except Europe and Eurasia. Emerging economies accounted for 97% of increase in the global consumption growth. China's primary energy consumption increased by 1.5% and still recorded the world's largest increment in primary energy consumption for the 15th consecutive year.

The energy structure gradually shifts from coal toward lower carbon fuels, but petroleum is still the world's most important fuel, accounting for 32.9% of global primary energy consumption. Coal, the second largest fuel, accounts for 29.2% of global primary energy consumption, the lowest share since 2005. The consumption of natural gas is still below the 10-year average, accounting for 23.8% of primary energy consumption. The primary energy consumption structure of some countries in the world in 2015 is shown in Tab. 1.3.

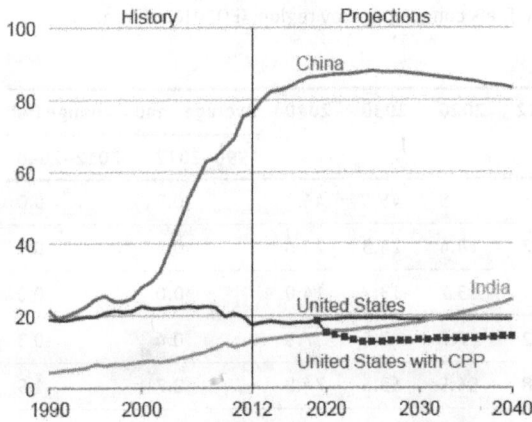

Fig. 1.1: Coal consumption in China, the USA, and India, 1990–2040 (quadrillion Btu).
Source: EIA International Energy Outlook 2016.

Tab. 1.3: Primary energy consumption in some countries in 2015 (million tons of oil equivalent).

Country	Crude oil	Natural gas	Coal	Nuclear power	Hydropower	Renewable energy	Total
USA	851.6	713.6	396.3	189.9	57.4	71.7	2280.6
Canada	100.3	92.2	19.8	23.6	86.7	7.3	329.9
Brazil	137.3	36.8	17.4	3.3	81.7	16.3	292.8
Russia	143.0	352.3	88	44.2	38.6	0.1	666.8
France	76.1	36.1	8.7	99.0	12.2	7.9	239.0
Germany	110.2	67.2	78.3	20.7	4.4	40.0	320.6
UK	71.6	61.4	23.4	15.9	1.4	17.4	191.2
South Africa	31.1	4.5	86.0	2.4	0.2	1.0	124.2
China	559.7	177.6	1920.4	38.6	254.9	62.7	3014.0
India	195.5	45.5	407.2	8.6	28.1	15.5	700.5
Japan	189.6	102.1	119.4	1.0	21.9	14.5	448.5

In 2015, the global proven petroleum reserves decreased from 2.4 billion barrels (0.1%) to 1.6976 trillion barrels. Global petroleum consumption reached 94.4 million barrels per day – nearly double the recent average growth rate in the past 10 years. China once again accounted for the largest increment to demand (+0.77 million barrels per day). The situation of petroleum dependence on imports has also become more and more serious in China. According to the statistics of the International

Energy Agency (IEA), the daily consumption of petroleum in China has increased to a record level of 10.32 million barrels in 2015. Based on the data reported by the PetroChina Research Institute of Economics and Technology, domestic petroleum in 2015 apparent consumption is estimated at 543 million tons. In the next 6 years, daily petroleum consumption of China will increase by 2.5 million barrels, and net imports will increase by 9.6% to 7.37 million barrels per day, which will break the highest historical record. The degree of external dependence in China will exceed 60% for the first time and is currently the largest petroleum importer and consumer in the world. In the long run, this trend will continue and it is expected that the dependence on petroleum import will increase from 59% in 2014 to 76% in 2035, which is higher than the peak of the USA in 2005.

The situation of increasing energy consumption is facing challenges. The reserves of fossil fuels on the Earth are limited. According to the *BP Statistical Review of World Energy 2016*, the global proven petroleum reserves could be sufficient to meet 50.7 years of global demand, natural gas is sufficient to meet 52.8 years, and coal is sufficient to meet 114 years, which is the largest ratio of reserve to production for any fossil fuel. According to the estimates by the World Health Organization, the global population will reach 10 billion to 11 billion by 2060. If the level of energy consumption per capita in all countries reaches today's level in developed countries, one-third of the world's 35 major minerals will be depleted in 40 years, including oil, natural gas, coal (assuming 2 trillion tons), and uranium, so the supply of fossil fuels in the world is facing serious shortage crisis situation.

To cope with the serious shortage of fossil fuels, we must gradually change the structure of energy consumption, vigorously develop renewable energy represented by solar energy, and take the road of sustainable development in the energy supply field, so as to ensure the prosperity of the economy and the continuous development of human society.

1.1.2 Protecting the ecological environment gets attention

Due to human activities mainly relying on the burning of fossil fuels, the environment was polluted, along with global warming, melting of icebergs, rising sea levels, and increasing desertification and more frequent occurrence of natural disasters. People gradually realized that reducing greenhouse gas (GHG) emissions, controlling the atmosphere environment, and preventing pollution have reached a moment of urgency.

A 2007 Synthesis Report issued by the United Nations Intergovernmental Panel on Climate Change (IPCC) stated that since 1750, concentrations of CO_2, methane (CH_4), and nitrous oxide (N_2O) have increased significantly in the global atmosphere due to human activities [2]. It has now far exceeded the concentration values of thousands of years before industrialization based on ice core records. In 2005, the concentration of CO_2 and CH_4 in the atmosphere far exceeded the natural range of the past 650,000 years. The increase in CO_2 concentration is mainly due to the use of fossil fuels. Since the era of industrialization, human activities have caused an increase in global GHG emissions, which increased by 70% from 1970 to 2004. Since the middle of the twentieth century, most of the increase in the global average temperature is likely due to the increase in GHG concentrations by human activities. Over the past 50 years, there has been a significant climate warming by human activities on all continents (except Antarctica). CO_2 is the most important anthropogenic GHG. In 1970–2004, the annual CO_2 emissions increased by about 80%, from 21 billion tons to 38 billion tons. In 2004, it accounted for 77% of the total anthropogenic GHG emissions. Emissions of GHG at current or higher rates will cause further increase in temperature during the twenty-first century and will induce many changes in the global climate system. These changes are likely to be greater than those observed during the twentieth century and may cause some irreversible effects. A medium confidence study has shown that if the global average temperature increases by more than 1.5–2.5 °C (compared with 1980–1999), the estimated 20–30% of species in this study may be at risk of extinction. If the global mean temperature increases by more than about 3.5 °C, model predictions show that there will be a large number of extinctions in the world (40–70% of the assessed species). Table 1.4 shows model-based estimates for the rise of global surface temperature and that of mean sea level for the end of the twenty-first century (2090–2099).

The United Nations IPCC concluded that the global average temperature increase should be controlled at 2.0–2.4 °C by 2050, which need the global CO_2 emissions be reduced by 50–85% based on 2000. Higher emission levels will lead to more significant climate change, as mentioned in Tab. 1.5. Current research finds that climate change is faster than previously expected, and even a 50% reduction in global CO_2 emissions by 2050 is not enough to avoid the danger of temperature rise.

In 2015, CO_2 emissions from energy consumption increased by only 0.1%. This was the lowest growth rate since 1992, except during the economic recession in 2009. The absolute value of emission reductions in the USA (−2.6%) and Russia (−4.2%) was the largest, but India (+5.3%) experienced the largest increase in emissions. The emissions of China have declined for the first time since 1998.

The Paris Climate Change Agreement, which entered into force in November 2016, is essentially an energy agreement. Energy consumption accounts for at least two-thirds of GHG emissions, and the revolutionary transformation of the energy industry is crucial to achieving the goals of the Paris Agreement. In 2015, energy-related CO_2 emission growth basically stagnated. This is mainly due to a 1.8% reduction in the energy demand of the global economy, an increase in energy efficiency, and an increase in the use of clean energy around the world. The annual investment of the energy industry is about US$ 1.8 trillion, of which more and more investment is attracted to clean energy. At the same time, the investment in the upstream petroleum and gas industry has been drastically reduced. The level of subsidies for fossil

Tab. 1.4: Projected global average surface warming and sea level rise at the end of the twenty-first century.

Case	Temperature change (°C at 2090–2099 relative to 1980–1999)[a,d]		Sea level rise (m at 2090–2099 relative to 1980–1999)
	Best estimate	Likely range	Model-based range excluding future rapid dynamical changes
Constant year 2000 concentrations[b,c]	0.6	0.3–0.9	Not available
B1 scenario	1.8	1.1–2.9	0.18–0.38
A1T scenario	2.4	1.4–3.8	0.20–0.45
B2 scenario	2.4	1.4–3.8	0.20–0.43
A1B scenario	2.8	1.7–4.4	0.21–0.48
A2 scenario	3.4	2.0–5.4	0.23–0.51
A1F1 scenario	4.0	2.4–6.4	0.26–0.59

[a]These estimates are assessed from a hierarchy of models that encompass a simple climate model, several earth models of intermediate complexity, and a large number of atmosphere-ocean general circulation models (AOGCMs) as well as observational constraints.
[b]Constant composition of year 2000 is derived from AOGCMs only.
[c]All scenarios above are six SRES marker scenarios. Approximate CO_2-eq concentrations corresponding to the computed radiative forcing due to anthropogenic GHGs and aerosols in 2100 for the SRES B1, AIT, B2, A1B, A2, and A1FI illustrative marker scenarios are about 600, 700, 800, 850, 1,250, and 1,550 ppm, respectively.
[d]Temperature changes are expressed as the difference from the period 1980 to 1999. To express the change relative to the period 1850–1899, add 0.5 °C.

fuels in 2015 decreased from nearly US$ 500 billion in the previous year to US$ 325 billion, which reflects the decline in the price of fossil fuels and the progress of the reform of subsidies for fossil fuels made in several countries.

IEO 2016 counts and predicts of CO_2 emissions in some countries and regions from 1990 to 2040 are shown in Tab. 1.6. The average annual growth rate of CO_2 emissions during 2012–2040 will be 1.0%.

The report also makes statistics and forecasts on CO_2 emissions from the combustion of different types of fuels. The results are shown in Fig. 1.2. In 1990, CO_2 emissions associated with the consumption of liquid fuels accounted for the largest portion (43%) of global emissions. In 2012, they had fallen to 36% of total emissions, and they remain at that level through 2040. Coal, which is the most carbon-intensive fossil fuel, became the leading source of world energy-related CO_2 emissions in 2006, and it remains the leading source through 2040. However,

Tab. 1.5: Characteristics of post-TAR stabilization scenarios and resulting long-term equilibrium global average temperature and the sea level rise component from thermal expansion only.

Global average temperature increase above pre-industrial at equilibrium, "best estimate" climate sensitivity[d,e]	CO_2 equivalent concentration at Stabilization including GHGs and aerosols (2005 = 375 ppm)[b]	CO_2 concentration at stabilization (2005 = 379 ppm)[b]	Change in global CO_2 emissions in 2050 (percent of 2000 emissions)[a,c]
°C	ppm	ppm	percent
2.0–2.4	445–490	350–400	−85 to −50
2.4–2.8	490–535	400–440	−60 to −30
2.8–3.2	535–590	440–485	−30 to +5
3.2–4.0	590–710	485–570	+10 to +60
4.0–4.9	710–885	570–660	+25 to +85
4.9–6.1	885–1130	660–790	+90 to +140

Source: *IPCC 2007.*

[a]The emission reductions to meet a particular stabilization level reported in the mitigation studies assessed here might be underestimated due to missing carbon cycle feedbacks.

[b]Atmospheric CO_2 concentrations were 379 ppm in 2005. The best estimate of total CO_2-eq concentration in 2005 for all long-lived GHGs is about 455 ppm, while the corresponding value including the net effect of all anthropogenic forcing agents is 375 ppm CO_2-eq.

[c]Ranges correspond to the 15th–85th percentile of the post-TAR (Third Assessment Report) scenario distribution. CO_2 emissions are shown, so multigas scenarios can be compared with CO_2-only scenarios.

[d]The best estimate of climate sensitivity is 3 °C.

[e]Note that the global average temperature at equilibrium is different from the expected global average temperature at the time of stabilization of GHG concentrations due to the inertia of the climate system. For the majority of scenarios assessed, stabilization of GHG concentrations occurs between 2100 and 2150.

[f]Equilibrium sea level rise is for the contribution from ocean thermal expansion only and does not reach equilibrium for at least many centuries. These values have been estimated using relatively simple climate models (one low-resolution AOGCM and several EMICs based on the best estimate of 3 °C climate sensitivity) and do not include contributions from melting ice sheets, glaciers, and ice caps. Long-term thermal expansion is projected to result in 0.2–0.6 m/°C of global average warming above preindustrial. (AOGCM refers to atmosphere-ocean general circulation model and EMICs to earth system models of intermediate complexity.)

although coal accounted for 39% of total emissions in 1990 and 43% in 2012, its share is projected to decline to 38% in 2040, only slightly higher than the liquid fuels share. The natural gas share of CO_2 emissions, which was relatively small at 19% of total GHG emissions in 1990 and 20% in 2012, will increase to 26% of total fossil fuel emissions in 2040.

Tab. 1.6: World energy-related carbon dioxide emissions by region and country in the reference case with and without the US Clean Power Plan (CPP), 1990–2040 (billion metric tons).

Region/country	1990	2012	2020	2030	2040	Average annual percent change, 1990–2012	Average annual percent change, 2012–40	Total change, 2012–40 (billion metric tons)	Percent change, 2012–40
USA	5.0	5.3	5.5	5.5	5.5	0.1	0.2	0.4	6.9
Canada	0.5	0.6	0.6	0.6	0.6	0.9	0.5	0.1	14.9
Mexico/Chile	0.3	0.5	0.5	0.6	0.7	1.9	1.1	0.2	35.7
Japan	1.0	1.2	1.2	1.2	1.1	0.8	-0.4	-0.1	-10.9
South Korea	0.2	0.6	0.7	0.8	0.8	4.5	1.0	0.2	32.9
Australia/New Zealand	0.3	0.4	0.5	0.5	0.6	1.8	0.8	0.1	26.7
Russia	2.4	1.8	1.8	1.9	1.9	-1.3	0.1	0.1	3.8
China	2.3	8.4	9.9	10.6	11.1	6.1	1.0	2.7	31.9
India	0.6	1.8	2.1	2.7	3.7	5.3	2.7	2.0	109.9
Middle East	0.7	1.9	2.4	2.9	3.4	4.8	2.2	1.6	82.0
Africa	0.7	1.2	1.4	1.8	2.2	2.7	2.3	1.1	89.2
Brazil	0.2	0.5	0.5	0.7	0.8	3.5	1.5	0.3	52.4
Total world	21.4	32.3	35.6	39.1	43.2	1.9	1.0	10.9	33.9

Source: International Energy Outlook 2016.

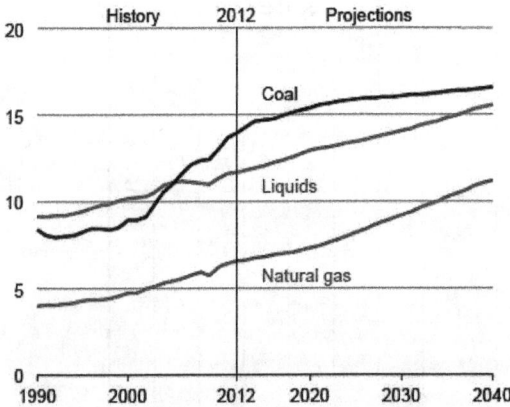

Fig. 1.2: World energy-related CO_2 emissions by fuel type, 1990–2040 (billion metric tons).
Source: International Energy Outlook 2016.

The *CO2 Emissions from Fuel Combustion Highlights (2016 Edition)* published by the IEA in May 2016 lists CO_2 emissions from fuel combustion in some of the countries from 1971 to 2014, as given in Tab. 1.7 [3]. And it also included CO_2 emission from different types of fuels in China in recent years, as shown in Tab. 1.8. It also pointed out that the top 10 emitting countries in 2014 were China, the USA, India, Russia, Japan, Germany, South Korea, Iran, Canada, and Saudi Arabia, as shown in Fig. 1.3.

Global CO_2 emissions in 2014 reached 32.4 Gt (10^9 t), while the total of the top 10 emitting countries was 21.8 Gt, accounting for two-thirds of the global CO_2 emissions. According to the industry classification statistics, electricity and heat generation, by far the largest, accounted for 42%, while transport accounted for 23%, industry accounted for 19%, residential accounted for 6%, and others accounted for 10%. It can be seen that it is very important to take measures to reduce CO_2 emission from electricity generation.

Table 1.7 shows that the average annual growth rate of global CO_2 emissions during the period from 1990 to 2014 was 57.9%, while the average annual growth rate in China during the same period reached 333.1%. On the one hand, this is due to the rapid economic growth that has led to an increase in emissions. On the other hand, it is related to the structure of energy consumption. The energy efficiency of China is not high, and energy consumption is dominated by coal. The harmful components such as sulfur contained in coal are high, so it has received widespread attention. The World Bank estimates that by 2020 the loss of environmental and health causing by air pollution in China will reach 13% of gross domestic products.

Reducing CO_2 emissions and protecting human ecological environment are urgent. Solar energy is a clean and pollution-free new energy. Solar photovoltaic (PV) system does not produce any waste during the power generation. The promotion of

Tab. 1.7: CO_2 emissions from fuel combustion (million tons of CO_2).

Country	1971	1975	1980	1985	1990	1995	2000	2005	2010	2013	2014	% change 1990–2014
USA	4,288.1	4,355.0	4,594.9	4,513.7	4,802.5	5,073.2	5,642.6	5,702.3	5,347.0	5,103.2	5,176.2	7.8%
Canada	340.1	377.0	422.2	393.8	419.5	448.9	516.2	535.1	525.8	549.7	554.8	32.2%
Japan	750.7	849.5	870.2	865.9	1,040.6	1,107.7	1,141.2	1,177.7	1,111.8	1,229.6	1,188.6	14.2%
France	423.2	422.9	455.1	351.7	345.5	343.5	364.5	370.4	340.1	317.1	285.7	17.3%
Russia					2,163.2	1,548.0	1,474.2	1,481.7	1,528.9	1,534.6	1,467.6	32.2%
Germany	978.2	973.4	1,048.4	1,004.6	940.3	856.7	812.4	786.8	758.9	763.9	723.3	23.1%
UK	621.0	575.9	570.5	543.4	547.7	513.7	521.2	531.2	476.8	449.7	407.8	25.5%
South Africa	157.1	203.0	208.4	222.9	243.8	259.8	280.5	372.3	406.7	423.3	437.4	79.4%
China	789.4	1,040.2	1,378.4	1,648.0	2,109.2	2,923.6	3,126.5	5,399.4	7,749.0	9,025.9	9,134.9	333.1%
India	181.0	217.1	262.0	375.8	530.4	707.7	890.4	1,079.6	1,594.3	1,852.5	2,019.7	280.8%
Brazil	87.5	129.6	167.7	156.2	184.3	227.7	292.3	310.5	370.5	451.3	476.0	158.4%
Total	13,942.2	15,484.1	17,706.3	18,246.5	20,502.5	21,362.0	23,144.5	27,037.7	30,450.4	32,129.4	32,381.0	57.9%

Source: CO_2 Emissions from Fuel Combustion Highlights (2016 Edition).

Tab. 1.8: CO_2 emissions of various fuels in China (million tons of CO_2).

Fuel	1971	1975	1980	1985	1990	1995	2000	2005	2010	2013	2014	% change 1990–2014
Coal	659.5	818.4	1,101.5	1,397.1	1,802.1	2,483.6	2,536.8	4,544.6	6,514.8	7,528.8	7,569.3	320.0%
Petroleum	122.5	204.3	248.6	234.3	286.5	412	546.9	776.8	1,016.0	1,016.9	1,195.7	317.3%
Natural gas	7.4	17.5	28.3	16.6	20.6	28.0	42.9	78.0	195.6	304.6	336.3	628.8%
Total	789.4	1,040.2	1,378.4	1,648	2,109.2	2,923.6	3,126.6	5,399.4	7,726.4	8,850.3	9,101.3	333.1%

Source: CO_2 Emissions from Fuel Combustion Highlights (2016 Edition).

GtCO$_2$

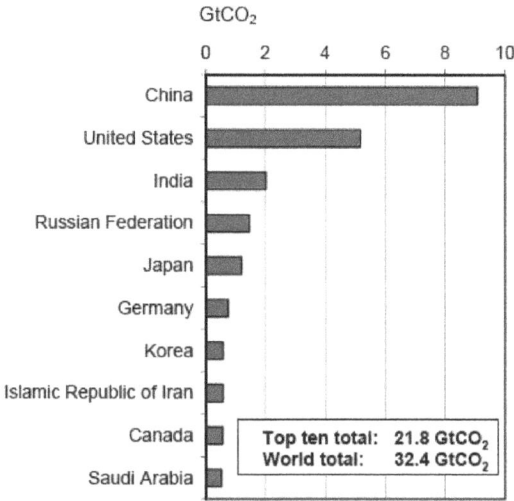

Fig. 1.3: Top ten emitting countries in 2014.
Source: CO$_2$ Emissions from Fuel Combustion Highlights (2016 Edition).

the application of PV will make an effective contribution to reducing atmospheric pollution and preventing global climate change.

1.1.3 Limitations of conventional power grids

According to the *World Energy Outlook 2016* published by the IEA, although many countries have stepped up energy supply network, there are still a large number of people unable to enjoy modern energy [4]. In 2014, nearly 1.2 billion people around the world could not obtain electricity, mainly in rural areas in sub-Saharan Africa. Latest estimates show that 2.7 billion people still rely on the traditional use of biomass for cooking, which means that they will continue to be exposed to the smoke-swept indoor environment, which will cause 3.5 million premature deaths each year. Most of the people live without electricity in remote areas where the economy is underdeveloped. Because of the decentralization of residence and inaccessibility of transportation, it is difficult to solve the problem of electricity consumption by extending the conventional electricity grid. The IEA estimates that by 2040 there will still be more than 500 million people in the world who do not have access to electricity as mentioned in Tab. 1.9.

According to the "2016 Global Renewable Energy Status Report" released by REN 21, there is 17% of the global area without grid coverage in 2013, and the global population without electricity is 1.201 billion, of which 80% is in the rural areas [5]. There are 635 million in Africa, 1 million in North America, 526 million

Tab. 1.9: Population without access to electricity in the world (million people).

Region	2014	2030	2040
Africa	634	619	489
Sub-Saharan Africa	633	619	489
Developing Asia	512	166	47
India	244	56	0
Latin America	22	0	0
Middle East	18	0	0
World	1,186	784	536

Source: International Energy Outlook 2016.

in developing countries such as Asia (1 million in China), 22 million in Latin America, and 17 million in the Middle East. In 2013, there were 20 countries with more than 15 million people without electricity (as given in Tab. 1.10). As the population of Africa has increased rapidly, the population of nonelectricity on the African continent has increased by 114 million since 2000. The IEA estimates that 950 million people will have access to electricity by 2040, but 530 million people still do not have access to electricity. In 2040, the coverage of the African continent's power grid should reach 70%, and it needs to invest US$ 7.5 billion annually.

Tab. 1.10: Countries with more than 15 million people without access to electricity in 2013.

Country	Number of people (10 million)	Country	Number of people (10 million)	Country	Number of people (10 million)
India	23.7	Tanzania	3.7	North Korea	1.8
Nigeria	9.5	Myanmar	3.6	Mozambique	1.6
Ethiopia	7.1	Kenya	3.5	Niger	1.5
The republic of Congo	6.1	Uganda	3.2	Malawi	1.5
Bengal	6.0	Sudan	2.5	Angola	1.5
Pakistan	5.0	Philippines	2.1	Côte d'Ivoire	1.5
Indonesia	4.9	Madagascar	2.0		

Source: 2016 Global Renewable Energy Status Report.

The lack of electricity supply has severely constrained the development of the local economy. These nonelectricity regions are often rich in solar energy resources, and the use of solar energy is an ideal choice.

The 2004 Bonn Declaration on the World Renewable Energy Conference proposed the goal of using solar energy to provide electricity to a billion people without electricity. For electricity supply in remote areas, PV systems will be of great use as an effective supplementary energy source.

1.2 Characteristics of solar energy electrical generation

1.2.1 Advantages of solar energy electrical generation

The main advantages of solar energy electrical generation are as follows:
(1) Solar energy is abundant. The solar radiation received by the Earth's surface is approximately 85,000 TW (1 TW = 1×10^{12} W), and the current global energy consumption is approximately 15 TW. Figure 1.4 shows a schematic representation of a comparison of solar energy versus fossil energy: the amount of solar radiation that shines on the Earth every year versus natural gas, petroleum, coal, nuclear energy, global proven energy reserves, and global annual energy consumption.
(2) Solar energy is inexhaustible. The solar radiation received by the Earth's surface among renewable energy sources, that is, solar energy is much larger than other energy sources, as shown in Tab. 1.11. Solar energy is safe and reliable, and it will not suffer from energy crisis or fuel market instability.

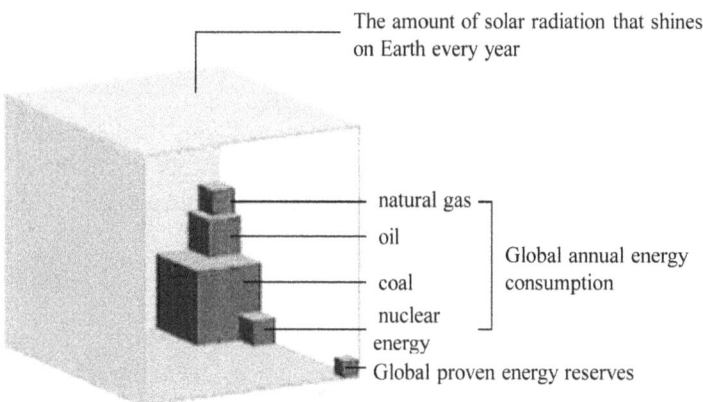

The amount of solar radiation that shines on Earth every year

natural gas
oil
Global annual energy consumption
coal
nuclear energy
Global proven energy reserves

Fig. 1.4: Comparison of solar energy and fossil energy.
Source: ECO Solar Equipment Ltd.

Tab. 1.11: Maximum power available for renewable energy.

Energy sources	Max power (TW)
Solar energy on the Earth	85,000
Solar energy on desert	7,650
Marine thermal energy	100
Wind energy	72
Geothermal energy	44
River water power	7
Biomass energy	7
Open ocean wave energy	7
Tidal energy	4
Coastal wave energy	3

Source: R. Winston et al. "Nonimaging Optics" 2005.

(3) Solar energy is available everywhere, and it can supply electricity nearby, without having to transport it over long distances, and avoid the loss of long-distance transmission lines.

(4) Solar energy does not use fuel, and its operating costs are low.

(5) Solar energy electrical generation has no moving parts (except tracker), is not easy to damage, and is easy to maintain. It is especially suitable for use under unattended conditions.

(6) Solar energy electrical generation does not produce any waste, no pollution, noise, and other public hazards, and no adverse impact on the environment is an ideal clean energy.

(7) The solar energy electrical generation system has a short construction period, is convenient and flexible, and can easily add or reduce the capacity of the PV array according to the increase or decrease of the load and avoid waste.

1.2.2 Disadvantages of solar energy electrical generation

The main disadvantages of solar energy electrical generation are as follows:

(a) Terrestrial applications are intermittent and random, and the amount of electrical energy generated is related to climatic conditions. It is impossible to generate electricity at night or on rainy days. If you need to power the load at any time, you need to have energy storage equipment.

(b) Low energy density: under standard conditions, the intensity of solar radiation received on the ground is 1,000 W/m². When used on a large scale, it takes up a large area.
(c) The current price is still high and the initial investment is large.

1.2.3 Types of solar energy electrical generation

There are two types of solar energy electrical generation.

1.2.3.1 Concentrator solar power generation

Concentrator solar power (CSP) refers to solar thermal power generation. It uses a large number of reflectors to focus the direct solar light, heat the working medium, generate high-temperature and high-pressure steam, and drive the turbine to generate electricity. CSP systems can be divided into the following three types according to solar energy collection methods.

1.2.3.1.1 Solar trough thermal power generation

The trough system uses a parabolic trough mirror to focus sunlight onto a tubular receiver (Fig. 1.5) and heats the heat transfer medium inside the tube to produce steam, which drives conventional turbines to generate electricity.

Fig. 1.5: Solar trough thermal power generation.

1.2.3.1.2 Solar tower thermal power generation

The tower system utilizes a large number of heliostats to reflect the solar heat radiation to a high-temperature collector (solar boiler) placed on the top of a tower (Fig. 1.6), and heat the working fluid to generate superheated steam, or the water in the collector is directly heated to generate superheated steam, which drives the turbine generator to generate electricity.

Fig. 1.6: Solar tower thermal power generation.

1.2.3.1.3 Solar dish-type thermal power generation

The dish system uses a curved condenser mirror to focus the incident sunlight at the focal point (Fig. 1.7) and place the Stirling engine directly at the focal point to generate electricity.

Solar thermal power generation has already had some practical applications, and its technology is still constantly improving and developing. At present, it has not yet reached the level of large-scale commercial application.

1.2.3.2 Solar PV electrical generation

At present, solar PV electrical generation has been widely used. This book mainly introduces the related information and knowledge of PV electrical generation.

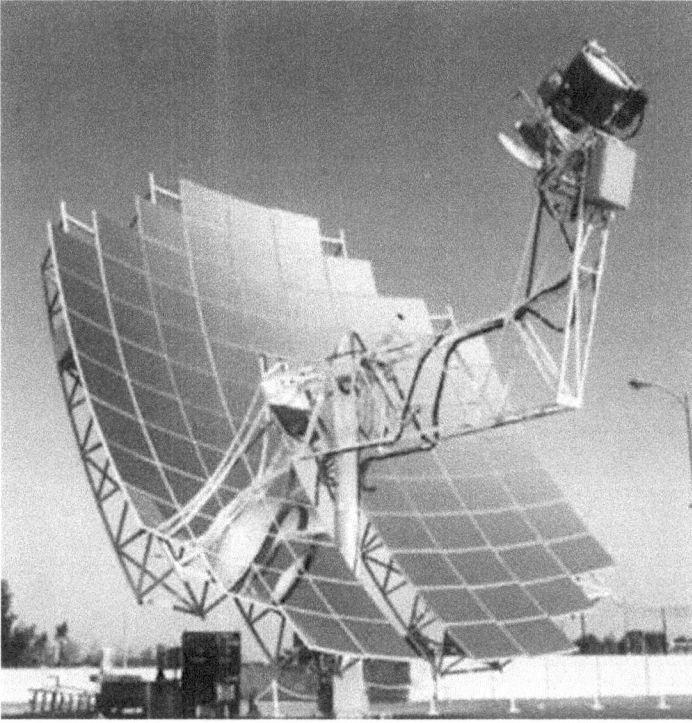

Fig. 1.7: Solar dish power generation.

1.3 The development of world's PV industry in recent years

1.3.1 PV cell production

In 1954, in the American Bell Laboratories, Daryl Chapin, Gerald Pearson, and Calvin Fuller made the first PV cell with the efficiency of 6%. In 1958, it was equipped on the American artificial satellite Pioneer No. 1 with the power of 0.1 W, the area of about 100 cm^2, and it worked for 8 years. Before the 1970s, PV power generation was mainly used in outer space. The vast majority of more than 6,000 spacecraft launched by human beings were powered by PV, which made important contributions to the aerospace industry. Due to the advances in technology, further improvement in the PV cell materials, structure, and manufacturing processes has reduced production cost, and PV power generation began to be applied on terrestrial and has been gradually applied to many fields after 1970s. However, due to the high prices, for a long period of time, they have fallen into a strange situation, where "to enable the market to expand, the cost of PV cells should be reduced; however, in order to further reduce the cost, we must mass produce, which relies on the expansion of the market, and there is no way for the expansion of the market to meet the demand for further price reduction." Until 1997, this situation began to be broken. Due to the promotion of lots of countries announced the implementation of "million solar roof project," the PV growth rate reached 42% in 1997. Before that, the average annual growth rate of PV cell production was about 12%. The annual PV cell production from 1977 to 1989 is given in Tab. 1.12, and the annual PV cell production from 1990 to 2006 is shown in Fig. 1.8.

Tab. 1.12: Annual production of solar cells from 1977 to 1989.

Year	1977	1982	1983	1984	1985	1986	1987	1988	1989
Production (MW)	0.5	9.3	21.6	25.0	24.4	27.5	29.1	35.0	42.2

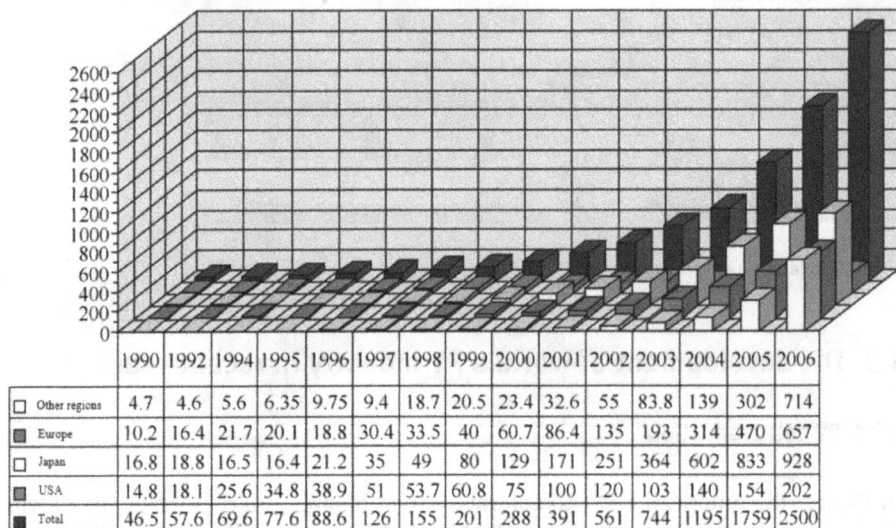

	1990	1992	1994	1995	1996	1997	1998	1999	2000	2001	2002	2003	2004	2005	2006
Other regions	4.7	4.6	5.6	6.35	9.75	9.4	18.7	20.5	23.4	32.6	55	83.8	139	302	714
Europe	10.2	16.4	21.7	20.1	18.8	30.4	33.5	40	60.7	86.4	135	193	314	470	657
Japan	16.8	18.8	16.5	16.4	21.2	35	49	80	129	171	251	364	602	833	928
USA	14.8	18.1	25.6	34.8	38.9	51	53.7	60.8	75	100	120	103	140	154	202
Total	46.5	57.6	69.6	77.6	88.6	126	155	201	288	391	561	744	1195	1759	2500

Fig. 1.8: Annual solar cell production (MW) worldwide 1990–2006.

For a long time, the USA ranked first in PV cell production, but it was surpassed by Japan in 1999. In the following 8 years, Japan has maintained its leading position. By 2007, China rapidly raised PV production and passed Japan as the number 1 in the world, and is far ahead now. The global PV cell production from 2006 to 2010 is given in Tab. 1.13. In the period from 1997 to 2007, the average annual growth rate of PV cell production was 41.3%, and it was more than 100% from 2007 to 2010.

Europe has always occupied a large share of the PV market. In recent years, it has been stagnant due to the factors such as policies and others. At the same time,

Tab. 1.13: Global solar cell production (MW) 2006–2010.

Region	2006	2007	2008	2009	2010
China	400	1,088	2,600	4,000	10,800
Taiwan (China)		450	900	1,000	3,400
Europe	657	1,062.8	2,000	2,800	3,127
Japan	928	920	1,300	1,800	2,182
USA	202	266.1	432	600	1,116
Rest of the world	314	213	668	500	3,274
Total	2,501	4,000	7,900	10,700	23,899

Note: Production statistics for the Taiwan region of China in the "Others of the World" in 2006.

the Asian PV market has risen rapidly since 2012 and has been catching up. According to the IEA–photovoltaic power systems program (PVPS), the global PV cell production in 2015 was approximately 63 GW, and China is still the largest PV cell producer with 41 GW of production, accounting for 65% of the world's total production, and compared with 33 GW of production in the previous year, increased by 24% [6]. The global PV cell production from 2011 to 2015 is given in Tab. 1.14. In 2015, the top three PV cell production plants produced more than 3 GW of PV cells, respectively, of which Hanwha Solar (bases located in China, Malaysia, and South Korea) produced 3.935 GW, Trina Solar produced 3.884 GW, and JA Solar produced 3.6 GW.

Tab. 1.14: Global solar cell production (GW) 2011–2015.

Year	2011	2012	2013	2014	2015
Production	37.2	35.97	40.3	52	63

In 2015, the global production of PV modules was approximately 63 GW. More than 90% of PV modules come from countries as members of IEA-PVPS. Due to the increase in market demand, the PV manufacturing capacity utilization rate has increased from 68% in 2014 to 80% in 2015. China is still the largest producer of PV modules with a production capacity of 45.8 GW, which accounts for 69% of the world's total production. Trina Solar is the world's largest manufacturer of PV modules, producing 5,873 MW modules. The production of thin-film PV modules was 3.6 GW, accounting for 6% of the total PV modules. In IEA-PVPS member countries, the production of PV modules in other countries was as follows: Malaysia 3.7 GW,

South Korea 3.4 GW, Japan 3.1 GW, Germany 2 GW (which is the largest PV module producer in Europe), and the USA 1.3 GW. The percentage of PV cells and PV modules produced in various countries is shown in Fig. 1.9.

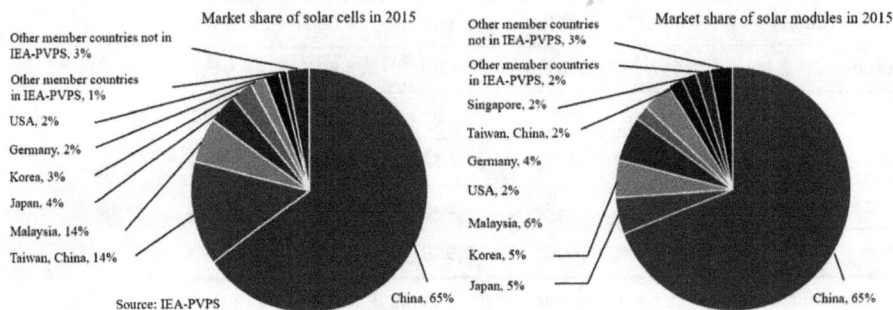

Fig. 1.9: Share of PV cells and PV modules production in 2015.

1.3.2 PV application market

Since PV power generation began to be applied on terrestrial in the 1970s, for a long period of time, it was mainly applied in off-grid areas, which has played a positive role in solving the basic livelihood of farmers and herdsmen in remote areas. At the same time, it provides safe and reliable power supply for navigation lights, microwave communication relay stations, railway signals, solar water pumps, and so on. The scale and field of application of off-grid PV systems are continuously expanding, and important contributions have been made to solve the special needs of industrial and agricultural electricity.

In 1990, Germany took the lead in putting forward a "1,000 Solar Roof Plan" to install a PV grid-connected system with a capacity of 1–5 kW on the roof of residential buildings. Due to some favorable policies, a total of 2,056 rooftop PV systems were installed at the end of the project. Taking this as an opportunity, Germany installed PV system capacity of 5 MW in 1995, doubled in 1996 to 10 MW, and further expanded to 15.6 MW in 1999. In January 1999, Germany began to implement the "100,000 Solar Roof Plan." In 2000, the installed capacity of PV power generation systems exceeded 40 MW. In 2006, a total of 850 MW was installed. In 2007, the installed capacity increased to 1,103 MW. In 2010, the cumulative installed capacity of German PV power generation systems has reached 17.37 GW, of which only 50 MW was off-grid PV systems. Germany's PV market has developed into a prosperous mature market from the exploration stage, and its installation capacity is far ahead of other countries in the initial stage.

In recent years, the installed capacity of global PV power generation has increased significantly. In 1994, the worldwide accumulated PV installation capacity was only 502 MW, 1,150 MW in 1999, 16 GW in 2008, and close to 100 GW in 2012, and in 2015, it reached 220 GW. According to the Renewable Capacity Statistic 2017 report released by the International Renewable Energy Agency (IRENA), by the end of 2016, the cumulative installed capacity of PV in the world reached 290 GW, of which approximately 77 GW was in China, accounting for more than one-fourth of the global installed capacity, followed by Japan, Germany, the USA, and Italy. PV installations in these five countries accounted for 73% of the global installed capacity [7]. Table 1.15 lists 10 countries with the most accumulated PV installations by the end of 2016. Installation capacity of Spain only fell behind Germany in a few years, and ranked second in the world. In 2010, PV installed capacity reached 3,921 MW in Spain, but due to slow development, in 2016, it was only 4,871 MW and was out of the top 10.

Tab. 1.15: Accumulated PV installation capacity (MW) in 2009–2016.

Country	2009	2010	2011	2012	2013	2014	2015	2016
China	300	800	3,300	6,800	17,450	28,050	43,180	77,420
Japan	2,627	3,618	4,914	6,632	13,643	23,300	33,300	41,600
Germany	110,564	17,552	25,037	32,641	36,335	38,234	39,786	40,986
USA	1,614	2,909	5,172	8,137	11,759	14,878	21,684	32,945
Italy	1,264	3,592	13,131	16,785	18,185	18,594	18,892	19,245
UK	27	96	995	1,756	2,873	5,424	9,187	11,250
India	12	37	563	1,277	2,269	3,144	5,271	9,658
France	277	1,044	2,796	3,965	4,652	5,654	6,755	6,767
Australia	105	399	1,394	2,434	3,255	4,004	5,031	5,626
Korea	524	650	730	959	1,467	2,481	3,615	5,500
Total world	22,578	38,903	69,746	99,347	135,426	172,289	220,132	290,791

Source: IRENA.

With the continuous reduction of PV cell prices, PV power generation has gradually approached or reached the level of grid parity in some areas. It is expected that the PV installation capacity will also grow rapidly. The industry expects that the global installed capacity of solar power generation in 2017 will reach 65–73 GW, and the global cumulative installed capacity can reach 600 GW by the end of 2020. In terms of installation capacity, PV is the third most important renewable energy source in the world followed by hydropower and wind power.

PV terrestrial application started in off-grid systems and has dominated for more than 20 years. Since the implementation of the solar roofing project in some countries, grid-connected PV systems have gradually become popular. In 2000, the amount of grid-connected PV systems exceeded that of off-grid PV systems. In recent years, due to the large number of PV power stations built in many countries, the share of grid-connected PV systems has rapidly expanded. Since then, the gap between grid-connected PV systems and off-grid PV systems has gradually widened. Today, the cumulative installed capacity of grid-connected PV systems accounts for more than 95% of the total capacity. It can be seen that PV power generation is playing an increasingly large role as an alternative energy source. The new installations and cumulative installations of off-grid and grid-connected PV systems in various regions of the world in 2015 are mentioned in Tab. 1.16.

Tab. 1.16: New installations and cumulative installations of off-grid and grid-connected PV systems in various regions of the world.

Country	Installed capacity (MW) in 2015				Cumulated capacity (MW) as of 2015			
	On-grid		Off-grid	Total	On-grid		Off-grid	Total
	Distributed	Centralized			Distributed	Centralized		
Australia	709	288	25	1,022	4,580	356	173	5,109
Canada	195	480	0	675	736	1,783	61	2,579
China	1,390	13,740	20	15,150	6,060	37,120	350	43,530
France	294	593	0	887	4,257	2,302	30	6,589
Germany	855	605	0	1,461	29,214	10,446	50	39,710
Italy	264	34	2	300	7,500	11,392	14	18,906
Japan	6,400	4,409	2	10,811	24,624	9,399	127	34,150
Korea	87	924	0	1,011	434	3,058	0	3,492
Spain	40	0	14	54	3,105	2,202	124	5,431
Turkey	0	208	0	208	12	254	0	266
USA	3,145	4,138	0	7,283	11,718	13,882	0	25,600
Other estimates				225				2,360
Total				50,655				227,736

According to the 2016 report issued by the Solar Energy Industries Association, USA has installed approximately 1 million PV power generation systems, of which 942,000 are household power generation systems, 56,500 are commercial power generation systems, and 1,500 are PV power plants [8]. PV power can be used by 6.5 million households, and CO_2 emissions can be reduced by 14 million tons each year, which is equivalent to the emissions of 9 coal-fired power plants. PV industry provided 209,000 jobs, which means one in every 83 jobs in the USA was from PV industry. The installation cost has been reduced by 70% in 10 years.

1.4 The development of China's PV industry

China began the research and development of solar PV power generation in 1958. In 1971, PV cells were successfully applied to the "The East Is Red 2" satellite for the first time. Since then, due to the development of technology, PV cells were applied terrestrial in 1973, and they were first used in the port of Tianjin for the navigation light power supply. In 1977, the national PV cell production was only 1.1 kW, and the price was about 200 CNY/W. Due to the price and production constraints, the market is slow to develop, and the terrestrial application is limited to low power system, which is typically between a few watts and several tens of watts.

In the 1970s, China established a number of PV companies, but the scale of production was small and lag in technology. In mid-1980s, five monocrystalline silicon and one amorphous silicon PV cell production lines were introduced successively to improve the product quality. The annual production capacity soared to 4.5 MW. The sales price dropped from 80 CNY/W in 1980 to around 50 CNY/W, but the actual production is only a few hundred kilowatts.

Since 2000, due to the impact of the international environment and the implementation of government projects, especially the "Song Dian Dao Xiang" project, electricity is delivered to every township, launched in 2002, involving a total of 4.7 billion yuan investment in the seven western provinces, which built 721 PV or PV–wind hybrid power stations and 268 small hydropower stations in 1,065 townships among 12 provinces (cities and districts) including Inner Mongolia, Qinghai, Xinjiang, Sichuan, Tibet, and Shaanxi provinces, to provide electricity to about 300,000 families and 1.3 million people. Among them, 15.5 MW PV–wind hybrid power stations were installed, and the investment total was 1.6 billion yuan. In 2002, the production of PV cells nationwide was 6 MW in China, and reached 12 MW in 2003. With the implementation of the "Song Dian Dao Xiang" project, which means electrification for all townships, a large number of PV companies were created in China, promoting development of PV industry and the talent cultivation and capacity building of China's PV industry, which played a tremendous role in the development of China's PV industry.

In recent years, due to the policies with active support adopted by the Chinese government and the booming European PV market, especially the German and

Spanish market, China has gradually become an important PV production and application market in the world. China's PV production capacity has grown rapidly. After 2003, China's annual growth in PV cells was doubled, and the growth rate has far exceeded that of other countries in the world. In 2001, the production was only 3 MW. In 2007, it reached 1,088 MW, and China became the largest producer of PV cells. After that, the production of PV cells was in a blowout-type development, and the production reached 41 GW by 2015. Most of China's PV cell products were exported. In 2015, exports exceeded US$ 12 billion. From 2013 to 2015, China ranked first in the world in terms of newly installed capacity for 3 consecutive years. In 2014, the PV installation was 10.6 GW, and the cumulative installed capacity reached 28 GW. In 2015, the installation reached 15.13 GW, and the cumulative installed capacity was 43.18 GW, which increased to about 42.74%. The annual installation of China's PV power generation system from 1976 to 2015 is mentioned in Tab. 1.17.

Tab. 1.17: Annual installation of China's PV power generation system (MW) from 1976 to 2015.

Year	Annual	Cumulative
1976	0.0005	0.0005
1980	0.008	0.0165
1985	0.07	0.2
1990	0.5	1.78
1995	1.55	6.63
2000	3.3	19
2005	5	70
2007	20	100
2010	513	800
2013	10,950	38,400
2015	15,130	43,180

1.5 PV power generation by some countries' organizations

1.5.1 Japan

Japan's New Energy Development Organization (NEDO), METI, PVTEC, and JPEA jointly developed and published an "Overview of the 2030 PV Roadmap" (PV2030 Plan)

in June 2004 [9]. Its overall goal is "to make PV power generation one of the key technologies by 2030." The specific target was that the cumulative installed PV capacity in
Japan should be 4.82 GW in 2010, and it should reach 100 GW by 2030. By that time,
50% of all residential electricity consumed in Japan will be provided by PV power generation, accounting for about 10% of the total power supply. It is estimated that the cost of
PV modules will be reduced to 75 JPY/W in 2020, and to less than 50 JPY/W in 2030. The
lifetime of PV modules will reach 30 years in 2020. The silicon consumption can be reduced to 1 g/W by 2030. The price of supporting components is also expected to continu-
ally decline. By 2020, the price of the inverter will be 15,000 JPY/kW, and the storage
battery will be reduced to 10 JPY/Wh. PV electricity generation prices will drop to 14
JPY/kWh by 2020, and to 7 JPY/kWh by 2030.

In order to adapt to the global need to reduce GHG emissions and to respond to
climate change and the rapid development of the PV industry, NEDO modified the
plan in 2009 and expanded to "PV2030 +." The overall goal was to make PV power
generation one of the key technologies by 2050, to play an important role in promoting the reduction of CO_2 emissions, not only for Japan but also for the global
community. The original PV2030 technical targets would be achieved in about 3–5
years, and the completion time for the new plan would be extended from 2030 to
2050. The concept of "realizing grid parity" remained unchanged, and the power
generation cost target in PV2030 remained the same. The target that will be
achieved in 2030 per "PV2030" should be achieved by 2025 per "PV2030 +." In addition, "PV2030 +" will achieve the goal that the cost of power generation is less
than 7 JPY/kWh. PV power generation should meet 5–10% of domestic primary energy demand by 2050, and provide about one-third of the components required by
the world's PV market. The key technical indicators are given in Tab. 1.18.

1.5.2 European Union

The EU Joint Research Center released a report of the PV Status 2016 in October
2016, which summarized the forecast of different sources for the global mid-term
(2040) PV market scenarios (Tab. 1.19). These scenarios' reference sources include
Greenpeace Organization research report, Bloomberg New Energy Finance (BNEF)
2016 New Energy Outlook, IEA 2014 Technology Roadmap of Solar PV Energy, *IEA
World Energy Outlook 2015 Edition* and *2016 Edition*, and some previous PV status
reports.

Greenpeace and the European PV Industry Association jointly published the
Solar Generation 6.2011 research report in February 2011, PV proposed three scenarios
for the future development of PV: paradigm shift scenario, formerly known as advanced scenario, accelerated scenario, formerly known as medium scenario, reference scenario: according to the name proposed in IEA 2009 World Energy Outlook
(WEO 2009), which means the basic conditions of the existing global energy markets

Tab. 1.18: Key indicators of PV2030 + PV development plan in Japan.

Goal	In 2010 or later	2020 (2017)	2030 (2025)	2050
Electricity cost	Equivalent to household retail electricity price 23 yen/kWh	Equivalent to commercial retail price 14 yen/kWh	It is equivalent to the general power generation price 7 yen/kWh	Equivalent to the general power generation price 7 yen/kWh or lower
Business component conversion efficiency (laboratory efficiency)	16% (20%)	20% (25%)	25% (30%)	Ultra-high performance components more than 40%
Used for Japanese domestic market production (GW/year)	0.5–1	2–3	6–12	25–35
Used for export market production (GW/year)	About 1	About 3	30–35	About 300
Main application	Single family home public utilities	Single/multiple family homes, public facilities, commercial buildings	Single/multiple family homes, public facilities, commercial use, rechargeable electric vehicles, etc.	Consumer use, industry, transportation, agriculture, etc., independent power supply

Tab. 1.19: World PV cumulative installation (GW) before 2040 estimated by different scenarios.

Year	2015	2020	2025	2030	2040
Actual installation amount	235				
Greenpeace (reference scenario)		332	413	494	635
Greenpeace (progress scenarios)		732	1,603	2,839	4,988
BNEF New Energy Outlook 2016	251	578	1,046	1,831	3,917
IEA PV Technology Roadmap (hiRen scenario)		450	790	1,721	4,130
IEA2015 current policy scenario		361	420	569	773
IEA2015 new policy scenario		397	560	728	1,066
IEA2015 450 ppm scenario		420	605	938	1,519
IEA2016 current policy scenario		424	592	708	991
IEA2016 new policy scenario		481	715	949	1,405
IEA2016 450 ppm scenario		517	833	1,278	2,108

Note: The data for 2013, 2030, and 2050 are obtained, and the data for 2020, 2025, and 2040 are extrapolated.

when the government does not change the existing policies and measures [10]. The average growth rate of the PV market in the first two scenarios is given in Tab. 1.20.

Tab. 1.20: Average growth rate of PV market under both scenarios.

Year	2011–2020	2021–2030	2031–2040	2041–2050
Paradigm shift scenario	42%	11% in the first 5 years and 9% later	7% in the first 5 years and 5% later	4%
Acceleration scenarios	26%	14% in the first 5 years, 10% after	7% in the first 5 years and 6% in the future	4%

Source: IEA 2009 World Energy Outlook (WEO 2009).

For the future development of the PV industry, three scenarios have made different predictions. The world cumulative PV installation capacity before 2050 is given in Tab. 1.21, and the ratio of PV power generation in the world's electricity consumption is given in Tab. 1.22.

Tab. 1.21: Historical and prediction of world PV installation by three scenarios.

Year	2007	2008	2009	2010	2015	2020	2030	2040	2050
Reference scenario									
PV installation (MW)	3	15,707	22,999	30,261	52,114	76,852	155,849	268,893	377,263
PV generation (TWh)	0	17	24	32	55	94	205	377	562
Acceleration scenario									
PV installation (MW)	3	15,707	22,999	34,986	125,802	345,232	1,081,147	2,013,434	2,988,095
PV generation (TWh)	0	17	24	37	132	423	1,421	2,822	4,450
Paradigm shift scenario									
PV installation (MW)	3	15,707	22,999	36,629	179,442	737,173	1,844,937	3,255,905	4,669,100
PV generation (TWh)	0	8	24	39	189	904	2,266	4,337	6,747

Editor's note: PV power generation data in 2008 paradigm shift scenario may be wrong.

Tab. 1.22: Predicting the proportion of PV power generation in world's electricity consumption by three scenarios.

Year		2010	2020	2030	2040	2050
Reference scenario						
The proportion of solar power accounts for electricity in the world reference (IEA Demand Projection)	%	0.2	0.4	0.7	1.1	1.4
The proportion of solar power accounts for electricity in the world energy revolution (Energy Efficiency)	%	0.2	0.4	0.8	1.3	1.8
Acceleration scenario						
The proportion of solar power accounts for electricity in the world reference (IEA Demand Projection)	%	0.2	1.9	4.9	8.2	11.3
The proportion of solar power accounts for electricity in the world's energy revolution (Energy Efficiency)	%	0.2	2.0	5.7	10.1	14.0
Paradigm shift scenario						
The proportion of solar power accounts for electricity in the world reference (IEA Demand Projection)	%	0.2	4.0	7.8	12.6	17.1
The proportion of solar power accounts for electricity in the world energy revolution (Energy Efficiency)	%	0.2	4.2	9.1	15.5	21.2

1.5.3 International PV technology roadmap steering committee

The International PV Technology Roadmap Steering Committee established by some PV companies and organizations published the International Technology Roadmap for PV, 8th Edition, in March 2017 [11]. It elaborated and forecasted the key technical indicators of the PV industry chain from materials to silicon wafers, PV cells, modules, and the system in 2016–2027, and analyzed the cost curve of module prices in the international market. It pointed out that due to the continuous decline in silicon material price and improvement in manufacturing process, the price of PV modules in the international market has dropped dramatically from 1.59 $/W in January 2011 to 0.58 $/W in January 2016, reduced by 64%. By January 2017, it dropped by another 36% to 0.37 $/W.

PV system LCOE (levelized cost of electricity) is related to factors such as solar insolation, system efficiency, and financial conditions. For large-scale PV power plants in the USA and Europe, the predicted PV system LCOEs are given in Tab. 1.23. The assumptions for this prediction include 80% debt with 18-year tenor, 20-year straight line depreciation, 25-year analysis period, 4% nominal debt, 5% nominal equity discount rates, and 2% inflation.

Tab. 1.23: Predicted LCOE of large PV system in the USA and EU.

System condition (kWhac/kWdc)	Years ($/kWh)					
	2016	2017	2019	2021	2024	2027
1,000	0.077	0.073	0.065	0.063	0.059	0.054
1,500	0.051	0.049	0.043	0.042	0.039	0.036
2,000	0.039	0.037	0.033	0.032	0.030	0.027
System cost ($/kW)	970	911.8	814.8	785.7	746.9	679

1.5.4 USA

In 2011, the capacity of PV installations in the USA was 1.2 GW, but the share of solar power in the USA's electricity supply was less than 0.1%. To speed up the development of renewable energy, the US Department of Energy launched a Sunshot Program with a goal that the cost of solar power can be competed with conventional power generation technologies in the absence of government subsidies in 2020. This means that the price of three types of PV power generation – the residential, commercial, and utility grid levels – should be reduced by approximately 75%. For the utility level, the goal is to reduce the cost of solar power to 6 ¢/kWh. Through industry efforts, the newly installed solar power generation capacity was approximately one-third of the total newly installed electricity generating capacity in the USA in 2014 and 2015. By December 2016, approximately 90% of the companies had achieved the goal of the cost of 6 ¢/kWh, 4 years ahead of the 2020 target in the plan. PV installation capacity has exceeded 30 GW, and the share of PV power in the US electricity supply has exceeded 1%, and the electricity generation cost at the utility grid level has been reduced to 7 ¢/kWh. As a result, the US solar energy plan made a modification on December 14, 2016, that the cost of solar power generation should be reduced by half, which means the cost should achieve 3 ¢/kWh (Tab. 1.24) by 2030. It is estimated that by 2050, 50% of the electricity supply in the USA will be provided by solar power.

Tab. 1.24: The US solar energy plan target (¢/kWh).

Year	2010	2016	2020	2030
Residential	42	18	9	5
Business	34	13	7	4
Public grid	27	7	6	3

Example analysis for the level of utility grid: PV power plant with capacity of 100 MW with single axis tracking, the power generation rate is 1,860 kWh/kW, under the condition of investment cost of 7% and inflation rate of 2.5%, using 5-year accelerated depreciation method and the value of the currency in 2016 to calculate the power generation cost was 7 ¢/kWh at that time. By using lower price modules of 0.65–0.30 $/W, the cost of power generation can be reduced by 1.2 ¢/kWh, with lower price balance system software and hardware about 0.85–0.55 $/W, the cost of power generation can be reduced by 1 ¢/kWh, and if the service life extends to 30–50 years, power degradation rate reduces to 0.75–0.2%/year, power generation cost can be reduced by 1.1 ¢/kWh. Reducing annual operation and maintenance costs to 14–4 $/kW, the power generation cost can be reduced by 0.7 ¢/kWh. In this way, the goal of generating electricity at a cost of 3 ¢/kWh can be achieved by 2030.

1.5.5 China

In December 2016, the National Energy Administration issued the "13th Five-Year Plan for Solar Energy Development" [12]. The plan points out that the "Thirteenth Five-Year Plan" will be a crucial period for the development of the solar energy industry. The basic tasks are upgrading the industry, reducing costs, and expanding applications to realize the goal of market-based self-sustaining development that does not rely on state subsidies and become an important force for achieving non-fossil fuel energy share of 15% and 20% of primary energy consumption in 2020 and 2030, respectively. The objectives regarding development and utilization proposed in the plan are as follows: By the end of 2020, the installed capacity of solar power generation will reach 110 GW or more; of which, the installed capacity of PV will reach more than 105 GW. By 2020, the annual utilization of solar energy will reach 140 million tons of standard coal or more. Cost target: the cost of PV power generation continues to decrease. By 2020, the PV power price level will decline by more than 50% on the basis of 2015, and the goal of grid parity will be realized on electricity use, the cost of solar thermal power generation will be lower than 0.8 CNY/kWh. The goal of technology progress: the conversion efficiency of commercialized advanced crystalline silicon PV cells reaches over 23%, and the conversion efficiency of commercialized thin-film PV cells should be significantly improved, and a number of new type PV cells start to be commercialized preliminarily. PV power generation system efficiency will be significantly improved, smart operation and maintenance will be realized, and the entire industrial chain integration capability will be established. Table 1.25 shows the main targets of solar energy utilization during the "Thirteenth Five-Year Plan" period.

In December 2016, China's Electronic Information Industry Development Institute and China's PV Industry Association released the "China's PV Industry Development Roadmap" covering the upstream and downstream links of the PV industry chain from

Tab. 1.25: Main targets of solar energy utilization during China's "Thirteenth Five-Year Plan" period.

Indicator target category	Main targets	2015	2020
Installation (10 MW)	PV power generation	4,318	10,500
	Solar thermal power generation	1.39	500
	Total	4,319	11,000
Power generation (100 million kWh)	Total power generation	396	1,500
Heat utilization (100 million m²)	Collecting area	4.42	8

2016 to 2025, including a total of 62 key indicators for polysilicon, silicon rods, silicon ingots, PV cells, modules, balance components, and systems, which can represent the level of development in the field.

With the implementation of the global climate agreement "Paris Agreement" and the continuous decline in the cost of PV power generation, the installation regions and application types of PV power generation will continue to expand, and the global PV market will increase year by year. Forecasting the future market size using the lowest value predicted by Bloomberg, Energy Trend, Gartner, and other institutions as a conservative and the highest values for future market size forecast, the newly global PV installed capacity in 2016 is expected to reach over 70 GW, and the estimated annual installed capacity of global PV in 2011 to 2025 is shown in Fig. 1.10. From 2016 to 2020, the global PV market will continue to expand its scale with a compound growth rate of 9%.

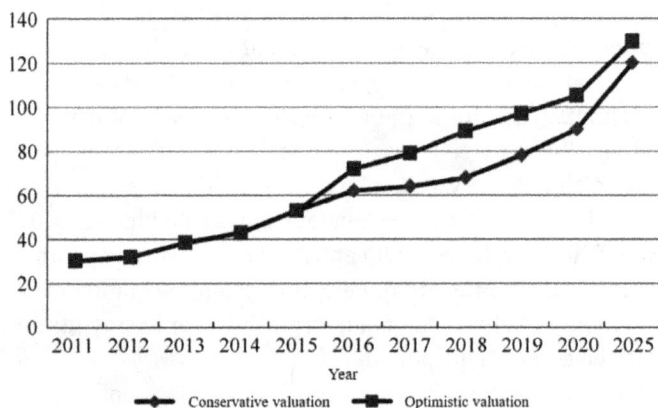

Fig. 1.10: World's installed and predicted PV installation 2011–2025.

Figure 1.11 shows the annual PV installation in China 2011–2015, and the prediction for 2016–2025.

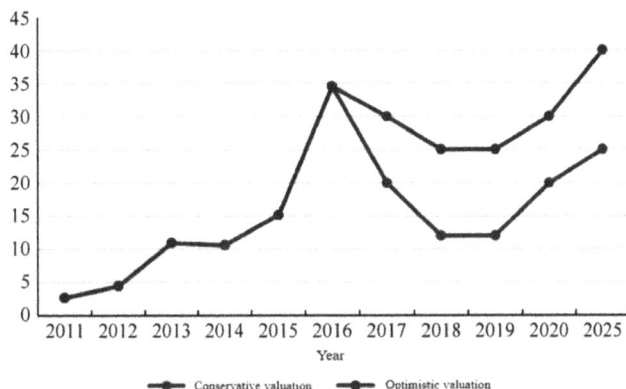

Fig. 1.11: China installed and predicted PV installation 2011–2025.

According to the "China Renewable Energy Development Roadmap 2050" jointly issued by the National Development and Reform Commission Energy Research Institute and other agencies, the installed capacity of China's PV power generation will reach 100, 400, and 1,000 GW by 2020, 2030 and 2050, respectively. In view of the industry, solar energy will transition from the current supplemental energy to alternative energy and gradually will become one of the main energy sources in the energy system.

1.5.6 International Energy Agency

The Energy Technology Perspectives 2014 Scenarios and Strategies to 2050 (ETP2014) published by the IEA and the Organization for Economic Cooperation and Development (OECD/IEA) is the latest edition after the 2010 edition [13]. It provides an in-depth analysis of the world's energy situation and proposes that there are three scenarios for the future of energy by 2050.

– 6 °C scenario (6DS): This scenario reflects the current situation to a large extent, and the potential destructive result the world is facing now has not changed. In 2050, energy consumption will increase by two-thirds over 2011, and GHG emissions will continue to rise. It is expected that the average global temperature will increase by at least 6 °C in the long term without taking any measures.

– 4 °C scenario (4DS): Considering the state's restriction of GHG emissions and efforts to improve energy efficiency, some major policy measures that are taken will keep the global average temperature rise within 4 °C for a long period of time, but by 2050 there will be a significant additional reduction in emissions to avoid a significant climate impact.

– 2 °C scenario (2DS): Through joint efforts of all countries in the world to combat climate change and take effective measures to reduce GHG emissions, CO_2 emissions in 2050 will be reduced by more than half from 2011 to establish a sustainable energy system. This will have at least a 50% chance of keeping the global average temperature increase below 2 °C.

Applying the principle to the electrical utility sector presented a 2DS hiRen scenario later based on the 2DS scenario, which accelerates the development of solar energy and wind energy in renewable energy, reducing or delaying the development of nuclear energy and carbon capture and storage technology deployment.

The development forecast of PV and Solar thermal power generation is shown in Table 1.26.

Tab. 1.26: Future development forecast of PV and CSP.

	6DS		4DS		2DS		2DS hiRen	
	2030	2050	2030	2050	2030	2050	2030	2050
PV power generation (TWh)	588	937	805	2,523	1,141	38,234	2,609	6,250
PV ratio (%)	1.6	1.9	2.3	5.6	3.5	9.5	8	16
PV (GW)	451	663	602	1,813	841	2,785	1,927	4,626
Thermal power generation (TWh)	92	359	147	796	554	2,835	986	4,186
Photothermal power generation ratio (%)	0.3	0.7	0.4	1.8	1.7	7.1	3	11
Photothermal power generation capacity (GW)	26	98	40	185	155	646	252	954

In 2013, total PV installation in the world reached 36 GW, with an average of 100 MW per day. The rapid growth of the application market has led to a rapid decline in the cost of PV modules and components, and increased the competitiveness of PV in the market.

Under the 2DS scenario, by 2050, solar (including PV and CSP) power generation will account for the fourth highest source, just behind wind power, hydropower, and nuclear energy in the global power supply, surpassing natural gas and bioenergy power generation, of which PV will account for 10%, and CSP for 7%. The predicted cost of new PV system in 2050 is given in Tab. 1.27.

In the 2DS hiRen scenario, solar power will become the largest source of electricity by 2040. By 2050, it will provide 27% of the world's electricity generation. PV

Tab. 1.27: New PV system cost by 2050 under 2DS scenario ($/MWh).

Year		2015	2020	2025	2030	2035	2040	2045	2050
Rooftop PV system	Min	135	108	94	83	72	62	58	53
	Max	539	427	359	312	265	225	208	191
	Mean	202	165	146	128	110	98	93	93
Utility-level PV system	Min	119	97	83	73	63	55	51	47
	Max	318	254	214	187	159	136	126	116
	Mean	181	137	113	97	91	79	71	71

(16%) and solar thermal power (11%) will take the third and fourth places respectively, falling behind only to wind power and hydropower.

The IEA has integrated PV development plans and roadmaps for various countries and regions, published the Technology Roadmap Solar PV Energy, and later released the 2014 edition [14]. It pointed out that since 2010, the installed PV capacity has exceeded the past 40 years, and it increased at a rate of 100 MW per day in 2013. At the beginning of 2014, the total global capacity has exceeded 150 GW. The distribution pattern has also undergone great changes. At the beginning, Europe (such as Germany and Italy) was the main force of the PV market. After 2013, China became the leader, followed by Japan and the USA. In the past 6 years, the price of PV power generation systems fell to one-third of the original price, while the module price was only one-fifth of the original price. The newly established system power generation cost is 90.300 $/MWh (Tab. 1.28). According to the roadmap, it is expected that by 2050, PV power will account for 16% of the global electricity supply, and increase by 11% from the target set out in the 2010 roadmap. PV power generation will account for 17% of clean power and account for 20% of all renewable energy power. China

Tab. 1.28: Predicted cost of PV power generation in the hiRen scenario ($/MWh).

Year		2013	2020	2025	2030	2035	2040	2045	2050
Public grid level	Min	119	96	71	56	48	45	42	40
	Mean	177	133	96	81	72	68	59	56
	Max	318	250	180	119	119	109	104	97
Rooftop PV system	Min	135	108	80	63	55	51	48	45
	Mean	201	157	121	102	96	91	82	78
	Max	539	422	301	231	197	180	171	159

still maintains the largest PV market and will account for 37% of global capacity by 2050. In order to realize the vision of the roadmap, capacity of 4,600 GW will be installed by 2050 and CO_2 emissions will be reduced by 4 Gt per year.

The roadmap predicted that with the development of the market, the average cost of PV power will be reduced by 25% by 2020, 45% by 2030, and 65% by 2050. Assuming that the cost of capital is 8%, the cost of PV power generation is in the range of 40–160 $/MWh. In order to achieve this goal, the PV installations will increase rapidly. In 2013, it will be 36 GW. After that, it will increase to 124 GW per year on average, and it will reach a peak of 200 GW in 2025–2050. It needs an average annual investment of US$ 225 billion, which is more than 2 times that of 2013.

It is expected that the grid level and rooftop PV systems will share the global market equally in the future. Although rooftop PV systems are currently more expensive, they can supply electricity nearby, saving on transmission costs and transmission line losses.

In addition to the current decline in the cost of PV power generation, it is necessary to strengthen the policy support mechanism for PV power generation. For a long time, the price of PV power generation does not reflect the impact of climate change and other environmental factors. The roadmap believes that the global CO_2 price will reach 46 $/t in 2020, 115 $/t by 2030, and 152 $/t by 2040.

With the gradual depletion of conventional energy sources, international petroleum prices have continued to soar. At the same time, people are increasingly paying attention to environmental pollution and climate warming. It is necessary to gradually change the energy consumption structure, vigorously develop renewable energy represented by solar energy, and take the road of sustainable development. And this has become a common consensus. Solar PV power generation will maintain a rapid development trend for a long period of time. After a certain period of time, it will gradually occupy a considerable share of the energy structure. It can be expected that solar power will become the main energy source by the end of this century, and a brilliant new solar energy era will come inevitably.

References

[1] John Conti. et al., International energy outlook 2016. DOE/EIA-0484(2016). May 2016. www. eia.gov/forecasts/ieo.

[2] R. K. Pachauri and A Reisinger. Climate change 2007: Comprehensive report. Switzerland: IPCC Geneva, 2007.

[3] Fatih Birol. CO_2 emissions from fuel combustion highlights (2016 Edition). IEA STATISTICS. https://www.iea.org/publications/freepublications/publication/ CO_2emissions from fuel combustion_highlights_2016.pdf.

[4] Tim Buckley. IEA. World Energy Outlook 2016[R] ieefa.org/wp-content/uploads/2016/11/ World-Energy-Outlook-2016-IEEFA.pdf

[5] Janet L. Sawin, et al., RENEWABLES 2016 Global Status Report[R] REN21 2016. ISBN978-3-9818107-0-7

[6] Stefan Nowak. Trends 2016 in photovoltaic applications. report IEA-PVPS T1-30:2016. ISBN978-3-906042-45-9 ISBN 978-3-906042-45-9.

[7] IRENA. RENEWABLE CAPACITY STATISTICS 2017 [R]. ISBN 978-92-9260-018-1 https://www. irena.org/ . . . /Publication/2017/Mar/IRENA_RE_Capacity_Statistics_2017.pdf

[8] Nat Kreamer and Tom Kimbis. SEIA 2016 ANNUAL REPORT [R] https://www.seia.org/research-resources/2016-annual-report

[9] Japanese NEDO. Overview of PV Roadmap Toward 2030. https://www.pvaustria.at/wp-content/uploads/2013/07/Roadmap_Nedo_2004.pdf

[10] EPIA, Greenpeace. Solar generation VI solar photovoltaic electricity empowering the world [R]. 2011. http://www.pv-era.net/doc_upload/documents/258_121_Solargeneration6.pdf.

[11] Giorgio Cellere, et al., International technology roadmap for photovoltaic 2016 results[R]. 8th edition, March 2017. ITRPV http://www.itrpv.net/Reports/Downloads/.

[12] China's National Energy Administration. "13th Five-Year Plan for Solar Energy Development". http://zfxxgk.nea.gov.cn/auto87/201612/t20161216_2358.htm

[13] Stephan Singer, et al., WWF the energy report -100 percent renewable energy by 2050. ISBN 978-2-940443-26-0.2011. http://assets.panda.org/downloads/101223_energy_report_final_print_2.pdf.

[14] Cédric Philibert. Technology roadmap solar photovoltaic energy 2014 Edition. International Energy Agency (IEA), © OECD/IEA, 2014. http://www.iea.org/publications/freepublications/publication/technology-roadmap-solar-photovoltaic-energy-2014-edition.html.

Exercises

1.1 Briefly explain the significance of developing and utilizing solar energy and the limitations of regular power grids.

1.2 What are the types of solar power generation? Briefly describe its working principle.

1.3 What are the advantages and disadvantages of solar PV power generation?

1.4 Give a brief description of current PV power production and market application.

1.5 What are the economic and technical problems in solar energy utilization?

1.6 Explain the application prospect of PV power generation.

Chapter 2
Solar radiation

2.1 Overview of the Sun

Looking out from the Earth upon which mankind lives and breeds, the most striking matter in the sky is the glorious Sun, a gas planet that glows and heats by itself. The interior of the Sun can be divided into three layers: the core area, the radiation zone, and the convective zone. The radius of the core area is about a quarter of the radius of the whole Sun but the mass of it accounts for more than half of the mass of the Sun. The temperature in this area is up to the range from 8×10^{16} to 40×10^{16} K, and the pressure in it is almost equal to 300 billion atm. Such extremely high temperature and pressure make 600 million tons of hydrogen into 5.96 million tons of helium through thermal fusion reaction at every second, and at the same time, release energy that is equal to burning 4 million tons of hydrogen. These energies radiate outward through the radiation zone and convective zone, which can last for 5 billion years.

The outside of the Sun comprises three layers: the photosphere, the chromosphere, and the corona. The surface of the Sun that people observe is the photosphere, and the layer thickness of it is about 500 km, all the visible light of the Sun is almost due to it. The surface of the photosphere has a granular structure – "rice shape tissue" – the bright area in the photosphere called the spot and the dark area called sunspots. The chromosphere is 2,000 km away from the surface of the photosphere, and there are plaques, dark stripes, and prominence, and intense flare also occurs frequently. The corona layer, which is outside the chromosphere, is extremely hot and extends to the Sun's radius several times. X-ray flares can be observed with the space telescope. Corona has a coronal hole, and the coronal hole is the source of the solar wind. The structure of the Sun is shown in Fig. 2.1.

Comparing the Sun with the Earth, the diameter of the Sun is about 1.39×10^9 m, which is 109.3 times larger than that of the Earth. The volume of the Sun is about 1.4122×10^7 km^3, which is 1.3 million times larger than that of the Earth. However, the average density of the Sun is about 1.41 g/cm^3, a little bit larger than that of water, but only about a quarter of that of the Earth, and the density is also changing with the altitude. There is difference between that of inside and outside. Most of the Sun's outer shell is gas and its density is very small, but its inside is denser. The density of the core area can be up to 160 g/cm^3, which is about 20 times larger than the density of steel. The total mass of the Sun is 1.9892×10^{27} ton, which is equal to 33.34 million Earths. The temperature on the surface is about 5,800 K. Sunlight consists of photons of different energy levels, that is, electromagnetic waves of different frequencies and wavelengths. Electromagnetic waves are usually distinguished by a range of bands, with different names, as mentioned in Tab. 2.1, where the visible light presents different colors due to the length of the wavelength, as given in Tab. 2.2.

https://doi.org/10.1515/9783110524833-002

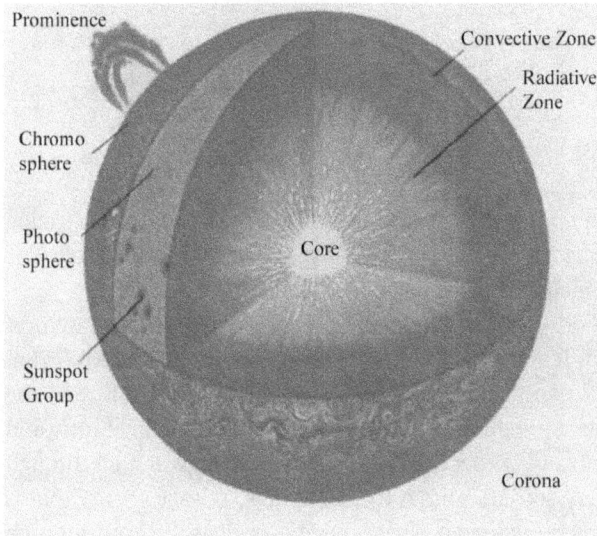

Fig. 2.1: The structure of the Sun.

Tab. 2.1: Wavelength range of electromagnetic waves.

Name	Wavelength	Name	Wavelength
Ultraviolet	100 Å to 0.4 μm	Ultra-far infrared light	15–1,000 μm
Visible light	0.4–0.76 μm	Millimeter wave	1–10 mm
Near-infrared light	0.76–3.0 μm	Centimeter wave	1–10 cm
Mid-infrared light	3.0–6.0 μm	Decimeter wave	10 cm to 1 m
Far-infrared light	6.0–15 μm		

Tab. 2.2: Wavelength range of visible light.

Color names	Wavelength range (μm)	Color names	Wavelength range (μm)
Purple	0.40–0.43	Yellow	0.56–0.59
Blue	0.43–0.47	Orange	0.59–0.62
Cyan	0.47–0.50	Red	0.62–0.76
Green	0.50–0.56		

The maximum value of the energy density in the solar spectrum is 0.475 µm, and it is sharply reduced toward the short-wavelength direction, but is slowly attenuated toward the long-wavelength direction as shown in Fig. 2.2. In the upper zone of atmosphere, about 7% of the total solar radiation energy is under the wavelength range of ultraviolet, 47% of the energy is in the visible light, and 46% of the energy is in the infrared light.

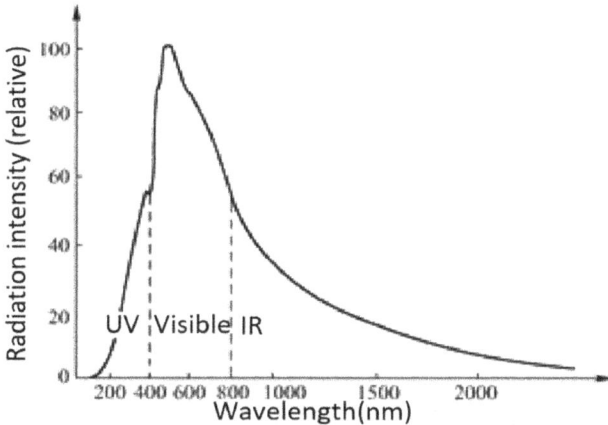

Fig. 2.2: Solar energy spectral distribution.

The amount of energy released by the Sun per second is 3.865×10^{26} J, which is equivalent to the energy generated by burning 1.32×10^{16} t of standard coal. The average distance between the Sun and the Earth is about 1.5×10^{11} m, and it is only about 1 over 22 billion of the energy of solar radiation to reach the extraterrestrial of the Earth, which is about 3.86×10^{23} kW. About 19% of it is absorbed by the atmosphere, about 30% is reflected back to the space by the air, dust particles, and the ground, and the solar radiation successfully reaching the Earth's surface through the atmosphere accounts for about 51%, as shown in Fig. 2.3. However, most of the surface of the Earth is covered by oceans, and the solar radiation reaching the land surface only accounts for 10% of the solar radiation energy reaching the Earth.

2.2 The relative movement of the Sun and the Earth

2.2.1 The overview of the Earth

The equatorial radius of the Earth is slightly longer, and the polar radius is slightly shorter. The polar axis is equivalent to the axis of rotation of an oblate sphere. According to the shape and size of the Earth published by the International Union of Geodesy and Geophysics in 1980, the main data are as follows:

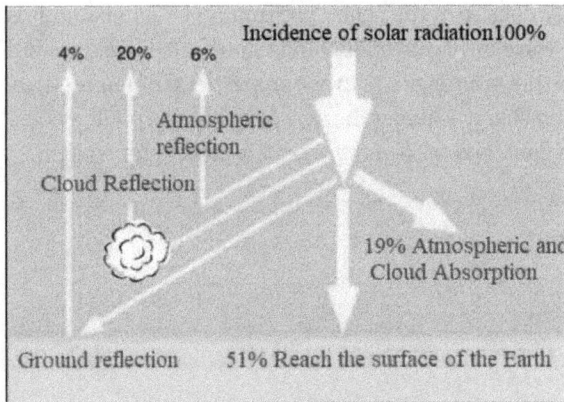

Fig. 2.3: Solar energy through the atmosphere.

- Equatorial radius   6378.137 km
- Polar radius   6,356.752 km
- Average radius   6,371.012 km
- Flat rate   1/298.257
- Equatorial perimeter   40,075.7 km
- Meridian perimeter   40,008.08 km
- Surface area 5.101×10^8 km^2
- Volume $10,832 \times 10^8$ km^3

In fact, the true shape of the Earth is slightly different from the above-mentioned oblate spheroids. The Southern Hemisphere is slightly thicker and shorter, with the South Pole sinking approximately 30 m inward, while the Northern Hemisphere is slightly thinner and longer, and the North Pole protrudes upward by about 10 m. So exaggeration to say, the true shape of the Earth is slightly like a pear. The weight of the Earth is about 5.98×10^{24} kg.

2.2.2 Apparent solar time

The Earth rotates from west to east around the Earth's south–north axis. One rotation takes 1 day, which is 24 h (actually one stellar day is 23 h 56 min and 04.0905 s). As the Earth rotates, it orbits the Sun following an oval orbit called the ecliptic (long axis 1.52×108 km, short axis 1.47×108 km, average Sun–Earth distance 1.496×10^8 km), this movement is revolution with a period of 1 year (actually 1 sidereal year is 365 days 06:06:9). Schematic of the relative movement of the Sun and the Earth is shown in Fig. 2.4.

The axis of rotation of the Earth is tilted at an angle of 23.45° from the normal line to the orbital plane (the ecliptic plane), and the axis of rotation always points

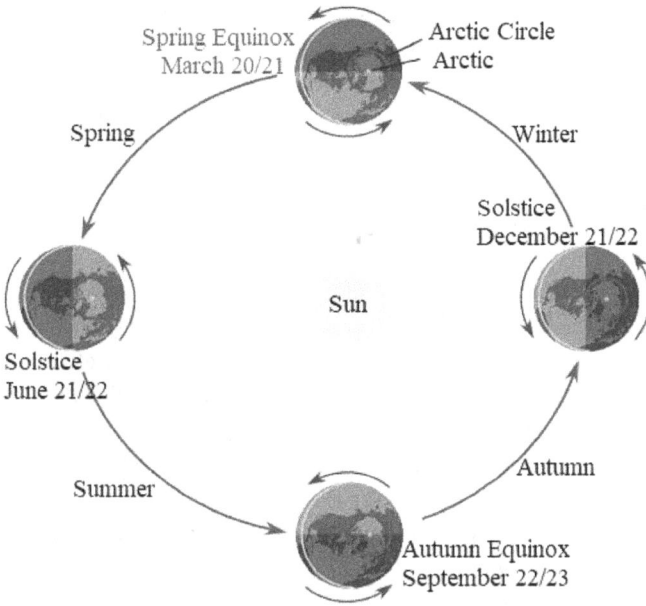

Fig. 2.4: Schematic of the Sun–Earth movement.

to the North Pole of the Earth during the Earth's revolution. This causes the Sun's direct rays to be sometimes on the equator, sometimes southerly, and sometimes northerly, forming a seasonal change on the Earth.

The 1884 International Conference established a method for dividing the time zone, which stipulates that every 15° of longitude is a time zone and the world is divided into 24 time zones. The longitude of the original warp through the Greenwich Observatory site in London, UK, is set as the 0° central meridian, and from 7.5° west longitude to 7.5° east longitude is the middle time zone, and it is divided into 12 time zones eastward and westward each.

Mean solar time is a time measurement system based mainly on the Earth's rotation cycle. The mean solar time assumes that the Earth's orbit around the Sun is a standard circle. Every day in the year, the rotation is uniform, and every day it rotates for 24 h, the rotated angle is $360°/24 = 15°$ per hour, and the time difference is $60/15 = 4$ min for every 1°. This is the basis for calculating the local time difference.

The mean solar time in a place is measured by the Sun's angle of time to the meridian circle of this place. The mean solar time is zero when this place is in lower culmination (0:00 midnight). In fact, the Beijing time we use is the mean solar time of 120°E longitude.

However, the Earth orbits around the Sun along an oval orbit. The Sun is located at a focal point of the ellipse. Therefore, the distance between the Sun and the Earth changes continuously throughout the year. Based on Kepler's second law of planetary motion, a line segment joining a planet and the Sun sweeps out equal areas during equal intervals of time. It can be seen that the Earth does not move at a constant speed in the orbit, so the rotation of the Earth relative to the Sun is not uniform. It is not always 24 h a day, sometimes it is less, and sometimes more. The apparent solar time take this factor into an account.

The time interval of the Sun successively passes twice in upper culmination (12:00 noon), known as the apparent solar day. One apparent solar day is divided into 24 apparent solar hours. The apparent solar time is measured in terms of hour angle in the center of the circle of the apparent solar. The starting point is the sky in the upper culmination of the apparent Sun. However, in our daily life, the starting point is usually the lower culmination, which is just 12 h apart. Therefore, in order to be consistent with people's habit of daily life, the apparent solar time is defined as the hour angle of the center of the circle plus 12 h.

Since the length of an apparent solar day is constantly changing within a year, so the length of 24 apparent solar hours during the day is constantly changing too. This is very inconvenient in practical applications. Therefore, it is not advisable to choose apparent solar time as timekeeping units.

The relationship between apparent solar time and mean solar time is

$$\text{Apparent solar time} = \text{mean solar time} + \text{time difference value}$$

The value of time difference is different for different days, reaching negative maximum as −14 min and 15 s on February 10 and positive maximum as 16 min and 25 s on November 2. It varies in this range on other days, the minimum difference happens on June 11 is 0 min and 1 s. So we can see that time difference is not large, and in practice, the difference between apparent solar time and mean solar time is often ignored.

The solar time in each place is related to the longitude of the place (not to the latitude). In case if the longitudes are different, the Sun rises and falls in different sequence. For example, 7:00 am in Beijing, China is 11:00 pm in London, UK, and between 4:00 am and 5:00 am in Xinjiang, China. They are all local solar time, but for convenience, different places in China all use the local time of Beijing, so while determining the apparent solar time for some other places in China, the local Beijing time should be calculated as the local mean solar time, and then convert to the local apparent solar time of other places. The method is as follows:

$$\text{Local mean solar time} = \text{mean solar time of Beijing} + 4\text{ min} \times (\text{local longitude} - 120°)$$

Therefore, in China, when calculating the mean solar time by local longitude, taking east longitude 120° as a benchmark, for every 1° decrease (westward), it decreases 4 min, and for every 1° increase (eastward), it increases 4 min. If a place is located at longitude 90°, then 90°−120° = −30°, −30 × 4 min = −120 min, then the local mean solar time is subtracted from Beijing time by 120 min.

After getting the local mean solar time, plus the time difference, you can get the local apparent solar time.

2.2.3 Sunrise and sunset patterns

In the Northern Hemisphere except the Arctic, only at vernal equinox and autumnal equinox, the Sun rises at east and falls at west. In the summer half of the year (vernal equinox – summer solstice – autumnal equinox), sunrise east biased to north and sunset west biased to north, and with more close to the summer solstice, the sunrise and sunset become more northward, on the summer solstice day, the sunrise and sunset become the most northward. In the winter half of the year (autumnal equinox – winter solstice – vernal equinox), sunrise east biased to south and sunset west biased to south, and with more close to the winter solstice, the sunrise and sunset become more southward, and on the winter solstice day, the sunrise and sunset become the most southward as shown in Fig. 2.5 [1].

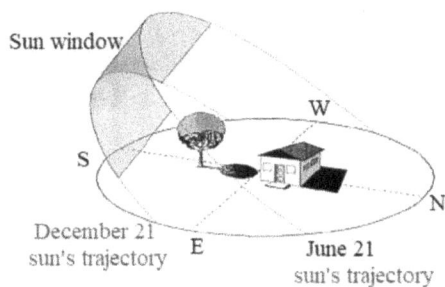

Fig. 2.5: Schematic of solar track.

In the Northern Hemisphere on the summer solstice (June 21 or 22), the Sun is directly at the zenith at 23.45°N. Therefore, the latitude 23.45°N is called the Tropic of Cancer. On the other hand, the Tropic of Capricorn (or the Southern Tropic) is the circle of latitude that contains the subsolar point on the winter solstice (December 21 or 22), which is 23.45°S.

On the vernal equinox day (March 20 or 21 in the Northern Hemisphere) and the autumnal equinox (September 22 or 23 in the Northern Hemisphere), the Sun just hits the Earth's equatorial plane.

2.3 Celestial coordinate system

The observer stands on the surface of the Earth and looks up at the sky like looking at the imaginary sphere around him/her, and according to the principle of relative motion, the Sun seems to move on this sphere from east to west every day. To determine the position of the Sun on the celestial sphere, the most convenient method is to use the celestial coordinate. Commonly used celestial coordinates are of two kinds: equatorial coordinate system and horizontal coordinate system [2].

2.3.1 Equatorial coordinate system

The equatorial coordinate system is the celestial coordinate system, which defines that the celestial equator QQ' is the basic circle and the intersection point O between the circle QQ' and every daily meridian is the origin. The P and P's are the northern celestial pole and the southern celestial pole, respectively, as shown in Fig. 2.6, and the great circles passing through PP' are all perpendicular to the celestial equator plane. Obviously, the semicircle passing through P and the position of the Sun (M) on the sphere is also perpendicular to the celestial equator plane, and the intersection is the point M'.

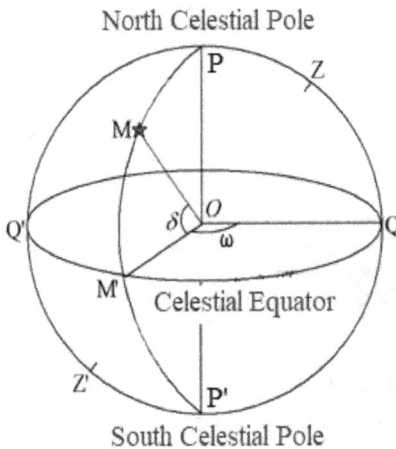

North Celestial Pole

South Celestial Pole

Fig. 2.6: Equatorial coordinate system.

In the equatorial coordinate system, the position of the Sun (M) is determined by two coordinates, the hour angle ω, and the Sun declination angle δ.

2.3.1.1 Hour angle ω

Equal to the arc QM', counting from the Q point (from the midday) on the daily meridian, the clockwise direction is positive and the counterclockwise direction is negative, that is, the morning is negative and the afternoon is positive. Usually it is expressed as ω, its value is equal to the time apart from the noon (hours) multiplied by 15°.

2.3.1.2 Sun declination angle δ

The intersect circle between the plane that is parallel to the equatorial plane and the Earth is called the latitude of the Earth. The latitude of the Sun's direct spot, which is the angle of the connection between the center of the Sun and the Earth, and the equatorial plane is often referred to as the Sun's declination angle, which is often denoted by δ. The change of the Sun declination angle on the Earth is shown in Fig. 2.7. For the Sun, $δ = 0°$ at the vernal equinox and the autumnal equinox day changes from $0°$ to $+23.45°$ till the summer solstice toward northern celestial pole and from $0°$ to $-23.45°$ at the winter solstice toward southern celestial pole. Sun's declination angle is a continuous function of time, and its rate of change is greatest at the vernal equinox and autumnal equinox day, around 0.5° a day. The declination angle is only related to which day of the year, and has nothing to do with the location, that is, anywhere on the Earth, the Sun's declination angle is the same.

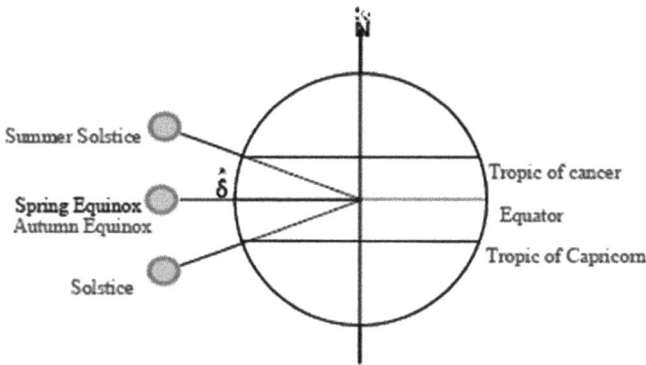

Fig. 2.7: Change of the solar declination angle on the Earth.

The solar declination can be approximated using the Cooper equation:

$$δ = 23.45 \sin\left(360 \times \frac{284 + n}{365}\right) \qquad (2.1)$$

where n is the date serial number in a year. For example, $n = 1$ is the New Year's Day, $n = 81$ is the vernal equinox, $n = 365$ is December 31 in common year, but in leap year, it is 366.

This is an approximated calculation formula. When the specific calculation is performed, the result of the δ value on the vernal equinox day and the equinox day equal to 0 cannot be obtained by this formula. More accurate calculations (error < 0.035°) can be obtained using another approximate formula derived by Iqbal [3]:

$$\delta = (180/\pi)(0.006918 - 0.399912 \cos B + 0.070257 \sin B - 0.006758$$
$$\cos 2B + 0.000907 \sin 2B° - 0.002697 \cos 3B + 0.00148 \sin 3B) \qquad (2.2)$$

where $B = (n-1) \times 360 / 365$; n is the date serial number in a year.

Example 2.1 Calculate the Sun's declination angle of September 22.
Solution $n = 265$ on September 22, and inserting into formula (2.1), the following is obtained:

$$\delta = 23.45 \sin\left(360 \times \frac{284 + 265}{365}\right) = -0.6°$$

The declination angles for different dates in a year are shown in Tab. 2.3.

Tab. 2.3: Sun declination angle δ.

Date	1	5	9	13	17	21	25	29
Jan	−23.1	−22.7	−22.2	−21.6	−20.9	−20.1	−19.2	−13.2
Feb	−17.3	−16.2	−14.9	−13.6	−12.3	−10.9	−9.4	
Mar	−7.9	−6.4	−4.8	−3.3	−1.7	−0.1	1.5	3
Apr	4.2	5.8	7.3	8.7	10.2	11.6	12.9	14.2
May	14.8	16	17.1	18.2	18.1	20	20.8	21.5
Jun	21.9	22.5	22.9	23.2	23.4	23.4	23.4	23.3
Jul	23.2	22.9	22.5	21.9	21.3	20.6	19.8	19
Aug	18.2	17.2	16.1	14.9	13.7	12.4	11.1	9.7
Sep	8.6	7.1	5.6	4.1	2.6	1	−0.5	−2.1
Oct	−2.9	−4.4	−5.9	−7.5	−8.9	−10.4	−11.8	−13.2
Nov	−14.2	−15.4	−16.6	−17.7	−18.8	−19.7	−20.6	−21.3
Dec	−21.7	−22.3	−22.7	−23.1	−23.3	−23.4	−23.4	−23.3

2.3.2 Horizontal coordinate system

When a person looks at the position of the Sun in the sky relative to the ground plane on the Earth, the position of the Sun relative to the Earth is relative to the ground plane and is usually determined by two coordinates, altitude angle and azimuth angle, as shown in Fig. 2.8.

At some point, because of the different locations on the Earth, the altitude angle and azimuth angle are not the same everywhere.

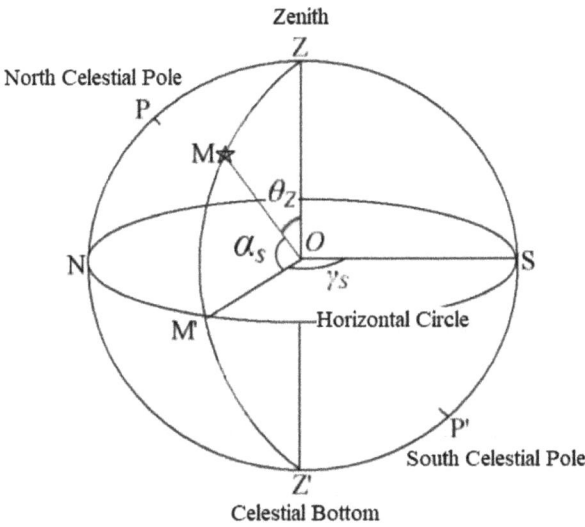

Fig. 2.8: Horizontal coordinate system.

2.3.2.1 Zenith angle θ_z

The zenith angle refers to the angle between the Sun's rays MO and the normal line OZ of the ground plane.

2.3.2.2 Altitude angle α_s

The altitude angle refers to the angle between the Sun's rays MO and the projection line OM' on its ground plane. It represents the angle of the Sun above the horizontal plane. The relationship between altitude angle and zenith angle is

$$\theta_z + \alpha_s = 90° \tag{2.3}$$

2.3.2.3 Azimuth angle γ_s

The azimuth angle refers to the angle y_s between the Sun's rays' projection OM' on the ground plane and south direction line on the ground plane OS. It indicates that the horizontal projection of the Sun's rays deviates from the just south direction. Take the south direction as the starting point (0°), westward (clockwise) as positive, and eastward (counterclockwise) as negative.

2.3.3 The calculation of solar parameters

2.3.3.1 The calculation of altitude angle

The relationship among altitude angle, zenith angle and latitude, declination and hour angle is

$$\sin a_s = \cos\theta_z = \sin\varphi\sin\delta + \cos\varphi\cos\delta\cos\omega \qquad (2.4a)$$

At the noon Sun, $\omega = 0$, then eq. (2.4a) can be simplified to

$$\sin a_s = \sin[90° \pm (\varphi - \delta)] \qquad (2.4b)$$

When the midday Sun is in the south of the zenith, which means $\phi > \delta$

$$\alpha_s = 90° - \phi + \delta \qquad (2.4c)$$

When the midday Sun is in the north of the zenith, which means $\phi < \delta$

$$\alpha_s = 90° + \phi - \delta \qquad (2.4d)$$

Example 2.2 Calculate the altitude and zenith angle at 12:00 am and 3:00 pm on September 22 in Shanghai, China.
Solution The latitude of the Shanghai area is 31.12°, we can obtain $\delta = -0.6°$ from example (2.1).
   At 12:00 am, the hour angle $\omega = 0$, at 3:00 am, and the hour angle $\omega = 3 \times 15 = 45°$.
   At 12:00 am, the altitude angle $\phi > \delta$ based on eq. (2.4c):

$$\alpha_s = 90° - \phi + \delta = 90° - 31.12° + (-0.6°) = 58.28°$$

The zenith angle at this time is $\theta_z = 90° - \alpha_s = 90° - 58.28° = 31.72°$.
   At 3:00 pm, the hour angle is $\omega = 3 \times 15 = 45°$, based on eq. (2.4a), the altitude angle at this time is

$$\sin\alpha_s = \sin 31.12\sin(-0.6) + \cos 31.12\cos(-0.6)\cos 45 = 0.5998$$

So, $\alpha_s = 36.86°$.
   The zenith angle at this time is $\theta_z = 90° - \alpha_s = 90° - 36.86° = 53.14°$.

2.3.3.2 The calculation of azimuth angle

The relationship between azimuth angle, declination and altitude angle, latitude and hour angle is

$$\sin\gamma_s = \frac{\cos\delta\sin\omega}{\cos\alpha_s} \qquad (2.5)$$

$$\cos\gamma_s = \frac{\sin\alpha_s\varphi - \sin\delta}{\cos\alpha_s\cos\varphi} \qquad (2.6)$$

Example 2.3 Calculate the azimuth angle γ_s at 14:00 on September 22 in Shanghai, China.
Solution From Example 2.1, we can know that
   On September 22, in Shanghai, China, $\delta = -0.6°$, $\phi = 31.12°$,

$$\omega = 2 \times 15° = 30°.$$

Using eq. (2.4a), we can obtain the altitude angle:

$$\sin\alpha_s = \sin 31.12 \times \sin(-0.6) + \cos 31.12 \times \cos(-0.6) \times \cos 30 = 0.7359$$

So $\alpha_s = 47.38°$.

Inserting into eq. (2.5), we obtain

$$\sin\gamma_s = \frac{\cos\delta\sin\omega}{\cos\alpha_s} = \frac{\cos-(0.6)\times\sin 30}{\cos 47.38} = 0.738$$

Therefore, $\gamma_s = 47.6°$.

2.3.3.3 The hour angle at sunrise and sunset ω_s

The altitude angle α_s at sunrise and sunset is $0°$, which can be obtained from eq. (2.4a)

$$\cos\omega_s = -\tan\phi\tan\delta \tag{2.7}$$

Because $\cos\omega_s = \cos(-\omega_s)$, so

$$\omega_{sr} = -\omega_s \; ; \; \omega_{ss} = \omega_s$$

where ω_{sr} is the sunrise hour angle and ω_{ss} is the sunset hour angle. Using the unit of degree, negative value sunrise hour angle and positive value for sunset hour angle, it can be seen that for a certain location, hour angles for sunrise and sunset are symmetrical with respect to the midday Sun.

2.3.3.4 Sunshine duration N

Sunshine duration is the local time interval from sunrise to sunset. Since the Earth rotates by $15°$ per hour, the sunshine duration N can be obtained by dividing the sum of the absolute values of the hour angle at sunrise and sunset by $15°$:

$$N = \frac{\omega_{ss} + |\omega_{sr}|}{15} = \frac{2}{15}\arccos(-\tan\varphi\tan\delta) \tag{2.8}$$

Example 2.4 Calculate the hour angle at sunrise and sunset and sunshine duration in Shanghai, China, at the winter solstice.

Solution In Shanghai, $\phi = 31.12°$. The solar declination angle at winter solstice $\delta = -23.45°$, and inserting into eq. (2.7),

$$\cos\omega_s = -\tan\phi\tan\delta = -\tan 31.12 \times \tan(-23.45) = 0.2619$$

Thus, the hour angle at sunrise is $\omega_{sr} = -74.82°$ and the hour angle at sunset is $\omega_{ss} = 74.82°$. In Shanghai, China, at winter solstice the sunshine duration is

$$N = 2 \times \frac{74.82°}{15°/h} = 9.98\,h$$

2.3.3.5 Azimuth angle at sunrise and sunset

The altitude angle at sunrise and sunset is $\alpha_{so} = 0°$; therefore, $\cos \alpha_s = 1$ and $\sin \alpha_s = 0$, and inserting into eq. (2.6):

$$\cos \gamma_{s,o} = -\sin\delta/\cos\phi \tag{2.9}$$

Azimuth angles at sunrise and sunset have two sets of solutions, so we must choose the correct set of solution. China can be roughly divided into two climate zones: the north tropical (0°–23.45°) and the north temperate zone (23.45°–66.55°). When the solar declination angle is $\delta > 0°$ (summer half-year), the sunrise and sunset will be in the northern quadrant (first and second quadrants in mathematical). When the solar declination angle is $\delta < 0°$ (winter half year), the sunrise and sunset will be in the southern quadrant (the third and fourth quadrants in mathematical).

Example 2.5 In search of the azimuth angle at the sunrise and sunset on September 22 in Shanghai, China.
Solution From Example 2.2, we can obtain that $\phi = 31.12°$; on the day of September 22, the declination angle $\delta = -0.6°$.
Inserting into eq. (2.9),

$$\cos \gamma_{s,o} = -\sin\delta/\cos\phi = -\sin(-0.6°)/\cos 31.12° = 0.01223$$

Thus, $\gamma_{s,o} = 89.30°$, or $\gamma_{s,o} = -89.30°$.
Therefore, the azimuth angle at sunrise is $\gamma_{s,or} = -89.30°$ and at sunset is $\gamma_{s,os} = 89.30°$.

2.3.3.6 Solar incidence angle

When the Sun irradiates on a tilt surface of the Earth, the angle between the solar beam radiation on a surface and the normal to that surface is defined as the solar incident angle θ_T. The relationship between the solar incidence angle and other angles is shown in Fig. 2.9. Thus, the geometric relationship between solar incidence angle and other angles is as follows:

$$\cos \theta_T = \sin \delta \sin \varphi \cos \beta - \sin \delta \cos \varphi \sin \beta \cos \gamma$$
$$+ \cos \delta \cos \varphi \cos \beta \cos \omega + \cos \delta \sin \varphi \sin \beta \cos \gamma \cos \omega \tag{2.10}$$
$$+ \cos \delta \sin \beta \sin \gamma \sin \omega$$

Besides,

$$\cos \theta_T = \cos \theta_Z \cos \beta + \sin \theta_Z \sin \beta \cos(\gamma_s - \gamma) \tag{2.11}$$

where θ_T is the solar incidence angle, δ is the solar declination angle, ϕ is the local latitude, β is the inclination of the tilt surface, y is the azimuth angle for the tilt surface, ω is the hour angle, θ_Z is solar zenith angle, and y_s is the solar azimuth angle.

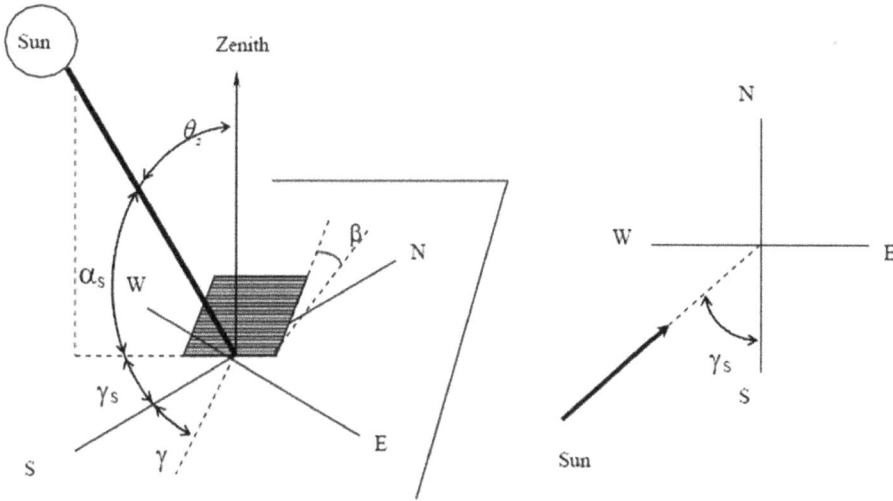

Fig. 2.9: Relationship between solar incident angle and other angles.

For the tilt surface of the Northern Hemisphere toward the equator ($y = 0°$), we can obtain that

$$\cos \theta_T = \sin \delta \sin \varphi \cos \beta - \sin \delta \cos \varphi \sin \beta$$

$$+ \cos \delta \cos \varphi \cos \beta \cos \omega + \cos \delta \sin \varphi \sin \beta \cos \omega \qquad (2.12a)$$

$$= \cos(\varphi - \beta)\cos \delta \cos \omega + \sin(\varphi - \beta)\sin \delta$$

For the tilted (also called inclined) surface of the Southern Hemisphere facing the equator ($y = 180°$), we can obtain that:

$$\cos \theta_T = \cos(\varphi + \beta)\cos \delta \cos \omega + \sin(\varphi + \beta) \sin \delta \qquad (2.12b)$$

If it is on the horizontal surface, that is, $\beta = 0°$, the solar incidence angle at this time is equal to the zenith angle, and its value is

$$\cos \theta_T = \cos \theta_z = \cos \varphi \cos \delta \cos \omega + \sin \varphi \sin \delta \qquad (2.12c)$$

If it is in the vertical surface, that is, $\beta = 90°$, then there is

$$\cos \theta_T = - \sin \delta \cos \varphi \cos y + \cos \delta \sin \varphi \cos y \cos \omega + \cos \delta \sin y \sin \omega \qquad (2.12d)$$

Example 2.6 Calculate the solar incident angle of a tilt surface with a tilt angle of 45° and an azimuth angle of 15° at 10:30 am on February 13 in Beijing.

Solution $N = 44$, $\delta = -14°$ on February 13, the hour angle in the morning of 10:30 am is $\omega = -22.5°$, $\beta = 45°$, and $y = 15°$. The latitude in Beijing, China, is $\phi = 39.48°$.

Substituting into eq. (2.10), we get

$$\cos\theta_T = \sin(-14)\sin 39.48 \cos 45 - \sin(-14)\cos 39.48 \sin 45 \cos 15$$
$$+ \cos(-14)\cos 39.48 \cos 45 \cos(-22.5)$$
$$+ \cos(-14)\sin 39.48 \sin 45 \cos 15 \cos(-22.5)$$
$$+ \cos(-14)\sin 45 \sin 15 \sin(-22.5)$$

which is

$$\cos\theta_T = (-0.2419)\times 0.6358 \times 0.707 - (-0.2419)\times 0.7718 \times 0.707 \times 0.9659$$
$$+ 0.9703 \times 0.7718 \times 0.707 \times 0.9239 + 0.9703 \times 0.6358 \times 0.707 \times 0.9659 \times 0.9239$$
$$+ 0.9703 \times 0.707 \times 0.2588 \times (-0.3827) = -0.1087 + 0.1275 + 0.4892 + 0.3892 - 0.068$$
$$= 0.8292$$

So the solar incidence angle at this time is $\theta_T = 34°$.

2.4 Angle of tracking surface

Some solar collectors track the Sun in a certain way in order to minimize the incidence angle of beam radiation and to maximize the amount of solar radiation received by the surface. For such motion of the surface, it is necessary to know the solar incidence angle and the azimuth of the surface [4].

The tracking system can be categorized according to its movement mode: one type is single-axis tracking, and the axis can be of any orientation. In fact, it is usually only horizontal east–west, horizontal south–north, and vertical or parallel to the direction of the Earth's axis, and another type is two-axis tracking.

(1) For a surface with daily tilt angle adjustment around the horizontal east–west axis, the beam radiation is normal to the surface at noon each day:

$$\cos\theta_T = \sin^2\delta + \cos^2\delta \cos\omega \tag{2.13a}$$

The slope of this surface will be fixed for each day:

$$\beta = |\varphi - \delta| \tag{2.13b}$$

The azimuth of the surface will be 0° or 180° throughout the day, depending on the latitude and the declination angle:

$$\text{If } (\phi - \delta) > 0, \text{so } \gamma = 0°$$
$$\text{If } (\phi - \delta) < 0, \text{so } \gamma = 180°$$

(2) For a surface continuously rotating around horizontal east–west axis, the solar incident angle:

$$\cos\theta_T = \left(1 - \cos^2\delta\sin^2 w\right)^{1/2} \tag{2.14a}$$

The slope of this surface can be determined by the following equation:

$$\tan\beta = \tan\theta_Z\left|\cos\gamma_s\right| \tag{2.14b}$$

If the solar azimuth angle passes through ±90°, the orientation of this kind of the azimuth angle between 0° and 180° for either hemisphere:

$$\text{If } \left|\gamma_s\right| < 90°, \text{so } \gamma = 0°$$

$$\text{If } \left|\gamma_s\right| > 90°, \text{so } \gamma = 180° \tag{2.14c}$$

(3) For the surface rotating around the axis of the horizontal north and south and can be continuously adjusted, when the solar incidence angle is minimized

$$\cos\theta_T = (\cos^2\theta_z + \cos^2\delta\sin^2 w)^{1/2} \tag{2.15a}$$

Its slope is given by the following formula:

$$\tan\beta = \tan\theta_Z\left|\cos(\gamma - \gamma_s)\right| \tag{2.15b}$$

The surface azimuth angle y will be 90° or –90° depending on the sign of the solar azimuth angle:

$$\text{If } \gamma_s > 0, \ \gamma = 90°$$

$$\text{If } \gamma_s \le 0, \ \gamma = -90° \tag{2.15c}$$

(4) For a surface that rotates around a vertical axis at a fixed inclination angle, the solar incidence angle is minimized when the surface azimuth angle is equal to the solar azimuth angle.

From eq. (2.11), the solar inclination angle θ_T can be obtained from the following equation:

$$\cos\theta_T = \cos\theta_Z\cos\beta + \sin\theta_Z\sin\beta \tag{2.16}$$

Since the inclination angle is fixed, β is a constant, and the azimuth angle $y = y_s$.

(5) For a surface rotating around a north–south axis parallel to the Earth's axis with continuous adjustment to minimizing θ_T:

$$\cos\theta_T = \cos\delta \tag{2.17a}$$

Since the inclination angle is changing continuously and is equal to

$$\tan\beta = \frac{\tan\varphi}{\cos\gamma} \tag{2.17b}$$

The surface azimuth angle is

$$\gamma = \arctan\frac{\sin\theta_Z \sin\gamma_s}{\cos\theta'\sin\varphi} + 180 C_1 C_2 \tag{2.17c}$$

where

$$\cos\theta' = \cos\theta_Z \cos\varphi + \sin\theta_Z \sin\varphi \tag{2.17d}$$

$$C_1 = \begin{cases} 0 & \text{if } \left(\arctan\frac{\sin\theta_Z \sin\gamma_s}{\cos\theta'\sin\varphi}\right)\gamma_s \geq 0 \\ +1 & \text{else} \end{cases} \tag{2.17e}$$

$$C_2 = \begin{cases} +1 & \text{if } \gamma_s \geq 0 \\ -1 & \text{if } \gamma_s < 0 \end{cases} \tag{2.17f}$$

(6) For a surface that is continuously tracking with two axes, its incidence angle must be minimized when:

$$\cos\theta_T = 1 \tag{2.18a}$$

$$\beta = \theta_Z \tag{2.18b}$$

$$\gamma = \gamma_s \tag{2.18c}$$

Example 2.7 For continuous rotation with the horizontal east and west axes to minimize θ_T, the surface is at:

(1) $\phi = 40°, \delta = 21°, \omega = 30°$ (2:00pm)

(2) $\phi = 40°, \delta = 21°, \omega = 100°$

Calculate the incidence angle of the beam solar radiation and the zenith angle.

Solution

(1) For the surface moving in this way, first calculate the incidence angle, according to eq. (2.14a), then we obtain

$$\theta_T = \arccos\left(1 - \cos^2 21 \times \sin^2 30\right)^{1/2} = 27.8°$$

Next, calculate the solar zenith angle θ_Z from eq. (2.4a),

$$\theta_Z = \arccos(\cos 40 \times \cos 21 \times \cos 30 + \sin 40 \times \sin 21) = 31.8°$$

(2) The same process as part (1),

$$\theta_T = \arccos \left(1 - \cos^2 21 \times \sin^2 100\right)^{1/2} = 66.8°$$

$$\theta_Z = \arccos(\cos 40 \times \cos 21 \times \cos 100 + \sin 40 \times \sin 21) = 83.9°$$

2.5 Solar radiation

The amount of energy emitted from the Sun in the form of radiation in 1 s is called the solar radiation power or radiant flux, and the unit is watts (W). The radiation power (radiant flux) projected by the Sun onto a unit area is called irradiance and the unit is watts per square meter (W/m^2). The radiant power projected by the Sun onto a unit area over a period of time (e.g., hourly, daily, monthly, and yearly) is called irradiation, and the unit is kilowatt hours per square meter per day (month, year) ($kWh/m^2/d$, $kWh/m^2/m$, etc.).

Due to historical reasons, different unit systems are sometimes used and unit conversions are needed:

$$1\,kWh = 3.6\,MJ$$

$$1\,cal = 4.1868\,J = 1.16278\,mWh$$

$$1\,MJ/m^2 = 23.889\,cal/cm^2 = 27.8\,mWh/cm^2$$

$$1\,kWh/m^2 = 85.98\,cal/cm^2 = 3.6\,MJ/m^2 = 100\,mWh/cm^2$$

$$1\,cal/cm^2 = 0.0116\,kWh/m^2$$

$$1\,MJ/m^2 = 0.2778\,kWh/m^2$$

2.5.1 Solar radiation outside the atmosphere

2.5.1.1 Solar constant

Outside the Earth's atmosphere, the solar radiation on unit area perpendicular to the direction of the Sun at an average Sun–Earth distance is essentially a constant. This radiation is called the solar constant, or irradiance in the case of air mass (AM) is equal to 0 (AM0).

In October 1981, the value of the solar constant adopted by the eighth session of the World Meteorological Organization for Instruments and Methods of Observation held in Mexico was

$$I_{sc} = 1367 \pm 7 \, \text{W/m}^2$$

According to a recent report by Wikipedia, the solar irradiance above the Earth's atmosphere was measured by satellites, and the solar constant was obtained after adjustment using the inverse square law. The result is

$$I_{sc} = 1.3608 \pm 0.0005 \, \text{kW/m}^2$$

Actually, the distance between the Sun and the Earth is changing during a year, so the value of I_{sc} also changes slightly.

2.5.1.2 Solar radiation reaching extraterrestrial surface

The average daily solar radiation H_0 at extraterrestrial atmosphere can be calculated by the following formula:

$$H_0 = \frac{24 \times 3,600}{\pi} y \cdot I_{sc} \left(\frac{\pi \omega_s}{180} \sin \varphi \sin \delta + \cos \varphi \cos \delta \sin \omega_s \right) \tag{2.19}$$

where I_{sc} is the Sun constant; ω_s is the angle at sunrise or sunset; δ is the solar declination angle; y is the correction value of the solar radiation flux of the upper atmosphere boundary caused by the change of the Sun–Earth distance, and is obtained by the following equation:

$$y = \left(1 + 0.034 \cos \frac{2\pi n}{365} \right)$$

where n is the number of days in a year; the unit of H_0 is MJ/m^2.

Similarly, the amount of hourly solar radiation on the horizontal surface above the atmosphere can also be obtained:

$$I_0 = \frac{12 \times 3,600}{\pi} I_{sc} \left(1 + 0.033 \cos \frac{360n}{365} \right)$$

$$\times \left[\frac{\pi(\omega_2 - \omega_1)}{180} \sin \varphi \sin \delta + \cos \varphi \cos \delta (\sin \omega_2 - \sin \omega_1) \right] \tag{2.20}$$

where ω_1 and ω_2 are the starting and ending hour angles.

When considering the average extraterrestrial solar radiation in each month, for example, if it is represented on the 16th of each month, there is a certain deviation, especially in June and December. So Klein proposed in 1977 that the day that is closest to the average declination angle in each month be the "monthly average day," as mentioned in Tab. 2.4, which also lists n the number of days in a year and the declination angle δ of that day [5].

Tab. 2.4: Declination angle for the average days of each month.

Month	Date	Number of days	$\delta(°)$
Jan	17	17	−20.92
Feb	16	47	−12.95
Mar	16	75	−2.42
Apr	15	105	9.41
May	15	135	18.79
Jun	11	162	23.09
Jul	17	198	21.18
Aug	16	228	13.45
Sep	15	258	2.22
Oct	15	288	−9.60
Nov	14	318	−18.91
Dec	10	344	−23.05

In eq. (2.19), we use the average value in a month as input of n and δ to get the total extraterrestrial solar radiation $\overline{H_0}$ in each month. From this, we can get the average irradiance for each month in the upper boundary of the atmosphere at different latitudes, which is given in Tab. 2.5.

Tab. 2.5: Monthly average daily extraterrestrial radiation ($MJ/m^2/day$).

Latitude (°)	Jan	Feb	Mar	Apr	May	Jun	Jul	Aug	Sep	Oct	Nov	Dec
90	0.0	0.0	1.2	19.3	37.2	44.8	41.2	26.5	5.4	0.0	0.0	0.0
85	0.0	0.0	2.2	19.2	37.0	44.7	41.0	26.4	6.4	0.0	0.0	0.0
80	0.0	0.0	4.7	19.6	36.6	44.2	40.5	26.1	9.0	0.6	0.0	0.0
75	0.0	0.7	7.8	21.0	35.9	43.3	39.8	26.3	11.9	2.2	0.0	0.0
70	0.1	2.7	10.9	23.1	35.3	42.1	38.7	27.5	14.8	4.9	0.3	0.0
65	1.2	5.4	13.9	25.4	35.7	41.0	38.3	29.2	17.7	7.8	2.0	0.4
60	3.5	8.3	16.9	27.6	36.6	41.0	38.8	30.9	20.5	10.8	4.5	2.3
55	6.2	11.3	19.8	29.6	37.6	41.3	39.4	32.6	23.1	13.8	7.3	4.8

Tab. 2.5 (continued)

Latitude (°)	Jan	Feb	Mar	Apr	May	Jun	Jul	Aug	Sep	Oct	Nov	Dec
50	9.1	14.4	22.5	31.5	38.5	41.5	40.0	34.1	25.5	16.7	10.3	7.7
45	12.2	17.4	25.1	33.2	39.2	41.7	40.4	35.3	27.8	19.6	13.3	10.7
40	15.3	20.3	27.4	34.6	39.7	41.7	40.6	36.4	29.8	22.4	16.4	13.7
35	18.3	23.1	29.6	35.8	40.0	41.5	40.6	37.3	31.7	25.0	19.3	16.8
30	21.3	25.7	31.5	36.8	40.0	41.1	40.4	37.8	33.2	27.4	22.2	19.9
25	24.2	28.2	33.2	37.5	39.8	40.4	40.0	38.2	34.6	296	25.0	22.9
20	27.0	30.5	34.7	37.9	39.3	39.5	39.3	38.2	35.6	31.6	27.7	25.8
15	29.6	32.6	35.9	38.0	38.5	38.4	38.3	38.0	36.4	33.4	30.1	28.5
10	32.0	34.4	36.8	37.9	37.5	37.0	37.1	37.5	37.0	35.0	32.4	31.1
5	34.2	36.0	37.5	37.4	36.3	35.3	35.6	36.7	37.2	36.3	34.5	33.5
0	36.2	37.4	37.8	36.7	34.8	33.5	34.0	35.7	37.2	37.3	36.3	35.7
−5	38.0	38.5	37.9	35.8	33.0	31.4	32.1	34.4	36.9	38.0	37.9	37.6
−10	39.5	39.3	37.7	34.5	31.1	29.2	29.9	32.9	36.3	38.5	39.3	39.4
−15	40.8	39.8	37.2	33.0	28.9	26.8	27.6	31.1	35.4	38.7	40.4	40.9
−20	41.8	40.0	36.4	31.3	26.6	24.2	25.2	29.1	34.3	38.6	41.2	42.1
−25	42.5	40.0	35.4	29.3	24.1	21.5	22.6	27.0	32.9	38.2	41.7	43.1
−30	43.0	39.7	34.0	27.2	21.4	18.7	19.9	24.6	31.2	37.6	42.0	43.8
−35	43.2	39.1	32.5	24.8	18.6	15.8	17.0	22.1	29.3	36.6	42.0	44.2
−40	43.1	38.2	30.6	22.3	15.8	12.9	14.2	19.4	27.2	35.5	41.7	44.5
−45	42.8	37.1	28.6	19.6	12.9	10.0	11.3	16.6	24.9	34.0	41.2	44.5
−50	42.3	35.7	26.3	16.8	10.0	7.2	8.4	13.8	22.4	32.4	40.5	44.3
−55	41.7	34.1	23.9	13.9	7.2	4.5	5.7	10.9	19.8	30.5	39.6	44.0
−60	41.0	32.4	21.2	10.9	10.0	4.5	2.2	8.0	17.0	28.4	38.7	43.7
−65	40.5	30.6	18.5	7.8	2.1	0.3	1.0	5.2	14.1	26.2	37.8	43.7
−70	40.8	28.8	15.6	5.0	0.4	0.0	0.0	2.6	11.1	24.0	37.4	44.9
−75	41.9	27.6	12.6	2.4	0.0	0.0	0.0	0.8	8.0	21.9	38.1	46.2
−80	42.7	27.4	9.7	0.6	0.0	0.0	0.0	0.0	5.0	20.6	38.8	47.1
−85	43.2	27.7	7.2	0.0	0.0	0.0	0.0	0.0	2.4	20.3	39.3	47.6
−90	43.3	27.8	6.2	0.0	0.0	0.0	0.0	0.0	1.4	20.4	39.4	47.8

Example 2.8 Calculate the total amount of solar radiation H_0 at the level of extraterrestrial on April 15 in Changchun, China.
Solution From Tab. 2.4, we can obtain the information that $n = 105$, $\delta = 9.41°$ on April 15, and the latitude of Changchun is 43.45°, so $\phi = 43.45°$.
   From eq. (2.7), we can get the hour angle at sunrise or sunset as follows:

$$\cos\omega_s = -\tan 43.45 \times \tan 9.41 = -0.9473 \times 0.1657 = -0.1570$$

Therefore, $\omega_s = 99°$.
   Inserting into eqs. (2.19) and (2.20), we get

$$H_0 = \frac{24 \times 3,600 \times 1367}{\pi}\left(1 + 0.033\cos\frac{360 \times 105}{365}\right)\left(\frac{\pi 99}{180}\sin 43 \sin 9.4 + \cos 43 \cos 9.4 \sin 99\right)$$

So, $H_0 = 33.78$ MJ/m^2.

Example 2.9 Calculate the amount of solar radiation I_0 at extraterrestrial between 10 am and 11 am on April 15 in Changchun.
Solution From last example, we can know that $n = 105$ and $\delta = 9.41°$ on April 15. The latitude of Changchun is 43.45°. The hour angle $\omega_1 = -30°$ at 10 am, and the hour angle $\omega_2 = -15°$ at 11 am, and inserting them to eq. (2.20):

$$I_0 = \frac{12 \times 3,600 \times 1367}{\pi}\left(1 + 0.033\cos\frac{360 \times 105}{365}\right)$$
$$\times \left(\frac{\pi[-15 - (-30)]}{180}\sin 43.45 \times \sin 9.41\right) + \cos 43.45 \times \cos 9.41[\sin(-15) - \sin(-30)]$$
$$= 3.79 \ MJ/m^2$$

2.5.1.3 Air mass

When the Sun coincides with the zenith axis, the Sun's rays pass through the thickness of Earth's atmosphere with the shortest distance. The ratio of the Sun's rays actual distance through the atmosphere to this shortest distance is called AM. It is assumed that the path of the vertical incidence of the solar rays to sea level in the condition of the standard atmospheric pressure and 0 °C is AM = 1. Therefore, the AM at the upper boundary of the atmosphere is AM = 0. When the Sun is at other locations, the AM is greater than 1. If the ratio is 1.5, it is written as AM1.5. The schematic diagram of AM is shown in Fig. 2.10.

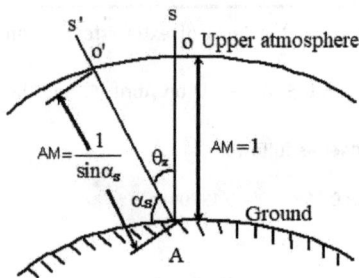

Fig. 2.10: Schematic diagram of air mass.

The equation for the air mas on the ground is

$$AM = \frac{1}{\cos \theta_z} \frac{p}{p_0} \tag{2.21}$$

where θ_z is the zenith angle, P is the local atmospheric pressure, and P_0 is the atmospheric pressure at sea level.

Equation (2.21) is derived from the trigonometric function relationship, ignoring the effects of refraction and curvature of the surface. When $\alpha_s < 30°$, there exists a large error. In the photovoltaic system engineering calculation, the following formula can be used to calculate:

$$AM(\alpha_s) = [1229 + (614\sin\alpha_s)^2]^{1/2} - 614\sin\alpha_s \tag{2.22}$$

Solar radiation passes through the Earth's atmosphere. Due to the absorption and scattering of the solar spectrum by the atmosphere, the range of solar spectral and energy distribution is going to change. When the solar altitude angle is 90°, the ultraviolet spectrum of the Sun reaching the ground accounts for about 4%, visible light accounts for 46%, and infrared ray accounts for 50%. When the solar altitude angle is as low as 30°, the corresponding ratio is 3%, 44%, and 53%. When the solar altitude angle is lower, the ultraviolet energy is almost equal to zero, the energy of visible light is reduced to 30%, and the energy of infrared rays is dominant. This is caused by the strong scattering of shortwave portions by air molecules.

The larger the AM, the longer the path that the light travels through the atmosphere and the more attenuation it experiences, the less energy that can reach the ground.

2.5.2 Solar irradiance at the ground surface

2.5.2.1 Atmospheric transparency

Atmospheric transparency is a parameter that characterizes the degree of light transmission through the atmospheric [6]. In clear, cloudless days, the transparency of the atmosphere is high, and more solar radiation reaches the ground. When there are a lot of clouds or dust in the sky, the transparency of the atmosphere is low, and less solar radiation reaches the ground. According to the Bouguer–Lambert law, the solar irradiance $I_{\lambda,0}$ at the wavelength of λ, the irradiance will be as below, after passing through the atmosphere with a thickness of dm, and the irradiance decays to:

$$dI_{\lambda,n} = -C_\lambda I_{\lambda,0} dm$$

Integrating the above equation, we obtain

$$I_{\lambda,n} = I_{\lambda,0} e^{-c_\lambda m} \tag{2.23}$$

where $I_{\lambda,n}$ is the normal solar irradiance reaching the ground surface, $I_{\lambda,0}$ is the solar irradiance at the upper boundary of the atmosphere, C_λ is the extinction coefficient of the atmosphere, and m is the AM.

The equation can also be transferred to:

$$I_{\lambda,n} = I_{\lambda,0} P_\lambda^m \tag{2.24}$$

In this equation, $P_\lambda = e^{-c_\lambda}$ is called the monochromatic spectral transparency.

The full-spectrum solar irradiance can be obtained by integrating eq. (2.24) from the wavelength band 0 to ∞:

$$I_n = \int_0^\infty I_{\lambda,0} P_\lambda^m d\lambda \tag{2.25}$$

Let the average value of the monochromatic transparency in the entire solar radiation spectrum range be P_m, and eq. (2.25) after integration is

$$I_n = y \cdot I_{sc} P_m^m \tag{2.26a}$$

Or

$$P_m = \sqrt[m]{\frac{I_n}{y \cdot I_{sc}}} \tag{2.26b}$$

where y is the daily Sun–Earth distance correction value; P_m is a composite transparent factor, which represents the degree of attenuation of solar radiation energy by the atmosphere.

2.5.2.2 Normal solar irradiance at the ground surface

In order to compare atmospheric transparency under different conditions of AM, atmospheric transparency must be corrected based on a given AM. For example, the atmospheric transparency P_m with AM m is corrected to atmospheric transparency P_2 with AM equal to 2,

$$I_n = \gamma . I_{sc} P_2^m \qquad (2.27)$$

where y is the Sun–Earth distance correction value, I_{sc} is the solar constant, and P_2^m is the P_m value when it is corrected to $m = 2$.

2.5.2.3 Direct solar irradiance on horizontal surface

The relationship between the Sun's direct irradiance and the solar altitude is shown in Fig. 2.11.

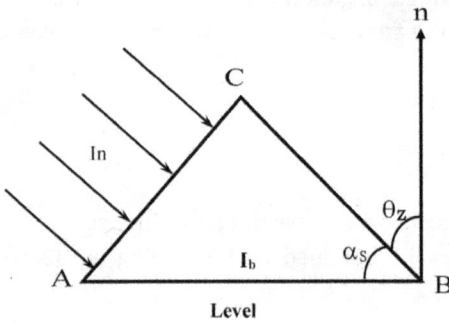

Fig. 2.11: Relationship between solar beam irradiance and altitude angle.

Since the direct sunlight is incident on the AC and AB surfaces with equal energy, there are

$$I_b = I_n \sin \alpha_s = I_n \cos \theta_z \qquad (2.28)$$

where I_b is the direct irradiance on the horizontal surface, α_s is the solar altitude angle, and θ_z is the zenith angle.

Taking eq. (2.27) into eq. (2.28), we obtain

$$I_b = \gamma . I_{sc} \, p_m^m \sin \alpha_s$$

Integrating the above equation from the time at sunrise to sunset, we obtain

$$H_b = \int_0^t \gamma . I_{sc} \, p_m^m \sin \alpha_s dt = \gamma . I_{sc} \int_0^t p_m^m \sin \alpha_s \, dt \qquad (2.29)$$

where H_b is the total amount of direct irradiation on the horizontal surface. When dt in the formula is changed to the hour angle ω, there is

$$H_b = \frac{T}{2\pi} \gamma . I_{sc} \int_{-\omega}^{+\omega} p_m^m (\sin \varphi \sin \delta + \cos \varphi \cos \delta \cos \omega) d\omega \qquad (2.30)$$

where T is the length of a whole day and night (1,440 min a day) and ω is the hour angle of sunrise and sunset.

2.5.2.4 Diffuse solar irradiance on the horizontal surface

On sunny days, the diffuse solar irradiance that reaches the horizontal surface of the ground surface depends mainly on the solar altitude angle and atmospheric transparency:

$$I_d = C_1 (\sin \alpha_s)^{C_2} \qquad (2.31)$$

where I_d is the diffuse radiation on a horizontal surface, α_s is the solar altitude angle, and C_1 and C_2 are the empirical coefficients.

2.5.2.5 Total solar irradiance on the horizontal surface

The total solar irradiance is the sum of the direct irradiance and the diffuse irradiance reaching the surface of the Earth, which is

$$I = I_b + I_d \qquad (2.32)$$

where I is the total solar irradiance on the horizontal surface, I_b is the direct irradiance on the horizontal surface, and I_d is the diffuse irradiance on the horizontal surface.

2.5.2.6 Clarity index

Sometimes the clarity index K_T can also be used as a measure of the attenuation of the irradiance as it passes through the atmosphere. It is defined as the ratio of the total solar irradiance reaching the Earth's horizontal surface to the extraterrestrial solar irradiance. The values are not the same at different time periods. The ratio of the monthly average solar radiation on the horizontal surface $\overline{H}$ to the monthly average extraterrestrial solar radiation $\overline{H}_0$ is the monthly average clarity index $\overline{K}_T$, which is expressed as

$$\overline{K}_T = \frac{\overline{H}}{\overline{H}_0} \qquad (2.33a)$$

Similarly, the ratio of the daily average solar radiation H on the horizontal surface to the daily average extraterrestrial solar radiation H_0 is the daily average clarity index K_T. The expression is

$$K_T = \frac{H}{H_0} \tag{2.33b}$$

At an hour, the ratio of the solar radiation I on the horizontal surface to the extraterrestrial solar radiation I_0 can be considered as the hourly clarity index k_T as follows:

$$k_T = \frac{I}{I_0} \tag{2.33c}$$

The greater the clarity index K_T, the more transparent the atmosphere, the less solar energy attenuates, and the greater the intensity of solar radiation reaching the ground.

2.5.2.7 The ratio of diffuse irradiation to total irradiation

The total amount of solar radiation received on the horizontal surface of the ground is composed of two parts: the beam solar radiation and the diffuse radiation. Even if the total solar radiation of the two places are the same, the proportion of the beam irradiation dose and the diffuse irradiation is usually not the same.

The factors affecting the proportion of solar beam irradiation and diffuse irradiation are very complicated. If there is no actual measured data, it can be estimated by approximate calculation formula. The following describes the approximate calculation methods for different time periods.

2.5.2.7.1 The ratio of hourly diffuse irradiation to total irradiation

In 1982, Erbs et al. proposed an approximate equation for calculating the ratio of hourly diffuse irradiation to total irradiation,

$$\frac{I_d}{I} = 1.0 - 0.09k_T, \text{ if } k_T \le 0.22$$

$$\frac{I_d}{I} = 0.9511 - 0.1604k_T + 4.388k_T^2 - 16.638k_T^3$$
$$+ 12.336k_T^4, \text{ if } 0.22 < k_T \le 0.80 \tag{2.34}$$

$$\frac{I_d}{I} = 0.165, \text{ if } k_T > 0.80$$

where k_T is the hourly clarity index.

2.5.2.7.2 The ratio of daily diffuse irradiation to total irradiation

Erbs et al. proposed the ratio of daily diffuse irradiation to total irradiance based on the ratio of hourly diffuse irradiance to total irradiance [7]. According to the two conditions when the sunset angle is greater than or less than 81.4°, the relationship is as follows:

For $w_s \leq 81.4°$

$$\frac{H_d}{H} = \begin{cases} 1.0 - 0.2727K_T + 2.4495K_T^2 - 11.9514K_T^3 + 9.3879K_T^4 & \text{if } K_T < 0.715 \\ 0.143 & \text{if } K_T \geq 0.715 \end{cases}$$

For $w_s > 81.4°$

$$\frac{H_d}{H} = \begin{cases} 1.0 + 0.2832K_T - 2.5557K_T^2 + 0.8448K_T^3 & \text{if } K_T < 0.722 \\ 0.175 & \text{if } K_T \geq 0.722 \end{cases} \tag{2.35}$$

2.5.2.7.3 Ratio of monthly diffuse irradiation to total irradiation

In the design of solar energy application systems, it is often necessary to know the local average monthly total solar irradiation and diffuse irradiation (or beam irradiation), but sometimes it may only have the monthly total solar irradiation data. Find out what percentage of beam and diffuse irradiation each month accounts for, that is, to make "beam-diffuse separation." The empirical formula proposed by Erbs et al. [7] can usually be used:

For $w_s \leq 81.4°$ and $0.3 \leq \bar{K}_T \leq 0.8$

$$\frac{\bar{H}_d}{\bar{H}} = 1.391 - 3.56\bar{K}_T + 4.189\bar{K}_T^2 - 2.137\bar{K}_T^3 \tag{2.36a}$$

For $w_s > 81.4°$ and $0.3 \leq \bar{K}_T \leq 0.8$

$$\frac{\bar{H}_d}{\bar{H}} = 1.311 - 3.022\bar{K}_T + 3.427\bar{K}_T^2 - 1.821\bar{K}_T$$

For the monthly average diffuse irradiance of all sky (including the effects of clouds in the sky), NASA published the Surface Meteorology and Solar Energy (SSE) Release 6.0 Methodology Version 3.2.0 on June 2, 2016 [8]. It is recommended to use the following approximate method:

For latitude between 45°S and 45°N:

$$\frac{\bar{\bar{H}}_d}{\bar{\bar{H}}} = 0.96268 - 1.45200\bar{K}_T + 0.27365\bar{K}_T^2 + 0.04279\bar{K}_T^3 + 0.000246w_s + 0.001189\alpha_s$$

For latitude, between 90°S and 45°S and between 45°N and 90°N:

If $0° \leq w_s \leq 81.4°$:

$$\frac{\bar{H}_d}{\bar{H}} = 1.441 - 3.6839K_T + 6.4927K_T^2 - 4.147K_T^3 + 0.0008w_s - 0.008175\alpha_s$$

If $81.4° < \omega_s \le 100°$:

$$\frac{\overline{H}_d}{\overline{H}} = 1.6821 - 2.5866K_T + 2.373K_T^2 - 0.529K_T^3 + 0.00277\omega_s - 0.004233\alpha_s$$

If $100° < \omega_s \le 125°$:

$$\frac{\overline{H}_d}{\overline{H}} = 0.3498 + 3.8035\overline{K}_T - 11.765\overline{K}_T^2 + 9.1748\overline{K}_T^3 + 0.001575\omega_s - 0.002837\alpha_s$$

If $125° < \omega_s \le 150°$:

$$\frac{\overline{H}_d}{\overline{H}} = 1.6586 - 4.412\overline{K}_T + 5.8\overline{K}_T^2 + 3.1223\overline{K}_T^3 + 0.000144\omega_s - 0.000829\alpha_s$$

If $150° < \omega_s \le 180°$:

$$\frac{\overline{H}_d}{\overline{H}} = 0.6563 - 2.893\overline{K}_T + 4.594\overline{K}_T^2 - 3.23\overline{K}_T^3 + 0.004\omega_s - 0.0023\alpha_s$$

where ω_s is the hour angle at sunset on the "monthly average day," α_s is the noon solar altitude angle, $\overline{K}_T$ is the monthly mean clarity index (see eq. (2.33a)).

2.5.3 Hourly solar irradiation on tilt surfaces

2.5.3.1 Hourly solar beam irradiation on tilt surfaces $I_{T,b}$

The meteorological observatory measures the solar irradiation on the horizontal surface. In practical applications, whether it is photovoltaic or solar heat utilization, the lighting surface is usually tilted; therefore, the amount of solar irradiation on the tilt plane needs to be calculated. The solar irradiation on the tilted surface consists of three parts: solar beam irradiation, diffuse irradiation, and ground reflection irradiation.

From Fig. 2.12, it can be seen that the total hourly solar irradiation on the tilt plane of the ground has the following relationship with the beam irradiation dose:

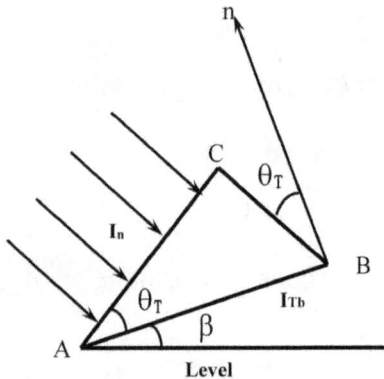

Fig. 2.12: Solar beam irradiation on tilted surfaces.

$$I_{T,b}/I_n = \cos\theta_T$$

So we have:

$$I_{T,b} = I_n\cos\theta_T \tag{2.37}$$

where θ_T is the incident angle of the Sun's rays on the tilt plane, so substituting eq. (2.10) into eq. (2.37), we can get beam radiation on the tilt plane:

$$\begin{aligned} I_{Tb} = I_n\,(&\sin\delta\,\sin\varphi\,\cos\beta - \sin\delta\cos\varphi\,\sin\beta\,\cos\gamma \\ &+ \cos\delta\cos\varphi\,\cos\beta\,\cos\omega + \cos\delta\,\sin\varphi\,\sin\beta\,\cos\gamma\cos\omega \\ &+ \cos\delta\,\sin\beta\,\sin\gamma\sin\omega) \end{aligned} \tag{2.38}$$

where β is the angle between the tilt plane and the horizontal surface, φ is the local latitude, δ is the solar declination angle, ω is the hour angle, and y is the azimuth angle of the tilt plane.

2.5.3.2 Ratio of hourly beam radiation on tilted to horizontal surface R_b

The ratio of the hourly beam irradiation of the tilt plane to that of the horizontal surface can be obtained from eqs. (2.37) and (2.28):

$$R_b = \frac{I_{T,b}}{I_b} = \frac{I_n\cos\theta_T}{I_n\cos\theta_z} = \frac{\cos\theta_T}{\cos\theta_z} \tag{2.39}$$

For a titled surface facing the equator, $y = 0$, and substituting eqs. (2.12a), (2.12b), and (2.4a) into eq. (2.39), we obtain as follows:

For the Northern Hemisphere:

$$R_b = \frac{\cos(\varphi - \beta)\cos\delta\cos\omega + \sin(\varphi - \beta)\sin\delta}{\sin\varphi\,\sin\delta + \cos\varphi\cos\delta\cos\omega} \tag{2.40a}$$

For the Southern Hemisphere:

$$R_b = \frac{\cos(\varphi + \beta)\cos\delta\cos\omega + \sin(\varphi + \beta)\sin\delta}{\sin\varphi\,\sin\delta + \cos\varphi\cos\delta\cos\omega} \tag{2.40b}$$

If at 12:00 noon, $\omega = 0$, substituting to eqs. (2.40a) and (2.40b), we obtain separately:

For the Northern Hemisphere:

$$R_{bn} = \frac{\cos|\varphi - \delta - \beta|}{\cos|\varphi - \delta|} \tag{2.41a}$$

For the Southern Hemisphere:

$$R_{bn} = \frac{\cos|-\varphi + \delta - \beta|}{\cos|-\varphi + \delta|} \tag{2.41b}$$

Example 2.10 Calculate the ratio of the beam irradiation on a tilt plane with an angle of 30° facing south to that on the horizontal surface in Beijing on February 13 at 10:30 am.

Solution From Example 2.6, we know that $n = 44$, $\delta = -14°$. At 10:30 am, the hour angle $\omega = -22.5°$ and $\beta = 30°$. The latitude of Beijing is $\phi = 39.48°$, and substituting to eq. (2.40a), we obtain that

$$R_b = \frac{\cos(39.48 - 30)\cos(-14)\cos(-22.5) + \sin(39.48 - 30)\sin(-14)}{\sin 39.48 \sin(-14) + \cos 39.48 \cos(-14)\cos(-22.5)}$$

which is equal to 1.57.

2.5.3.3 Hourly diffuse irradiation on tilt planes

The hourly diffuse irradiation on the titled surface can be obtained from the following equation:

$$I_{T,d} = \frac{1 + \cos \beta}{2} I_d \tag{2.42}$$

In the above equation, $I_{T,d}$ is the hourly diffuse irradiation on the titled surface, I_d is the hourly diffuse irradiance on the horizontal surface, and β is the angle (tilt) between the tilted surface and the horizontal surface.

2.5.3.4 Ground reflected irradiation

Assuming that the ground reflection is isotropic, the following equation can be obtained using the law of exchange of the angle coefficients:

$$I_{T,\theta} = \rho \frac{1 - \cos \beta}{2}(I_d + I_b) = I\rho \frac{(1 - \cos \beta)}{2} \tag{2.43}$$

where ρ is the ground reflectance, which is related to the coverage of the ground surface. The reflectivity of different ground conditions is given in Tab. 2.6.

Tab. 2.6: The reflectivity of different ground surface.

Surface condition	Reflectivity ρ	Surface condition	Reflectivity ρ	Surface condition	Reflectivity ρ
Desert	0.24–0.28	Dry grass	0.15–0.25	New snow	0.81
Dry ground	0.10–0.20	Wet grass	0.14–0.26	Left snow	0.46–0.7
Wet and bare ground	0.08–0.09	Forest	0.04–0.10	Water surface	0.69

In general, $\rho = 0.2$ is preferable.

2.5.3.5 Total hourly solar irradiation on tilt planes – sky isotropic model

The idea that the solar diffuse irradiation in the sky is isotropic is first proposed by Liu and Jordan in 1963 [9]. The total hourly solar irradiation on a tilt plane consists of three parts: direct solar irradiation, diffuse irradiation, and ground reflection irradiation

$$I_T = I_b R_b + I_d \left(\frac{1 + \cos \beta}{2}\right) + I\rho \left(\frac{1 - \cos \beta}{2}\right) \tag{2.44a}$$

The above equation can also be written as

$$R = \frac{I_b}{I} R_b + \frac{I_d}{I} \left(\frac{1 + \cos \beta}{2}\right) + \rho \left(\frac{1 - \cos \beta}{2}\right) \tag{2.44b}$$

where R is the ratio of the hourly total solar irradiation I_T on the tilted surface to the hourly total irradiation I on the horizontal surface.

2.5.3.6 Total hourly solar irradiation on tilt planes – sky anisotropy model

2.5.3.6.1 HDKR model

Although the isotropic model of solar irradiation is easy to understand with convenient calculation, it is not very accurate. The diffuse irradiation that surrounds the Sun is not the same in all directions. In the Northern Hemisphere, because the Sun is basically operating in the southern sky, the average diffuse irradiation in the southern sky is obviously larger than in the northern sky. The study pointed out that the average amount of diffuse irradiation in the southern sky in June was 63%. Later, Hay, Davies, Klucher, Rcindl (HDKR) et al. proposed improved sky diffuse anisotropy models, respectively [10]. Finally integrating into the HDKR model, total solar irradiation on a tilt plane can be calculated using the following equation:

$$I_T = (I_b + I_d A_i) R_b + I_d (1 - A_i) \left(\frac{1 + \cos \beta}{2}\right) \left[1 + f \sin^3 \left(\frac{\beta}{2}\right)\right] + I\rho \left(\frac{1 - \cos \beta}{2}\right) \tag{2.45}$$

where

$$A_i = \frac{I_{bn}}{I_{0n}} = \frac{I_b}{I_0}$$

$$f = \sqrt{\frac{I_b}{I}}$$

I_b is hourly direct solar irradiation on the horizontal surface, I_d is the hourly solar diffuse irradiation on the horizontal surface, I_0 is the hourly solar total irradiation outside the atmosphere, R_b is the ratio of hourly direct solar irradiation on the tilt plane to that on the horizontal surface, β is the angle between the tilt plane and the horizontal surface (tilted angle), ρ is the ground reflectivity, and I is the hour total solar irradiation on the horizontal surface.

2.5.3.6.2 Perez model

Perez et al. analyzed the diffuse irradiation components on the tilt plane of the ground in detail, and proposed that the hourly solar diffuse irradiation on the tilt plane can be calculated by the following equation [11]:

$$I_{d,T} = I_d \left[(1 - F_1) \left(\frac{1 + \cos \beta}{2} \right) + F_1 \frac{a}{b} + F_2 \sin \beta \right] \tag{2.46}$$

where F_1 and F_2 are circumsolar and horizon brightness coefficients. They are functions of three parameters that describe the sky conditions, including the zenith angle θ_z, the clarity ξ, and the brightness $\triangle$, and are determined by the following equations:

$$F_1 = \max\left\{ 0, \left(f_{11} + f_{12} \triangle + \frac{\pi \theta_z}{180} f_{13} \right) \right\}$$

$$F_2 = \left(f_{21} + f_{22} \triangle + \frac{\pi \theta_z}{180} f_{23} \right)$$

In the above equation, the brightness $\triangle$ is equal to $\triangle = m I_d / I_{on}$, m is the AM of the atmosphere; I_{on} is the solar irradiation on the vertical surface of sunlight outside the atmosphere.

The clarity ξ is a function of the hourly diffuse irradiation I_d and the direct irradiation I_n on the vertical surface of sunlight, and the relationship is

$$\xi = \frac{\frac{I_d + I_n}{I_d} + 5.535 \times 10^{-6} \theta_z^3}{1 + 5.535 \times 10^{-6} \theta_z^3}$$

The brightness coefficients $f_{11}, f_{12}, ..., f_{23}$ are given in Tab. 2.7.

In eq. (2.46), a and b take into account the effect of incident angle around the Sun on tilted and horizontal surfaces. The irradiation around the Sun is emitted as if the Sun was a point light source:

$$a = \max[0, \cos \theta_z]$$

$$b = \max[\cos 85°, \cos \theta_z]$$

In this way, the total radiation on the tilted surface includes five terms: the beam, the isotropic diffuse, the circumsolar diffuse, the diffuse from the horizon, and the ground-reflected term. The relationship is as follows:

Tab. 2.7: Brightness coefficients in the Perez model.

Range of ξ	f_{11}	f_{12}	f_{13}	f_{21}	f_{22}	f_{23}
0–1.065	−0.196	1.084	−0.006	−0.114	0.180	−0.019
1.065–1.230	0.236	0.519	−0.180	−0.011	0.020	−0.038
1.230–1.500	0.454	0.321	−0.255	0.072	−0.098	−0.046
1.500–1.950	0.866	−0.381	−0.375	0.203	−0.403	−0.049
1.950–2.800	1.026	−0.711	−0.426	0.273	−0.602	−0.061
2.800–4.500	0.978	−0.986	−0.350	0.280	−0.915	−0.024
4.500–6.200	0.748	−0.913	−0.236	0.173	−1.045	0.065
6.200–	0.318	−0.757	0.103	0.062	−1.698	0.236

$$I_T = I_b R_b + I_d (1-F_1) \left(\frac{1+\cos\beta}{2} \right) + I_d F_1 \frac{a}{b} + I_d F_2 \sin\beta + I\rho \left(\frac{1-\cos\beta}{2} \right) \qquad (2.47)$$

2.5.4 Monthly mean solar irradiation on tilt plane

In the design of solar energy application system, energy balance calculation is needed. Due to the random nature of solar irradiation, it is neither meaningful nor too cumbersome to calculate the energy balance according to days and more impossible to calculate by hours. However, calculating annually is too rough. The most reasonable calculation should be based on monthly energy balance calculations. Meteorological observatories generally provide solar irradiation on the horizontal surface. Therefore, how to obtain the monthly mean solar irradiation on tilt plane from the solar irradiation on the horizontal surface is the basis for the design of solar energy application systems.

2.5.4.1 Sky isotropic model
For a long time, the commonly used method of sky isotropic model was first proposed by Liu and Jordan in 1963, and then improved by Klein in 1977. It is believed that solar diffuse and ground reflection are isotropic, and the equation for calculating the monthly average solar irradiation on a tilt plane is as follows:

$$\bar{H}_T = \bar{H} \left(1 - \frac{\overline{H_d}}{\overline{H}} \right) \bar{R}_b + \bar{H}_d \left(\frac{1+\cos\beta}{2} \right) + \bar{H} \left(\frac{1-\cos\beta}{2} \right) \rho \qquad (2.48a)$$

or:

$$\bar{R} = \frac{\bar{H}_T}{\bar{H}} = \left(1 - \frac{\bar{H}_d}{\bar{H}}\right)\bar{R}_b + \frac{\bar{H}_d}{\bar{H}}\left(\frac{1+\cos\beta}{2}\right) + \rho\left(\frac{1-\cos\beta}{2}\right) \tag{2.48b}$$

where $\bar{H}_T$ is the monthly mean solar irradiation on the tilt plane, $\bar{H}$ is the monthly mean solar irradiation on the horizontal surface, $\bar{H}_d$ is the monthly mean solar diffuse irradiance on the horizontal surface, and $\bar{R}_b$ is the ratio of the direct solar irradiation on the tilt plane and that on the horizontal surface.

For the tilt plane of the Northern Hemisphere toward the equator ($y_s = 0°$), it can be simplified as

$$\bar{R}_b = \frac{\cos(\varphi - \beta)\cos\delta\sin\omega_s' + (\pi/180)\omega_s'\sin(\varphi-\beta)\sin\delta}{\cos\varphi\cos\delta\sin\omega_s + (\pi/180\omega_s)\sin\varphi\sin\delta} \tag{2.49a}$$

where ω_s' is the time angle at sunset of the "monthly average day," determined by the following equation:

$$\omega_s' = \min\begin{bmatrix} \arccos(-\tan\varphi\tan\delta) \\ \arccos(-\tan(\varphi-\beta)\tan\delta) \end{bmatrix}$$

For a tilt plane facing the equator ($y_s = 180°$) in the Southern Hemisphere, it can also be simplified as

$$\bar{R}_b = \frac{\cos(\varphi + \beta)\cos\delta\sin\omega_s' + (\pi/180)\omega_s'\sin(\varphi+\beta)\sin\delta}{\cos\varphi\cos\delta\sin\omega_s + (\pi/180\omega_s)\sin\varphi\sin\delta} \tag{2.49b}$$

In the above equation,

$$\omega_s' = \min\begin{bmatrix} \arccos(-\tan\varphi\tan\delta) \\ \arccos(-\tan(\varphi+\beta)\tan\delta) \end{bmatrix}$$

2.5.4.2 Sky anisotropy model
Similarly, although the calculation of the isotropic model of solar irradiation is more convenient, it is not very accurate. In particular, the monthly average value of solar irradiation is much different from the actual situation.

2.5.4.2.1 Klein and Theilacker's method
In 1981, Klein and Theilacker proposed a calculation method based on the sky anisotropy model, starting with the special case of the Northern Hemisphere tilted to the equator (azimuth $y = 0°$) [12]. The ratio $\bar{R}$ of the monthly mean total solar irradiation

on the tilt plane to the monthly average total solar irradiation on the horizontal surface can be calculated by the following equation:

$$\bar{R} = \frac{\sum_1^N \int_{t_{sr}}^{t_{ss}} G_T dt}{\sum_1^N \int_{t_{sr}}^{t_{ss}} G dt} \tag{2.50}$$

where G_T is the solar irradiance on the tilt plane, G is the solar irradiance on the horizontal surface, t_{ss} is the sunset time on the tilt plane, and t_{sr} is the sunrise time on the tilt plane.

Applying eq. (2.44a), we obtain

$$N\bar{I}_T = N\left[(\bar{I} - \bar{I}_d)R_b + \bar{I}_d\left(\frac{1+\cos\beta}{2}\right) + \bar{I}\rho\left(\frac{1-\cos\beta}{2}\right)\right]$$

where $\bar{I}$ and $\bar{I}_d$ are the long-term average of the total and diffuse irradiation on the horizontal surface, which can be obtained by summing the hourly total irradiation I and the hourly diffuse irradiation I_d for each hour in N days and then divided by N. Substituting to eq. (2.50), we obtain

$$\bar{R} = \frac{\int_{t_{sr}}^{t_{ss}} \left[(\bar{I} - \bar{I}_d)R_b + \bar{I}_d\left(\frac{1+\cos\beta}{2}\right) + \bar{I}\rho\left(\frac{1-\cos\beta}{2}\right)\right] dt}{\bar{H}} \tag{2.51}$$

Collares-Pereira and Rabl proposed in 1979 that the ratio of the hourly total solar irradiation to the daily solar irradiation on the horizontal surface can be approximated as follows:

$$\frac{I}{H} = \frac{\pi}{24}(a + b\cos\omega)\frac{\cos\omega - \cos\omega_s}{\sin\omega_s - \frac{\pi\omega_s}{180}\cos\omega_s} \tag{2.52}$$

where $a = 0.4090 + 0.5016\sin(\omega_s - 60)$, $b = 0.6609 - 0.4767\sin(\omega_s - 60)$, ω_s is the time angle at sunset, and ω is the time angle.

According to the method proposed by Liu and Jordan (1960), the ratio of hourly solar diffuse irradiation to daily solar diffuse irradiation is

$$\frac{I_d}{H_d} = \frac{\pi}{24}\frac{\cos\omega - \cos\omega_s}{\sin\omega_s - \frac{\pi\omega_s}{180}\cos\omega_s} \tag{2.53}$$

Substituting eqs. (2.52) and (2.53) into eq. (2.51), we can get the ratio of the monthly mean solar irradiation on the tiled surface located in the Northern Hemisphere toward the equator (bearing angle $y = 0°$) to the monthly mean total irradiation on the horizontal surface:

$$\bar{R} = \frac{\cos(\varphi - \beta)}{d \cos \varphi} \left\{ \left(a - \frac{\bar{H}_d}{\bar{H}} \right) \left(\sin \omega_s' - \frac{\pi}{180} \omega_s' \cos \omega''_s \right) \right.$$

$$\left. + \frac{b}{2} \left[\frac{\pi}{180} \omega_s' + \sin \omega_s' (\cos \omega_s' - 2 \cos \omega''_s) \right] \right\} \tag{2.54}$$

$$+ \frac{\bar{H}_d}{2\bar{H}} (1 + \cos \beta) + \frac{\rho}{2} (1 - \cos \beta)$$

In the above equation,

$$\omega_s' = \min \left[\begin{matrix} \arccos(- \tan \varphi \tan \delta) \\ \arccos(- \tan(\varphi - \beta) \tan \delta) \end{matrix} \right]$$

$$\omega''_s = \arccos[- \tan(\varphi - \beta) \tan \delta]$$

$$d = \sin \omega_s - \frac{\pi}{180} \omega_s \cos \omega_s$$

Finally, Klein and Theilacker generalized the above conclusions to general cases of arbitrary azimuths, considering some factors such as the sunrise and sunset times are still symmetric with respect to the solar noon on the tilt plane facing the equator (azimuth $y = 0°$). However, at any azimuth angle, the sunrise and sunset times are not symmetric with respect to the midday Sun. $\bar{R}$ can still be expressed as

$$\bar{R} = D + \frac{\bar{H}_d}{2\bar{H}} (1 + \cos \beta) + \frac{\rho}{2} (1 - \cos \beta) \tag{2.55}$$

where $\overline{H_d}$ is the monthly mean solar diffuse irradiation on the horizontal surface, $\overline{H}$ is monthly mean total solar irradiation on horizontal surface, β is the square tilt angle, and ρ is the ground reflectivity:

$$D = \begin{cases} \max \{0, G(\omega_{ss}, \omega_{sr})\} & (\omega_{ss} \geq \omega_{sr}) \\ \max \{0, [G(\omega_{ss}, -\omega_s) + G(\omega_s, \omega_{sr})]\} & (\omega_{sr} > \omega_{ss}) \end{cases} \tag{2.56}$$

where the function G can be obtained by the following method:

$$G(\omega_1, \omega_2) = \frac{1}{2d} \left[\left(\frac{bA}{2} - a'B \right) (\omega_1 - \omega_2) \frac{\pi}{180} \right.$$

$$+ (a'A - bB)(\sin \omega_1 - \sin \omega_2) - a'C(\cos \omega_1 - \cos \omega_2)$$

$$\left. + \frac{bA}{2} (\sin \omega_1 \cos \omega_1 - \sin \omega_2 \cos \omega_2) + \frac{bC}{2} (\sin^2 \omega_1 - \sin^2 \omega_2) \right]$$

In the above equation,

$$A = \cos\beta + \tan\varphi\cos\gamma\sin\beta$$

$$B = \cos\omega_s\cos\beta + \tan\delta\sin\beta\cos\gamma$$

$$C = \frac{\sin\beta\sin\gamma}{\cos\varphi}$$

$$a = 0.409 + 0.5016\sin(\omega_s - 60°)$$

$$b = 0.6609 - 0.4767\sin(\omega_s - 60°)$$

$$d = \sin\omega_s - \frac{\pi}{180}\,\omega_s\cos\omega_s$$

$$a' = a - \frac{\bar{H}_d}{\bar{H}}$$

where y is the azimuth of the tilt plane which is 0° toward the south, 180° toward the north, negative toward the east, and positive toward the west. δ is a solar declination angle, ω_s is the time angle at sunset on the horizontal surface, $\cos\omega_s = -\tan\varphi\tan\delta$, and ω_{sr} is the time angle at sunrise on the tilt plane, which can be expressed as

$$|\omega_{sr}| = \min\left[\omega_s,\ \arccos\frac{AB + C\sqrt{A^2 - B^2 + C^2}}{A^2 + C^2}\right]$$

$$\omega_{sr} = \begin{cases} -|\omega_{sr}| & \text{if}(A > 0 \text{ and } B > 0) \text{ or } (A \geq B) \\ +|\omega_{sr}| & \text{else} \end{cases}$$

ω_{ss} is the time angle at sunset on the tilt plane, which can be expressed as

$$|\omega_{ss}| = \min\left[\omega_s,\ \arccos\frac{AB - C\sqrt{A^2 - B^2 + C^2}}{A^2 + C^2}\right]$$

$$\omega_{ss} = \begin{cases} -|\omega_{ss}| & \text{if}(A > 0 \text{ and } B > 0) \text{ or } (A \geq B) \\ +|\omega_{ss}| & \text{else} \end{cases}$$

In this way, the monthly average solar irradiation can be calculated for arbitrary orientations and different tilt planes. However, when it is actually used, the calculation is very complicated and it is usually necessary to compile special calculation software to easily calculate the monthly average total solar irradiation of different orientations and various tilt planes.

2.5.4.2.2 RETScreen method

The RETScreen method is basically the same as that of Klein and Theilacker, except taking into account that the tilted angle of the array will change continuously throughout the day in the tracking system in order to be able to extend the application into the tracking system, and in some places the simplification is made [13]. There are three steps in total:

(1) Assuming that each day in a month has the same total solar irradiation as that of the "monthly average day," the specific dates called "monthly average day" in each month are shown in Tab. 2.4. Using the method of Collares-Pereira and Rabl (1979) (eq. (2.52)), the total solar irradiation I on the horizontal surface was calculated from the hourly solar irradiation between 30 min after sunrise and 30 min before sunset during a day.

Then, according to the method of Liu and Jordan (1960) (formula (2.53)), hourly solar diffuse irradiation I_d on the horizontal surface is calculated from that of the horizontal surface.

(2) Calculate all hourly total solar irradiation on the tilt plane (or tracking surface). Total solar irradiation on tilt plane = direct solar irradiation + diffuse irradiation + ground reflection

$$I_{Th} = (I_h - I_{dh}) \frac{\cos \theta_{Th}}{\cos \theta_{Zh}} + I_{dh} \frac{1 + \cos \beta_h}{2} + I_h \rho_s \frac{1 - \cos \beta_h}{2} \tag{2.57}$$

The subscript h in the equation above is to consider that some parameters will change in each hour during tracking, and β_h is the angle between squares to the horizontal surface in each hour. β_h is a constant in a fixed-square or vertical axis tracking system. For two-axis tracking systems, $\beta_h = \theta_Z$. ρ_s is the ground reflection coefficient. When the monthly average temperature is above 0 °C, take 0.2; when it is lower than –5 °C, take 0.7; and when the temperature is in the range from –5 to 0 °C, take a linear relationship to calculate it.

The cosine of the zenith angle can be obtained from eq. (2.4a):

$$\cos \theta_{Zh} = \sin \varphi \sin \delta + \cos \varphi \cos \delta \cos \omega$$

The cosine of the incident angle can be obtained from eq. (2.11):

$$\cos \theta_{Th} = \cos \theta_{Zh} \cos \beta_h + (1 - \cos \theta_{Zh})(1 - \cos \beta_h) \cos(\gamma_{sh} - \gamma_h)$$

In the above equation, γ_{sh} is the azimuth angle in each hour. The azimuth is zero when facing the equator, positive toward the west, and negative toward the east. γ_h is the azimuth angle of the tilt plane in each hour. When facing the equator, the azimuth is zero, positive toward the west, and negative toward the east. For the fixed tilt plane, γ_h is a constant; for vertical axis or two-axis tracking systems, $\gamma_h = \gamma_{sh}$.

(3) Adding all hourly solar irradiation on the "average day of the month" is total solar irradiation of the "average day of the month." Considering the number of days of that month, you can get the average monthly total solar irradiation H_T of that month.

There is no doubt that this method is less accurate than the on-site measurement of hour-by-hour data. RETScreen's research shows that although this method is not very accurate, the error compared with data measured hour by day is between 3.9% and 8.9%. So it can meet the requirements of the general conditions of use but it is more complicated to calculate the monthly average solar irradiation in arbitrary orientations and tilted surfaces.

References

[1]   S. R. Wenham, M. A Green and M. E. Watt. Applied Photovoltaics[M]. ISBN0 86758 909 4
[2]   Rongsheng Fang, Licheng Xiang. et al.., Solar application technology [M], Beijing: China Agricultural Machinery Press, 1985.
[3]   M Iqbal. An introduction to solar radiation[M], Toronto : Academic Press, 1983, https://www.scirp.org/reference/ReferencesPapers.aspx?ReferenceID=986369
[4]   J. A Duffie and W. A. Backman. Solar engineering of thermal processes, Fourth Edition[M] New York: John Wiley & Sons. Inc, Hoboken, New Jersey 2013.
[5]   S. A. Klein. Calculation of monthly average insolation on tilted surfaces[J]. Solar Energy, 1977, 19(4), 325–329.
[6]   A. A. M. Sayigh. Solar energy engineering [m], Translated by Renxue Xu, Jianmin Liu, Beijing: Science Press, 1984.
[7]   D. G Erbs, S. A Klein and J. A. Duffie. Estimation of the diffuse radiation fraction for hourly, daily and monthly average global radiation[J]. Solar Energy, 1982, 28(4), 293–304.
[8]   Paul W. Stackhouse, et al. Surface meteorology and Solar Energy (SSE) Release 6.0 Methodology Version 3.2.0 http://power.larc.nasa.gov/documents/SSE_Methodology.pdf.
[9]   Y. H Liu and R. C. Jordan. The interrelationship and characteristic distribution of direct, diffuse and total solar radiation[J]. Solar Energy, 1960, 4, 1–19.
[10]   J. E. Hay. Calculating solar radiation for inclined surfaces: practical approaches[J]. Solar Energy, 1993, 3(4, 5), 373–380.
[11]   R Perez. et al.., An anisotropic hourly diffuse radiation model for sloping surfaces: Description, performance validation, site dependency evaluation[J]. Solar Energy, 1986, 36 (6), 481–497.
[12]   S A Klien and J C. Theilacker. An algorithm for calculating monthly-average radiation on inclined surfaces[J]. Journal of Solar Energy Engineering, 1981, 103, 29–33.
[13]   Canada clean energy decision support centre. Retscreen®Engineering & Cases Textbook[C]. ISBN: 0-662-35672-1. http://www.retscreen.net.

Exercises

2.1   Briefly describe the difference between the apparent solar time and the mean solar time.

2.2   What are the time angles at 9:30 am and 16:00 pm, respectively?

2.3   What is the difference between the solar azimuth angle and the azimuth angle of tilt plane?

2.4 The latitude of the Shanghai City is 31.14°N, and what is the altitude, azimuth, and zenith angles at 10:00 am on October 1?

2.5 The latitude of the Beijing City is 39.56°N. Please calculate the time angle at sunrise and sunset and azimuth angles at the winter solstice, and the sunshine hours of the whole day.

2.6 The latitude of the Lanzhou region is 36.03°N. Calculate the solar incident angle on the tilt plane with inclination angle of 45° and an azimuth angle of 15° at 3:00 pm on February 13.

2.7 Summarize the meaning of radiation flux, irradiance, and radiation.

2.8 What is the solar constant and what is the commonly used solar constant value?

2.9 What is the meaning of atmospheric mass and AM1.5?

2.10 When the solar zenith angle is 0°, the air quality is 1, so when the zenith angle is 48.2°, what is the atmospheric mass? What is the zenith angle when the atmospheric mass is 2?

2.11 What are the three components of solar radiation on a tilted surface?

2.12 Calculate the ratio of the hourly direct radiation on the tilted surface with a 30° tilted angle toward the south to the hourly total solar radiation on the horizontal surface in Shanghai on April 1.

Chapter 3
Basic principles of crystalline silicon PV cells

Photovoltaic (PV) cell (also called solar cell) is a device that directly converts solar radiation energy into electrical energy. The requirements of ideal PV cell materials are as follows: (1) high efficiency; (2) element abundance; (3) nontoxicity; (4) performance stability, good weather resistance, and long service lifetime; and (5) good mechanical properties for easy processing and preparation, especially for large-scale production.

3.1 Classification of PV cells

3.1.1 Classification according to the substrate materials

3.1.1.1 Crystalline silicon PV cells

Crystalline silicon is an indirect bandgap semiconductor material, and its bandgap width (1.12 eV) is significantly different from 1.4 eV [1]. Strictly speaking, silicon is not the ideal PV cell material. However, silicon is the second most abundant element on the Earth besides oxygen. It is easy to mine and extract, and exists mainly in the form of sand and quartz. With the development of the semiconductor device industry, the technology of growth and process of crystalline silicon has become increasingly mature, so crystalline silicon has become the main material of PV cells.

Crystalline silicon PV cells are made on crystalline silicon wafers. Crystalline silicon is the most widely used material for PV cells, including monocrystalline silicon (mono c-Si) PV cells, polycrystalline silicon (poly c-Si) PV cells, and quasi-monocrystalline silicon PV cells.

3.1.1.2 Monocrystalline silicon PV cells

Mono c-Si PV cells are made on mono c-Si wafers, and this kind of PV cell is the earliest developed PV cell with the most mature technology. Compared with other kinds of PV cells, mono c-Si PV cells have stable performance with higher conversion efficiency. Currently, the efficiency of this kind of PV cells for large-scale production has reached 19.5–23%. Due to technology advances, the prices have continued to decline and have occupied the largest market share for a long period of time. However, for the higher cost of production, the annual output has been gradually exceeded by poly c-Si PV cells after 1998. But with the development of technology, mono c-Si PV cells are further developed toward thinner wafer with higher efficiency. It can be expected that mono c-Si PV cells will have the lower costs and the higher market share in the future.

https://doi.org/10.1515/9783110524833-003

3.1.1.3 Polycrystalline silicon PV cells

Poly c-Si PV cells are made on poly c-Si wafers. Different from mono c-Si, high-purity silicon as a raw material is melted and cast into square silicon ingots, then cut into silicon wafers. Since the silicon wafer is composed of a plurality of crystal grains with different sizes and different orientations, the conversion efficiency of the poly c-Si PV cells is lower than that of the mono c-Si PV cells, and the efficiency of this kind of PV cells for large-scale production has reached 18.5–20.5%. Because of its relatively low manufacturing cost, it has developed rapidly with the highest marker share in recent years.

3.1.1.4 Quasi-monocrystalline silicon PV cell

Quasi-monocrystalline also known as mono-like crystalline has been developed in the recent years. This technology combines the advantages of Czochralski mono c-Si and casted poly c-Si by using casting technology with some seeds putting on the bottom of crucible. It has the characteristics of square, single crystal, low oxygen concentration, low light-induced degradation, and low defect density. Compared to poly c-Si, quasi-mono c-Si wafers have fewer grain boundaries and lower dislocation density. The efficiency of PV cells is 0.7–1%, which is higher than that of ordinary poly c-Si. The quasi-monocrystalline technology cannot grow perfect mono c-Si ingots, with only about 90% of the area in the middle is monocrystalline. The quality of the monocrystalline in this region is not as good as that of ordinary monocrystalline due to higher dislocation density, which results about 0.5% lower efficiency than that of ordinary mono c-Si PV cells. Although quasi-monocrystalline has certain advantages, there are still many technical difficulties and more technologies need to breakthrough for long-term development.

3.1.1.5 Silicon-based thin-film PV cells

Silicon-based thin-film PV cells are based on rigid or flexible materials as substrates. By chemical vapor deposition, N-type amorphous silicon (a-Si) or P-type a-Si is obtained by doping phosphorus (P) or boron (B). Silicon-based thin-film PV cells have the advantages of low deposition temperature ($\approx$200 °C), large area, continuous production, and flexible. Compared with crystalline silicon PV cells, its application range is more wide, but the lower conversion efficiency is still the greatest weakness of silicon-based thin-film PV cells. How to improve the conversion efficiency, stability, and cost performance of silicon-based thin-film PV cells was the research focus in the past few years.

3.1.1.5.1 Amorphous silicon PV cells

The bandgap of a-Si) is 1.7 eV. In the visible light range of the solar spectrum, the absorption coefficient of a-Si is nearly one order of magnitude higher than that of crystalline silicon. For a-Si PV cells, the peak of spectral response is close to the peak of solar spectrum. a-Si material has larger intrinsic absorption coefficient, so its thickness only need about 1 μm to fully absorb sunlight, which is less than 1/100 to that of the crystalline silicon. Therefore, a-Si PV cells have higher power generation capability than crystalline silicon PV cells under weak light condition. Since the commercialization of a-Si PV cells in 1980, Sanyo Electric Company (Japan) first applied as the power source for calculators. Thereafter, the application scope is gradually expanded from a variety of electronic consumer products such as watches, calculators, and toys to household power and PV power stations. a-Si PV cells have a certain market potential for low cost, easy production on a large scale, and building integration.

However, the efficiency of a-Si PV cells is relatively low, with only about 6–10% for large-scale production. Due to the effect of light-induced degradation, the stability of this kind of PV cells is poor, especially for the single-junction a-Si PV cells. With the research and development of the last 10 years, the efficiency of a-Si single-junction and tandem-junction PV cells has been significantly improved. The stability problem has also been improved, but not completely resolved. Therefore, a-Si PV cells cannot be widely used in power stations.

3.1.1.5.2 Microcrystalline silicon (μc-Si) PV cells

In order to obtain silicon-based thin-film PV cells with high efficiency and high stability, microcrystalline silicon (μc-Si) PV cells have developed for the past few years. μc-Si can be prepared at near room temperature. μc-Si films with the grain size of 10 nm and the thickness of 2–3 μm can be produced by using silane and hydrogen. In the middle of 1990s, the highest efficiency of μc-Si PV cells had already exceeded 10%. Compared with a-Si PV cells, the efficiency of μc-Si PV cells was higher and the effect of light-induced degradation was relatively smaller. However, this kind of PV cells still cannot reach the level of large-scale industrial production. Tandem (a-Si/μc-Si) PV cells with a-Si ($E_g = 1.7$ eV) as the top layer and μc-Si ($E_g = 1.1$ eV) as the bottom layer had been applied, and the efficiency had exceeded 14% showing a good prospect of application. However, μc-Si thin film cannot prepare P–N junction directly like mono c-Si for higher defect density by containing a large amount of a-Si, so P–I–N junction is used on μc-Si thin-film cells. How to prepare intrinsic layers with lower defect density and to prepare μc-Si films with lower content of a-Si at relatively low temperature is the key to further improve the efficiency of μc-Si PV cells.

3.1.1.5.3 Compound PV cells

Compound PV cells are made by compound semiconductor materials. The main types are as follows:

3.1.1.5.3.1 Monocrystalline compound PV cells

Gallium arsenide (GaAs) PV cells are a kind of monocrystalline compound PV cells. The bandgap of GaAs is 1.4 eV, which is an ideal PV cell material. This kind of PV cell has the highest efficiency in single-junction PV cells. The efficiency of multijunction GaAs PV cells has exceeded 40% under highly concentrated sunlight. Because of its high efficiency, it has been mainly applied in space applications. GaAs PV cells are rarely used in the terrestrial PV system due to expensive cost and toxic arsenic.

3.1.1.5.3.2 Polycrystalline compound PV cells

There are many kinds of polycrystalline compound PV cells, such as cadmium telluride (CdTe) PV cells and copper indium gallium selenide (CIGS) PV cells.

In addition, there are organic semiconductor PV cells, dye-sensitized PV cells, perovskite PV cells, and so on. Details will be given in Chapter 4.

3.1.2 Classification according to the structure of PV cells

3.1.2.1 Homojunction PV cells

A P–N junction formed by the same semiconductor material is called a homojunction, and a PV cell made by a homojunction structure is called a homojunction PV cell.

3.1.2.2 Heterojunction PV cells

A P–N junction formed by two semiconductor materials with different bandgaps is called a heterojunction, and a PV cell made by a heterojunction is called a heterojunction PV cell.

3.1.2.3 Schottky junction PV cells

A P–N junction formed by using a Schottky potential barrier on a metal–semiconductor (MS) interface is called a Schottky junction PV cell, or MS cell for short, including metal–oxide–semiconductor and metal–insulator–semiconductor (MIS) PV cells.

3.1.2.4 Composite junction PV cells

Composite junction PV cells have two or more P–N junctions. This kind of PV cells can be divided into vertical multijunction PV cells and horizontal multijunction PV cells, such as an MIS PV cell with a P–N junction silicon PV cell tandem together to form a high-efficiency MIS NP composite junction silicon PV cell, whose efficiency has reached 22%. Composite junction PV cells are often made of tandem to broaden the spectral response of the entire PV cells. The wide bandgap materials are used as the top layers to absorb high-energy photons, and the narrow bandgap materials are used as the bottom layers to absorb low-energy photons. The efficiency of the aluminum gallium arsenide–gallium arsenide–silicon PV cell has reached 31%.

3.1.3 Classification according to the application of PV cells

3.1.3.1 Space PV cells
Space PV cells are mainly applied to spacecraft such as satellites and spaceships. Due to the special application environment, PV cells are required to have high-efficiency, lightweight, high- and low-temperature shock resistance, and strong antihigh-energy particle radiation ability. Therefore, space PV cells have fine production process and expensive cost.

3.1.3.2 Terrestrial PV cells
Terrestrial PV cells are widely used in terrestrial PV systems. This kind of PV cells needs abundant substrate material, easy to large-scale production, good weather resistance, and high performance to cost ratio.

3.2 Working principle of PV cells

PV cell is a semiconductor device that converts light energy directly into electrical energy. Its core structure is semiconductor P–N junction. Taking crystalline silicon PV cell as an example, this chapter discusses the conversion of light energy into electric energy in detail. The principle and process of thin-film PV cell are introduced in Chapter 4.

3.2.1 Semiconductor

As we all know, a conductor is an object with a large number of charged particles that can move freely and can conduct current easily. Usually, metals are conductors. For example, the conductivity of copper is about $10^6/\Omega \cdot cm$. If 1 V is applied to two opposite surfaces of a 1 cm × 1 cm × 1 cm copper cube, a current of 10^6 A flows between the two surfaces.

An insulator is an object that is extremely difficult to conduct current, such as ceramic, mica, grease, and rubber. For example, the conductivity of quartz (SiO_2) is about $10^{-6}/\Omega \cdot cm$.

Conductive properties of an object between the conductor and the insulator are called semiconductor [2]. The conductivity of semiconductor is between 10^{-4} and $10^4/\Omega \cdot$ cm, and its conductivity can be adjusted in the range by adding a small amount of impurities. The conductivity of high-purity semiconductors increases rapidly with increasing temperature, which is the most important characteristic of semiconductor.

A semiconductor can be an element such as silicon (Si), germanium (Ge), and selenium (Se). It can also be a compound such as cadmium sulfide and GaAs, or an alloy such as $Ga_xAl_{1-x}As$ (where x is any number between 0 and 1). Many organic compounds are also semiconductors.

Many electrical properties of semiconductors can be explained by a simple model. The atomic number of silicon is 14, so there are 14 electrons outside of the nucleus, in which 10 electrons in the inner layers are tightly bound by the nucleus and four electrons in the outer layer where electrons are less bound by the nucleus. If the electrons in the outer layer can obtain enough energy, they will be out of the bondage of the nucleus and become free electrons. At the same time, a hole is left in the original position. The electrons are negatively charged and the holes are positively charged. The four electrons in the outer layer of silicon nucleus are also called valence electrons. The schematic of silicon atom is shown in Fig. 3.1.

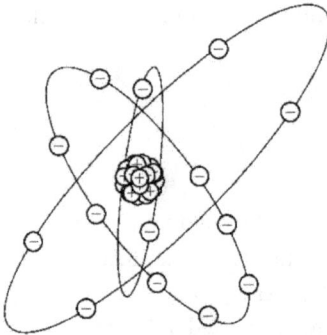

Fig. 3.1: Schematic of silicon atom.

There are four adjacent atoms around each atom in the silicon crystal and share two valence electrons with each neighboring atom to form a stable eight-atomic shell. These two valence electrons are called covalent electrons. The covalent bond structure of silicon crystal is shown in Fig. 3.2. Separating an electron from silicon atoms requires 1.12 eV, which is called the bandgap of silicon. The separated electrons are free to move and to transmit current. A vacancy that is left after an electron escapes from an atom is called hole. The electron coming from the neighboring atom can fill the hole, so the hole moves from one location to another new location, then creating an electrical current. The current generated by the flow of electrons is equivalent to the current generated when positively charged holes move in the opposite direction.

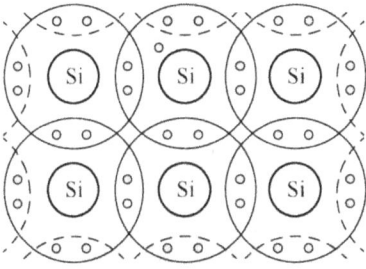

Fig. 3.2: Covalent structure of silicon crystals.

3.2.2 Bandgap structure

The properties of semiconductors can be explained by the bandgap structure. Silicon is a tetravalent element and each atom has four electrons on its outermost shell. In silicon crystals, each atom has four adjacent atoms and shares two valence electrons with each adjacent atom, forming a stable eight-electron shell.

The energy value of electrons in free space is basically continuous, but it is entirely different in crystals that the electrons in isolated atoms occupy a fixed set of discrete energy levels. Due to the interactions between extranuclear electrons of each atom in a crystal arranged in regular order, when the isolated atoms are close to each other, the separation energy levels in the isolated atom would expand and overlap with each other to become a band as shown in Fig. 3.3. The bandgap that allows electrons to occupy is called the allowed band, and the range between two allowed bands where electrons are forbidden to occupy is called the forbidden band.

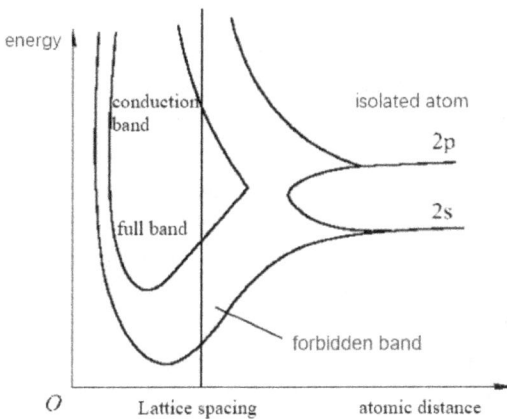

Fig. 3.3: Atom spacing and electron energy level.

At low temperature, electrons of crystals occupy the lowest possible energy state. However, the equilibrium state of the crystal is not a state in which the electrons are all at the lowest allowed energy level. According to the Pauli exclusion principle, each allowed energy level can only be occupied by at most two electrons that spin in the opposite direction. This means that at low temperature, all possible energy states below a certain energy level will be occupied by two electrons, which is called the Fermi energy (E_f). As temperature rises, some electrons get energy beyond the Fermi energy. Considering the limit of Pauli's exclusion principle, the occupation probability of an allowed electron energy state for any given energy E can be calculated according to statistical laws. This is the Fermi–Dirac distribution function $f(E)$ as follows:

$$f(E) = \frac{1}{1 + e^{(E - E_f)/KT}} \tag{3.1}$$

where E_f is the Fermi energy. Its physical meaning is that the probability of one state in the energy level of E_f occupied by electrons is 1/2. Therefore, the probability of unoccupied by electrons is larger at the state higher than the Fermi energy, that is, there are many vacant states (occupied probability is approximated to 0). On the contrary, the state lower than the Fermi energy has higher probability occupied by electrons, which can be approximated as being substantially occupied by electrons (occupied probability is approximated to 1).

Conductivity differs depending on the way the electrons filling in the allowed band. The allowed band that is completely occupied by electrons is called a conduction band, and electrons in full band cannot move even if applied on an electric field, so this kind of material is called an insulator. In the allowed band, the electrons can move to a little higher energy level affected by a small electric field and become free electrons, and the electrical conductivity of the material becomes very large, so this kind of material is called a conductor. Semiconductor means that the material has a band structure similar to an insulator, but with a small bandgap. In this case, the electrons in the full band obtained thermal energy at room temperature. They are possible to jump over the bandgap to the conduction band, and to become free electrons, which will contribute to the conduction of the material. The full band energy level involved in this conduction phenomenon is in most cases at the highest energy level of the full band, so the band structure can be simplified as shown in Fig. 3.4. In addition, since the electrons in this full band are located at the outermost layer of each atom, they are the valence electrons that participate in the bonding between atoms, so this kind of full band is called a valence band. In Fig. 3.4, the upper part of the conduction band and the lower part of the valence band are omitted.

Once the covalent bond is destroyed by absorbing external energy, the electrons will jump from the valence band into the conduction band, meanwhile leaving a vacancy in the valence band. This vacancy can be occupied by electrons from

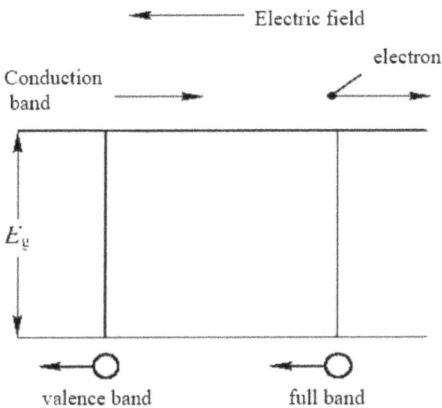

Fig. 3.4: Semiconductor band structure and carrier movement.

adjacent position in the valence band, and the new vacancy left by this electron movement can be filled by other electrons, so this phenomenon can be seen as the vacancy moves in sequence. It is equivalent to the positively charged particles that move in the opposite direction to the movement of electrons in the valence band, and the vacancy is called hole. In a semiconductor, holes and free electrons in a conduction band become charged particles (carriers). Electrons and holes move in the opposite direction under an external electric field. They have the same current direction for the opposite charged symbols and the superposition effects for the conductivity.

3.2.3 Intrinsic semiconductors and doped semiconductors

When the forbidden bandgap E_g is small, as the temperature rises, the number of electrons that move from the valence band to the conduction band will increase, and at the same time, the same number of holes are generated in the valence band. This process is called the generation of electron–hole pairs. The semiconductor with a certain conductivity that can produce electron–hole pairs at room temperature is called an intrinsic semiconductor [3]. It is an ultrapurified and defect-free semiconductor. Normally, due to impurities or lattice defects in semiconductors, there is an increase in free carriers (electrons or holes), which are called doped semiconductors. N-type semiconductors have redundant electrons, while P-type semiconductors have redundant holes.

Impurity atoms are entered (called doping) into the crystal structure in two ways. One way is that the impurity atoms are located in the interstice between the crystal atoms, the so-called interstitial impurities. Another way is that the impurity atoms replace the atoms in the crystal with the regular atomic arrangement of the crystal structure still maintained, the so-called substitutional impurities.

Atoms in groups III and V in the periodic table are substitutional impurities to silicon, such as a group V impurity atom replaces one atom of the silicon lattice, and four valence electrons of it can form covalent bonds with the surrounding silicon atoms, but the fifth valence electron is in different position. This electron cannot form a covalent bond, so it is not in the valence band. It is also bound to the group V atom and cannot move freely through the lattice, so it is not in the conduction band. It can be expected that releasing this excess electron requires less energy than releasing free electrons bound to the covalent bond, which means the energy required is much less than the silicon bandgap energy of 1.1 eV. The free electrons are in the conduction band, so the excess electron bound to the group V atom is located below the conduction band, as shown in Fig. 3.5.

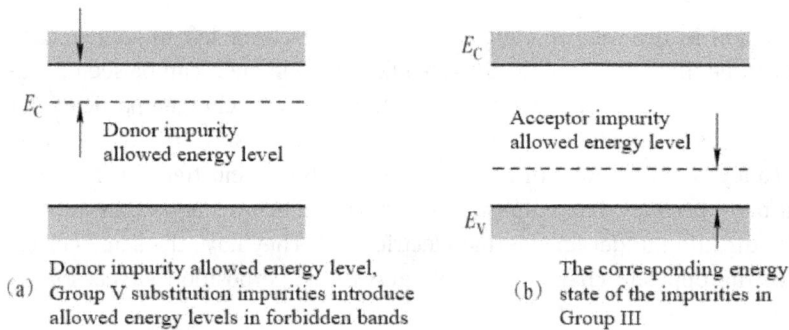

(a) Donor impurity allowed energy level, Group V substitution impurities introduce allowed energy levels in forbidden bands

(b) The corresponding energy state of the impurities in Group III

Fig. 3.5: Corresponding energy state of impurity in groups III and V.

An allowed energy level is placed in the "forbidden" bandgap. For example, when group V elements (Sb, As, P) are doping into semiconductor silicon crystals, these impurity atoms replace the positions of the silicon atoms into lattices. Among five valence electrons, there is one redundant valence electron left except to forming covalent bonds with adjacent silicon atoms. Compared to the covalent bond, the binding force between the redundant valence electron and the impurity atom is weak. Therefore, as long as the impurity atoms get a little weak energy, they can release electrons at room temperature to form free electrons, and the impurity atoms themselves become monovalent positive ions, but they cannot move because they are bound by lattices. In this case, the group V-doped silicon forms N-type semiconductor with redundant electrons. Such impurities that can provide free electrons to semiconductors are called donor impurities and the bandgap structure is shown in Fig. 3.6.

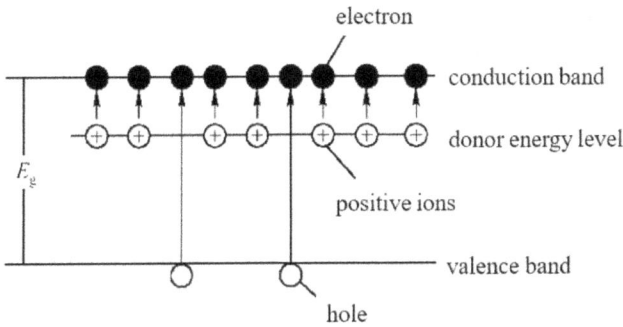

Fig. 3.6: Bandgap structure of N-type semiconductors.

In addition to the electrons generated from these donor levels, there are electrons excited from the valence band to the conduction band. Since this process is produced by electron–hole pairs, there are also the same number of holes. In an N-type semiconductor, there are more electrons than holes, electrons are referred to as majority carriers, and holes are referred to as minority carriers.

The analysis of impurity in group III is similar to the above. For example, when B, Al, Ga, and In are doping as impurities, one valence electron is taken from the adjacent silicon atom to form a complete covalent bond, because one electron is short to forming a complete covalent bond. The atoms that have taken away the electron leave a vacancy and become a hole. As a result, the impurity atoms become monovalent negative ions and provide holes with weak bonding. This bonding can be easily destroyed with a small amount of energy, then forming free holes. P-type semiconductor is the semiconductor with hole-excess. The impurity atom accepting electrons is called the acceptor impurity, and its bandgap structure is shown in Fig. 3.7. In this case, the majority carriers are holes and the minority carriers are electrons. In addition, there are some cases that the N and P conductivity types cannot be

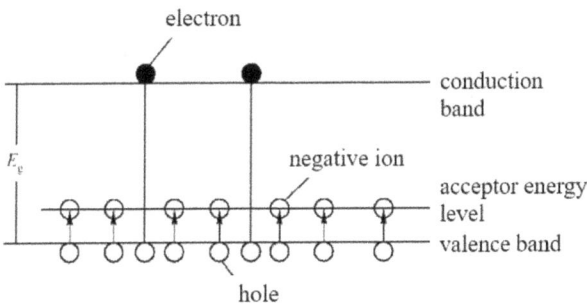

Fig. 3.7: Bandgap structure of P-type semiconductor.

obtained even if impurities are doping for the reason of a large vapor pressure difference between the constituent elements.

3.2.4 N-type and P-type semiconductors

3.2.4.1 N-type semiconductors

If pure silicon crystals are doping with a small amount of phosphorus (or arsenic, antimony, etc.), since the number of phosphorus atoms is much less than the number of silicon atoms, the overall structure remains basically unchanged, except a few silicon atoms at certain positions be replaced by phosphorus atoms. Since the phosphorous atom has five valence electrons (pentavalent), one phosphorous atom is covalently bonded to four adjacent silicon atoms, then one valence electron is left. This valence electron is not bonded in the covalent bond, and it is only affected by phosphorus nucleus attraction which is much weaker. Therefore, this valence electron easily breaks away from the attraction of phosphorus nucleus and becomes a free electron, which leads to the number of electron carriers in the silicon crystal being greatly increased. Since a pentavalent impurity atom can provide a free electron, this kind of impurity atom is called a donor. A tetravalent semiconductor (with four valence electrons) doped with a pentavalent impurity becomes an electronically conductive-type semiconductor, also called an N-type semiconductor. The schematic diagram of such semiconductor is shown in Fig. 3.8. In this kind of N-type semiconductor material, there are not only a large amount of free electrons generated by doping impurities, but also a small number of electron–hole pairs generated by thermal excitation. The number of hole is extremely small relative to the number of electrons, so the holes are called minority carriers and electrons are called majority carriers.

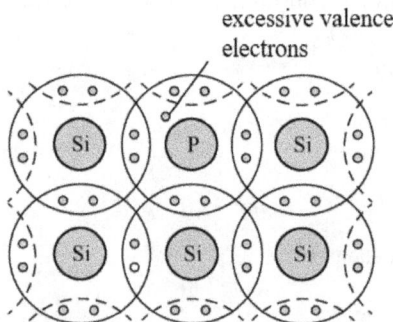

Fig. 3.8: Schematic of N-type semiconductor.

3.2.4.2 P-type semiconductors

Similarly, if a pure silicon crystal is doped with a trivalent impurity which has three valent electrons in its outermost layer, such as boron, or aluminum, gallium, and indium, it is capable of trapping electrons. When it forms a covalent bond with the neighboring silicon atoms, it still lacks one valence electron; thus, a hole appears in a covalent bond, and this hole can accept the filling of foreign electrons. The valence electrons of the nearby silicon atom are easily transferred to this position under thermal excitation. A hole appears in the covalent bond of the silicon atom, and the boron atom also becomes boron ion with one negative charge after receiving a valence electron. In this way, each boron atom can accept one valence electron, and at the same time generates a hole nearby, so the number of vacancy carriers in the silicon crystal greatly increases. Since a trivalent impurity atom can accept an electron and is called an acceptor impurity, a tetravalent semiconductor doped with a trivalent impurity is also called a P-type semiconductor. Of course, in the P-type semiconductor, there are not only a large number of vacancies generated by the doping of impurities, but also a small amount of electron–hole pairs generated by thermal excitation, but the number of electrons is much smaller. In contrast to N-type semiconductors, for P-type semiconductors, vacancies are majority carriers and electrons are minority carriers. A schematic diagram of a P-type semiconductor is shown in Fig. 3.9.

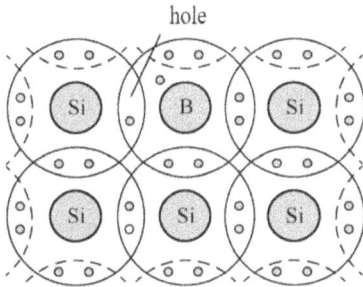

Fig. 3.9: Schematic of P-type semiconductor.

However, for pure semiconductors, no matter N-type or P-type, they are all electrically neutral with the equal number of electrons and holes. This is because the monocrystalline semiconductor and the impurity dopants are all electrically neutral. In the doping process, neither lose charge nor obtain charge from the outside, but the number of valence electrons of doped impurity atoms is one more or less than the atoms of the substrate material. There are a large number of removable electrons or holes in the semiconductor, but this does not destroy the balance of positive and negative charges in the entire semiconductor.

3.2.5 P–N junctions

3.2.5.1 The majority carrier diffusion motion

If one P-type and one N-type semiconductors are closely combined and integrated, the transition region between these two semiconductors with the opposite conductivity type is called a P–N junction. In two sides of the P–N junction, there are many holes and few electrons in the P region, while in the N region, there are many electrons and few holes. Therefore, the concentration of electrons and holes is not equal in these two sides of the interface between the P-type and N-type semiconductors, so that the diffusion motion of the majority carriers occurs [4].

In the P region near the interface, the holes diffuse from the high concentrated P region toward the low concentrated N region, and recombine with the electrons there, and then lots of positive ions of doped impurity appear in the N region. At the same time, in the P region, negative ions of doped impurity present as a result of the movement of the holes.

In the N region near the interface, the electrons diffuse from the high concentrated N region toward the low concentrated P region, and recombine with the holes there, and then lots of negative ions of doped impurity appear at the P region. At the same time, in the N region, positive ions of doped impurity present as a result of the movement of electrons.

The result of the diffusion is formed by a thin region in the interface, in which the side near the N region is accumulated positive charges and the other side of P region is accumulated negative charges. This thin region is referred to as a space charge region (also referred to as a depletion region), which is the P–N junction, as shown in Fig. 3.10. In the P–N junction, due to the accumulation of positive and negative charges on both sides, there will be a reverse electric field from the N region to the P region, which is called a built-in electric field (or barrier electric field).

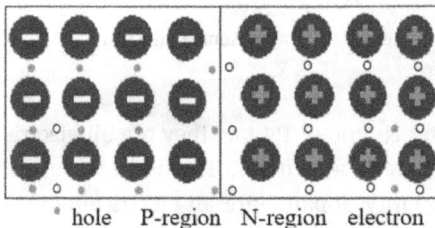

* hole   P-region   N-region   electron

Fig. 3.10: P–N junction.

3.2.5.2 Minority carrier drift motion

Because of the built-in electric field, there is a force acting on charges. The electric field will push the positive charges to move in the direction of the electric field and prevent it from moving in the reverse direction of the electric field. At the same time, the electric field will attract the negative charge to move against the direction of the electric field and prevent it from moving in the direction of the electric field. Due to the opposite direction between movement and built-in electric field, when the holes in the P region attempt to continue to diffuse into the N region by passing through the space charge region, they are even pulled back into P region by resistance of the built-in electric field. At the same time, the electrons in the N region attempt to continue to diffuse into the P region by passing through the space charge region. They are also even pulled back into the N region by the resistance of built-in electric field. In short, the presence of a built-in electric field hinders the diffusion motion of majority carriers. However, the electrons in the P region and the holes in the N region can be moved to the other side of the P–N junction by the built-in electric field. The movement of minority carriers under the built-in electric field is called drift motion. The direction of this movement is opposite to the direction of diffusion. The minority carrier drift motion caused by the P–N junction is finally balanced with the diffusion motion of the majority carrier. At this time, the number of carriers with diffusion and drift is equal, the direction of both motions is reversed, the total current is zero, and the diffusion is no longer going on. As a result, the thickness of the space charge zone no longer increases and reaches an equilibrium state. If the conditions and the environment are unchanged, this equilibrium state will not be destroyed, and the thickness of the space charge region will be constant. This thickness is related to the doping concentration.

Due to the existence of an electric field in the space charge region, the electric potentials of each points in the electric fields are different, and the direction of the electric field points to the direction of potential drop. Therefore, in the space charge region, the positive ion has a high potential, and the negative ion has a low potential, so a potential difference exists on the two sides in space charge region, called a potential barrier, also called contact potential difference, which can be expressed as

$$V_d = \frac{kT}{q} \ln \frac{n_n}{n_p} = \frac{kT}{q} \ln \frac{p_p}{p_n} \tag{3.2}$$

where q is the elementary charge (-1.6×10^{-19} C), T is the absolute temperature, k is the Boltzmann constant, n_n, n_p are electron concentrations in N-type and P-type semiconductor materials, and p_n, p_p are hole concentrations in N-type and P-type semiconductor materials.

3.2.6 Photovoltaic effect

When the semiconductor surface is exposed to sunlight, if energy of some photons is greater than or equal to the bandgap of the semiconductor, electrons can break the bondage of the nuclei and produce a large number of electron–hole pairs in the semiconductor [5]. This phenomenon is called the internal photoelectric effect (the phenomenon that atom hits electrons and makes them get out of the metal is the external photoelectric effect). Semiconductor materials rely on the internal photo-electric effect to convert light energy into electrical energy. Therefore, the condition for achieving the internal photoelectric effect is that the absorbed photon energy is greater than the bandgap of semiconductor materials, which is

$$hv \geq E_g \qquad (3.3)$$

where hv is the photon energy, h is Planck's constant, v is the frequency of the light wave, and E_g is the bandgap of the semiconductor material.

Since $C = v\lambda$, where C is the speed of light and λ is the wavelength of the light. Equation (3.3) can be rewritten as

$$\lambda \leq \frac{hC}{E_g} \qquad (3.4)$$

This means that electron–hole pairs can be produced if only the wavelength of photons satisfies the requirement of eq. (3.4). This wavelength is usually referred to as the cutoff wavelength and is expressed as λ_g. Photons with a wavelength greater than λ_g cannot generate carriers.

Due to the different bandgap of different semiconductor materials, the photon energy required to excite electron–hole pairs is also different. In a semiconductor material, photons exceeding the bandgap are absorbed and transfer to electrical energy, and photons with energy less than the bandgap are absorbed by the semiconductor and then converted into heat and rise the temperature of semiconductor instead of generating the electron–hole pairs. Obviously, the bandgap is very important for PV cells. The larger the bandgap, the less solar energy available for use, and it makes each kind of PV cells to have a certain limit on the wavelength of the absorbed light.

Sunlight shining on the PV cells is partly reflected off the surface of the PV cell, another part is absorbed by the PV cell, and a small amount passes through the PV cell. For photons absorbed by PV cells, photons with energy greater than the bandgap of semiconductor can excite the valence electrons of atoms in the semiconductor. Photon-generated electron–hole pairs, which are also called photogenerated carriers, are generated in the P region, space charge region, and N region. The electron–hole pairs migrate in all directions due to thermal motion. After the photogenerated electron–hole pairs are generated in the space charge region, they are immediately separated by the built-in electric field. Photogenerated electrons are pushed into the N region, and photogenerated holes are pushed into the P region. The total carrier concentration at the boundary of the space charge zone is approximately zero. In the N region, once the photogenerated electron–hole pairs are generated, the photogenerated holes diffuse toward the P–N junction boundary.

Once they reach the P–N junction boundary, they are immediately affected by the built-in electric field, make a drift motion under the action of the electric field force, and cross the space charge region into the P region; thus, photogenerated electrons (majority carriers) are left in the N region. The photogenerated electrons in the P region also diffuse toward the P–N junction boundary, and after reaching the P–N junction boundary, also due to the effect of the built-in electric field, they perform a drift motion under the action of the electric field force, entering the N region, and the photogenerated holes (majority carriers) are left in the P region. Therefore, positive and negative charges are accumulated on both sides of the P–N junction to form a photogenerated electric field with opposite direction to that of the built-in electric field. In addition to offsetting part of the built-in electric field, this electric field also makes the P-type layer positively charged and the N-type layer negatively charged, thus generating a photoelectromotive force, which is the "photogenerated voltage effect" (referred to as the PV effect).

3.2.7 PV cell photoelectric conversion principle

PV cells are semiconductor PV devices that convert light energy into electrical energy [6]. With light, there will be a certain potential difference between the upper and lower electrodes of the PV cells. When they are connected to the load by the wires, it will generate a direct current as shown in Fig. 3.11, so the PV cell can be used as a power source.

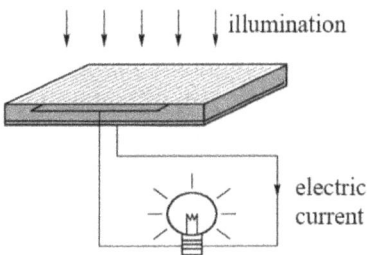

Fig. 3.11: PV cell working principle.

The physical process of photoelectric conversion is as follows.
- The photons are absorbed so that electron–hole pairs are generated on both sides of the P–N junction, as shown in Fig. 3.12a.
- Electrons and holes generated within one diffusion length away from the P–N junction reach the space charge region by diffusion, as shown in Fig. 3.12b.
- The electron–hole pairs are separated by the electric field. Therefore, the electrons with the high potential in P side slide to the N side, and the holes move in the opposite direction, as shown in Fig. 3.12c.

- If the P–N junction is open circuit, electrons and holes accumulated on both sides of the junction produce an open-circuit voltage, as shown in Fig. 3.12d. If there is a load connected to the cell, there will be current conduction in the circuit, as shown in Fig. 3.12a.

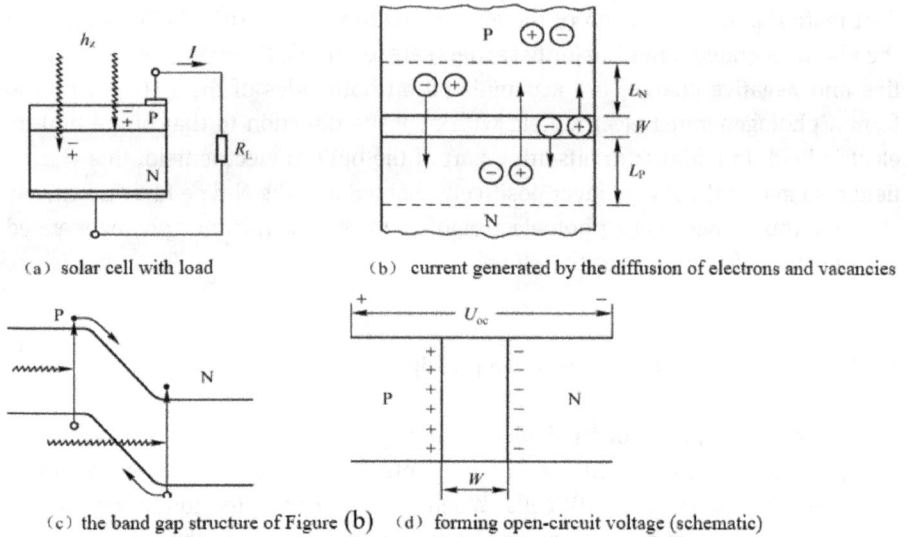

(a) solar cell with load   (b) current generated by the diffusion of electrons and vacancies

(c) the band gap structure of Figure (b)   (d) forming open-circuit voltage (schematic)

Fig. 3.12: Photoelectric conversion physical process.

3.2.8 Structure of crystalline silicon PV cells

The structure of typical crystalline silicon PV cells is shown in Fig. 3.13. The substrate material is P-type crystalline silicon with a thickness of about 0.18 mm. An N-type semiconductor of about 0.25 μm is formed by diffusion to form a P–N

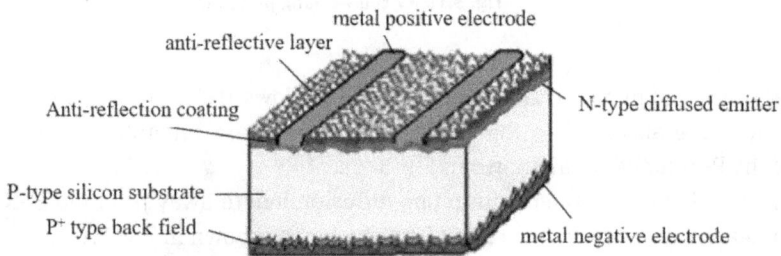

Fig. 3.13: Crystal silicon PV cell structure diagram.

junction. On the light-receiving surface of the PV cell that is the surface of the N-type semiconductor, there is a pyramidal textured structure and an antireflection layer. The upper surface has dense finger lines and several grid lines crossing the finger lines to form metal positive electrode for the current output. On the back of the PV cell that is the P-type substrate, there is a layer of P^+ back-surface field with a higher doping concentration, usually an aluminum back-surface field, or a boron back-surface field. Underneath the back-surface field is a metal negative electrode for current flow, thus constituting a typical single-junction (N–P–P^+) crystalline silicon PV cell.

The operating voltage of each crystalline silicon PV cell is approximately 0. 50–0.65 V. The magnitude of this value is independent of the size of the cell. The output current of the PV cell is related to the size of its area, the intensity of sunlight, and the temperature. When other conditions are the same, a larger area of the cell can generate a larger current, so the power is also greater.

PV cells are generally made into P^+/N type or N^+/P type structures. The first symbol, P^+ or N^+, indicates the conductivity type of the semiconductor material on the front side of the PV cell. The second symbol, N or P, indicates the conductivity type of the PV cell substrate semiconductor material. Under the sunshine, the output voltage of the PV cell is positive from the P-type side and negative from the N-type side.

3.3 PV cell electrical characteristics

3.3.1 Standard test conditions

The photogenerated electricity of PV cells is related to the irradiance of the light source, the temperature of the cell, and the spectral distribution of the irradiated light. Therefore, when testing the power of the PV cell, standard test conditions (STC) must be specified. At present, the STC for terrestrial PV cells are universally specified as follows:
- Light source irradiance: 1,000 W/m^2
- The temperature of the cell: 25 °C
- AM1.5 terrestrial solar spectral irradiance distribution

The specific distribution of the solar spectral irradiance of AM0 and AM1.5 is shown in Fig. 3.14.

3.3.2 PV cell equivalent circuit

If a load resistance R is connected to the positive and negative electrodes of the PV cell that is illuminated, the PV cell is in the working condition. The equivalent

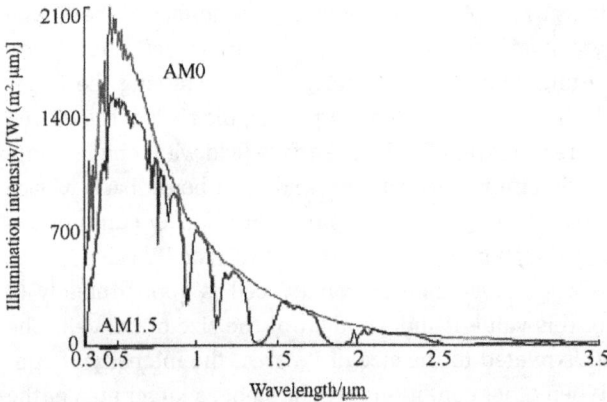

Fig. 3.14: Spectral irradiance distribution of AM0 and AM1.5.

circuit is shown in Fig. 3.15. It is equivalent to a constant current I_{ph} connected in parallel with a forward diode. The forward current flowing through the diode is called the dark current I_D in the PV cell. From the two sides of the load R, the forward voltage V which generates dark current can be measured. In addition to the current flowing through the load (I), the equivalent circuit of an ideal PV cell is formed. The actual used PV cells, that is, the actual PV cell's equivalent circuit is shown in Fig. 3.16, which is a little different due to its own resistance. R_{sh} is called shunt resistance, which is mainly formed by the following factors: surface leakage along the edge of the cell caused by surface contamination, leakage currents induced by the irregular diffusion along dislocations and grain boundaries, and leakage currents induced by the tiny bridges formed by microscopic cracks, the boundaries, and crystal defects after electrode metallization. R_s is called series resistance and consists of the surface resistance of the top diffusion region, the bulk resistance of the cell, the ohmic resistance between the front and back electrodes, and the PV cell and the resistance of the metal conductor.

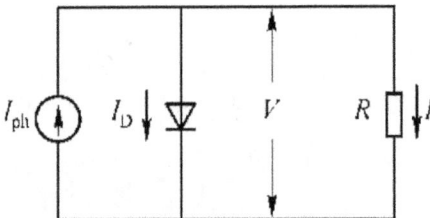

Fig. 3.15: Ideal PV cell equivalent circuit.

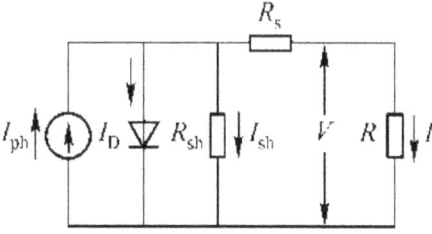

Fig. 3.16: Actual PV cell equivalent circuit.

As shown in Fig. 3.16, the voltage across the load is V, so the voltage across R_{sh} is $(V + IR_s)$, so there is

$$I_{sh} = (V + IR_s)/R_{sh}$$

The current passing through the load is $I = I_{ph} - I_D - I_{sh}$.

After changing the above equation, we obtain

$$I(1 + R_s/R_{sh}) = I_{ph} - (V/R_{sh}) - I_D \tag{3.5}$$

Among them, the dark current I_D is the sum of the injection current, the recombination current, and the tunnel current. In general, the tunnel current can be ignored so that the dark current I_D is the sum of the injection current and the recombination current.

The external voltage applied to the P–N junction is $V_j = V + IR_s$.

In order to estimate the output and efficiency of a PV cell by using the equivalent circuit, the injection current and the recombination current can be simplified to a single exponential form:

$$I = I_{ph} - I_0 \left[e^{qV_j/(A_0 kT)} - 1 \right]$$

where I_0 is the new exponential prefactor and A_0 is the structural factor of the junction, which reflects the influence of the structural integrity of the P–N junction on the performance. In the ideal case ($R_{sh} \to \infty$, $R_s \to 0$), we can get I_D from eq. (3.5):

$$I_D = I_0 \left[e^{qV_j/(A_0 kT)} - 1 \right] \tag{3.6}$$

Equation (3.6) is the current–voltage relationship of the PV cell under light irradiation. From eq. (3.6), we can see that when the load R is short-circuited, that is $V_j = 0$ (ignoring series resistance), the value of the short-circuit current I_{sc} is exactly equal to the photocurrent, that is $I_{sc} = I_{ph}$. When the load $R \to \infty$, the output current approaches 0, and the open-circuit voltage V_{oc} is determined by the following equation:

$$V_{oc} = (A_0 kT/q) \ln \left(I_{ph}/I_0 + 1 \right) \tag{3.7}$$

In the absence of light, the current–voltage relationship at the P–N junction is shown by the curve in Fig. 3.17, which is the dark current–voltage curve of the PV cell.

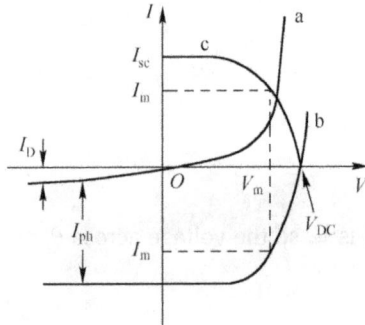

Fig. 3.17: Current–voltage characteristics of PV cells: (a) arc characteristic curve; (b) characteristic curve in light condition; and (c) transform coordinates to obtain PV cell characteristics curve.

The photocurrent I_{ph} generated under illumination causes the curve to shift I_{ph} in the negative direction of the current axis, resulting in the curve b in Fig. 3.17. For convenience, we transform the coordinate direction to get the curve c in Fig. 3.17. The curve c is the current–voltage curve ($I–V$) of the PV cell in the light condition, and its relation is shown in eq. (3.6).

3.3.3 The main technical parameters of PV cells

3.3.3.1 Current–voltage curve
When the load R changes from 0 to infinity, the relationship between the voltage V and the current I across the load R is the load characteristic curve of the PV cell, which is usually called the current–voltage curve of the PV cell, which is conventionally called $I–V$ curve.

In fact, $I–V$ curve is not usually calculated but experimentally measured. A variable resistor R is connected to the positive and negative electrodes of the PV cell. Under a certain solar irradiance and temperature, as the resistance value changes from 0 (short circuit) to infinity (open circuit), the current and the voltage across the resistance is measured at the same time. On the Cartesian graph, the ordinate represents the current and the abscissa represents the voltage. The connection of each point is achieved, which is the $I–V$ curve of the PV cell under the irradiance and temperature, as shown in Fig. 3.18.

3.3.3.2 Maximum power point
Under a certain condition of solar irradiance and operating temperature, any point on the $I–V$ curve is the operating point [7]. The connection between the working

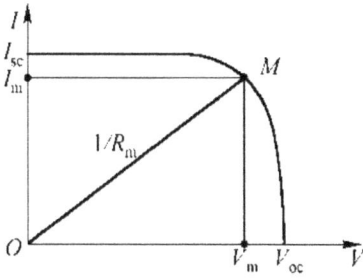

Fig. 3.18: *I–V* characteristics of PV cells.

point and the origin is called the load line, and the reciprocal of the load line slope is the load resistance R_L. The abscissa corresponding to the operating point is the operating voltage V, and the ordinate is the operating current I. The product of voltage V and current I is the output power. When the load resistance R_L is adjusted to a certain value R_m, the point M is obtained on the curve, and the product of the corresponding operating current I_m and the working voltage V_m is the maximum, that is,

$$P_m = I_m V_m = P_{max} \qquad (3.8)$$

The M point is said to be the best working point (or maximum power point) for the PV cell. I_m is the operating current at maximum power point, V_m is the operating voltage at maximum power point, R_m is the operating load resistance at maximum power point, and P_m is the maximum output power. It is also good to make a horizontal line through a working point on the $I–V$ curve, and the intersection point with the ordinate is I, and make a vertical line again, intersecting the abscissa at V. The area of the rectangle enclosed by these two lines is equal to the product of voltage V and current I in the numerical value, that is, the output power. Any operating point on the $I–V$ curve corresponds to a certain output power. Usually, the output power of different working points is also different, but a working point that surrounds a maximum area rectangle ($OI_m M V_m$) can always be found. That is, the product of the operating voltage V and the current I corresponding to this point is the largest; therefore, the output power is also the largest and this point is the best operating point, that is,

$$P = VI = V\left[I_{ph} - I_0\left(e^{qV/AkT} - 1\right)\right]$$

At this maximum power point, it has $dP_m/dV = 0$, so there is

$$\left(1 + \frac{qV_m}{AkT}\right)e^{\frac{qV_m}{AkT}} = \left(\frac{I_{ph}}{I_0}\right) + 1$$

After tidying the above equation, we obtain

$$I_m = \frac{(I_{ph} + I_0)qV_m/AkT}{1 + (qV_m/AkT)} \tag{3.9}$$

$$V_m = \frac{AkT}{q} \ln \left[\frac{1 + (I_{ph}/I_0)}{1 + qV_m/AkT} \right] \approx V_{oc} - \frac{AkT}{q} \ln \left(1 + \frac{qV_m}{AkT} \right) \tag{3.10}$$

Finally,

$$P_m = I_m V_m \approx I_{ph} \left[V_{oc} - \frac{AkT}{q} \ln \left(1 + \frac{qV_m}{AkT} \right) - \frac{AkT}{q} \right] \tag{3.11}$$

As shown in Fig. 3.18, if the PV cell operates at the left of the maximum power point, the voltage drops from the optimal operating voltage, and the output power will decrease; and when the voltage exceeds the optimal operating voltage, the power will also drop with increasing voltage.

Figure 3.19 shows the voltage–power curve of a 200 W solar module. When the voltage applied on the module increases, the power gradually increases, but there is a maximum value. If the voltage continues to rise, the power will decrease. And when the voltage eventually reaches to the open-circuit voltage, the power becomes zero. The V–P characteristic curve will decrease with the solar irradiance decreases.

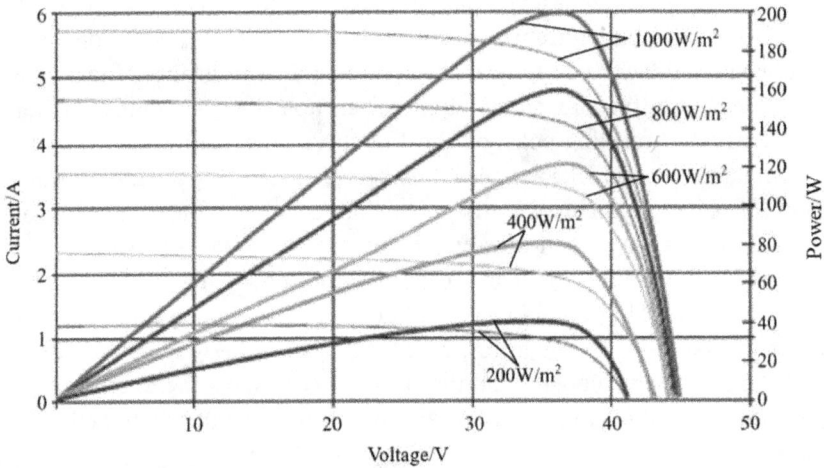

Fig. 3.19: Voltage–power curve for a component.

The power indicated by a PV cell usually refers to the power corresponding to the maximum power point under standard operating conditions. In the actual operation, it is not usually performed under STC, and generally does not necessarily meet the conditions of the best load. In addition to the fact that the solar irradiance and temperature are constantly changing during the day, there is very little time to actually achieve the rated output power. Some PV systems use the "maximum power tracker" to increase the output power to a certain extent.

3.3.3.3 Open-circuit voltage

Under certain temperature and irradiance conditions, the terminal voltage of a PV cell under no-load (open-circuit) conditions, that is, the voltage at which the I–V curve intersects with the abscissa, is usually represented by V_{oc}.

For a typical PV cell, it can be approximately regarded as an ideal PV cell, that is, the series resistance of the PV cell is zero and the shunt resistance is infinite. In the open-circuit condition, $I = 0$, the voltage V is the open-circuit voltage V_{oc}, we can see from eq. (3.7):

$$V_{oc} = \frac{AkT}{q} \ln\left(\frac{I_{ph}}{I_0} + 1\right) \approx \frac{Akt}{q} \ln\left(\frac{I_{ph}}{I_0}\right) \tag{3.12}$$

The open-circuit voltage V_{oc} of the PV cell is independent of the cell area, and the open-circuit voltage of crystalline silicon PV cell is general about 600–650 mV.

3.3.3.4 Short-circuit current

Under certain temperature and irradiance conditions, the output current of the PV cell when the terminal voltage is zero corresponds to the point where the I–V curve intersects with the ordinate, which is usually represented by I_{sc}.

From eq. (3.5), $I_{sc} = I_{ph}$, when $V = 0$.

The short-circuit current I_{sc} of the PV cell is related to the area of the PV cell, and the larger the area, the larger I_{sc}. The I_{sc} value for a typical 1 cm^2 crystalline silicon PV cell is approximately 35–38 mA.

3.3.3.5 Fill factor (curve factor)

The fill factor is an important parameter to characterize the performance of a PV cell. It is defined as the ratio of the maximum power of a PV cell to the product of the open-circuit voltage and the short-circuit current. It is usually expressed in terms of FF (or CF):

$$FF = \frac{I_m V_m}{I_{sc} V_{oc}} = 1 - \frac{AkT}{qV_{oc}} \ln\left(1 + \frac{qV_m}{AkT}\right) - \frac{AkT}{qV_{oc}} \tag{3.13}$$

where $I_{sc} V_{oc}$ is the limit output power of the PV cell, and $I_m V_m$ is the maximum output power of the PV cell.

In Fig. 3.18, the rectangular area A is surrounded by the vertical line through the open-circuit voltage V_{oc} and the horizontal line through the short-circuit current I_{sc} as well as the ordinate and the abscissa are the limit output power that the PV cell can reach. The rectangular area B is surrounded by the vertical and horizontal lines of the maximum power point, and the ordinate and abscissa are the maximum output power of the PV cell. The ratio between the two is the FF of the PV cell, that is

$$FF = B/A$$

The smaller the series resistance of the PV cell and the greater the shunt resistance, the larger the FF, and the larger the area enclosed by the I–V curve of the PV cell, indicating that the I–V curve is closer to the square, which means that the maximum output power of the PV cell is closer to the limit output power, the better the performance.

3.3.3.6 PV cell conversion efficiency

The percentage of the maximum power of a PV cell and the total amount of radiation power incident on the PV cell is referred to as the conversion efficiency of the PV cell:

$$\eta = V_m I_m / A_t \cdot P_{in} \tag{3.14}$$

Among them, V_m and I_m are, respectively, the voltage and current at the maximum output power point, A_t is the total area of the PV cell including the area of the gate line (also called full area), and P_{in} is the power of incident light per unit area.

Sometimes the aperture area A_a instead of A_t is used, that is, the area occupied by the gate lines is subtracted from the total area, so that the calculated efficiency is a little bit higher.

Example 3.1 For a PV cell with an area of 100 cm^2 and maximum power of 1.5 W, what is the conversion efficiency of the PV cell?

Solution. Based on eq. (3.14),

$$\eta = V_m I_m / A_t \cdot P_{in} = 1.5/100 \times 10^{-4} \times 1,000 = 15\%$$

According to the statistics from the National Renewable Energy Laboratory (NREL), by the end of 2018, the historical development of the highest conversion efficiency from various types of PV cell laboratories is shown in Fig. 3.20 [8]. As shown in the figure, the highest laboratory conversion efficiency records for single junction PV cells and submodules are mentioned in Tab. 3.1, and the maximum conversion efficiency for laboratories of multijunction terrestrial PV cells and submodules is given in Tab. 3.2.

3.3.3.7 Current temperature coefficient

When the temperature changes, the output current of the PV cell will change. Under the specified test conditions, when the temperature changes by 1 °C, the variation value of the short-circuit current of the PV cell is called the current temperature coefficient, which is usually expressed by α:

$$I_{sc} = I_0(1 + \alpha \, \Delta T) \tag{3.15}$$

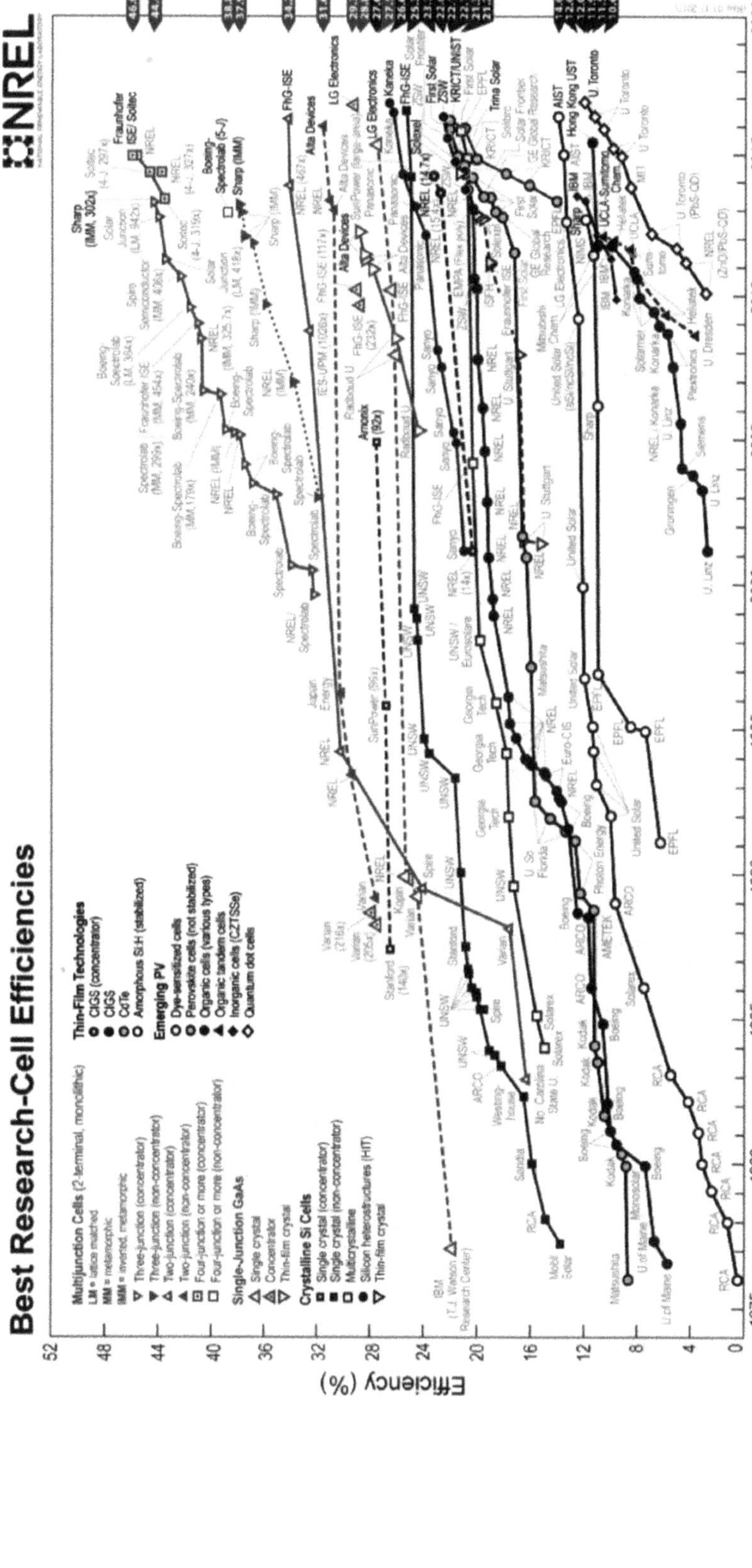

Fig. 3.20: Best research-cell efficiencies.
Source: https://www.nrel.gov/pv/assets/pdfs/pv-efficiency-chart.20181221.pdf

Tab. 3.1: Confirmed single-junction terrestrial cell and submodule efficiencies measured under the global AM 1.5 spectrum (1,000 W/m^2) at 25 °C.

Classification	Efficiency (%)	Area (cm^2)	V_{oc} (V)	J_{sc} (mA/cm^2)	FF (%)	Test center	Description
Silicon							
Si (crystalline cell)	26.3±0.5	180.43	0.7438	42.26	83.8	FhG-ISE	Kaneka, rear junction
Si (multicrystalline cell)	21.3±0.4	242.74	0.6678	39.80	80.0	FhG-ISE	Trina solar
Si (thin transfer submodule)	21.2±0.4	239.7	0.687	38.50	80.3	NREL	Solexel (35 μm thick)
Si (thin-film minimodule)	10.5±0.3	94.0	0.492	29.7	72.1	FhG-ISE	CSG solar (<2 μm on glass)
III–V cells							
GaAs (thin-film cell)	28.8±0.9	0.9927	1.122	29.68	86.5	NREL	Alta devices
GaAs (multicrystalline)	18.4±0.5	4.011	0.994	23.2	79.7	NREL	RTI, Ge substrate
InP (crystalline cell)	22.1±0.7	4.02(t)	0.878	29.5	85.4	NREL	Spire, epitaxial
Thin-film chalcogenide							
CIGS (cell)	21.0±0.6	0.9927	0.757	35.70	77.6	FhG-ISE	Solibro, on glass
CIGS (minimodule)	18.7±0.6	15.892	0.701	35.29	75.6	FhG-ISE	Solibro,4 serial cells
CdTe (cell)	21.0±0.4	1.0623	0.8759	30.25	79.4	Newport	First solar, on glass
CZTSSe (cell)	9.8±0.2	1.115	0.5073	31.95	60.2	Newport	IMRA Europe
CZTS (cell)	7.6±0.1	1.067	0.6585	20.43	56.7	NREL	UNSW
Amorphous/microcrystalline							
Si (amorphous cell)	10.2±0.3	1.001	0.896	16.36	69.8	AIST	AIST
Si (microcrystalline cell)	11.8±0.3	1.044	0.548	29.39	73.1	AIST	AIST
Perovskite							
Perovskite (cell)	19.7±0.6	0.9917	1.104	24.67	72.3	Newport	KRICT/UNIST
Perovskite (minimodule)	12.1±0.6	36.13	0.836	20.20	71.5	AIST	SJTU/NIMS, 10 serial cells

Tab. 3.1 (continued)

Classification	Efficiency (%)	Area (cm²)	V_{oc} (V)	J_{sc} (mA/ cm²)	FF (%)	Test center	Description
Dye sensitized							
Dye (cell)	11.9±0.4	1.005	0.744	22.47	71.2	AIST	Sharp
Dye (minimodule)	10.7±0.4	26.55	0.754	20.19	69.9	AIST	Sharp, 7 serial cells
Dye (submodule)	8.8±0.3	398.8	0.697	18.42	68.7	AIST	Sharp, 26 serial cells
Organic							
Organic (cell)	11.2±0.3	0.992	0.780	19.30	74.2	AIST	Toshiba
Organic (minimodule)	9.7±0.3	26.14	0.808	16.47	73.2	AIST	Toshiba (8 series cells)

Generally, for crystalline silicon PV cells: $\alpha = +(0.06–0.1)\%/$ °C, which means that the short-circuit current rises slightly when the temperature rises.

3.3.3.8 Voltage temperature coefficient

When the temperature changes, the output voltage of the PV cell also changes. Under the specified test conditions, the change value of the open-circuit voltage of the PV cell is called the voltage temperature coefficient for each temperature change of 1 °C. It is usually expressed by β:

$$V_{oc} = V_0(1 + \beta \Delta T) \tag{3.16}$$

Generally, for crystalline silicon PV cells: $\beta = -(0.3–0.4)\%/$ °C, which means that the open-circuit voltage will drop when the temperature rises.

3.3.3.9 Power temperature coefficient

When the temperature changes, the output power of the PV cell also changes under STC. The change value of the output power of the PV cell when the temperature changes to 1 °C is called the power temperature coefficient, which is usually expressed by y. Since $I_{sc} = I_0(1 + \alpha \Delta T)$, $V_{oc} = V_0(1 + \beta \Delta T)$, I_0 is the short-circuit current at 25 °C, V_0 is the open-circuit voltage at 25 °C, so the theoretical maximum power is

Tab. 3.2: Confirmed multiple-junction terrestrial cell and submodule efficiencies measured under the global AM 1.5 spectrum (1,000 W/m^2) at 25 °C.

Classification	Efficiency (%)	Area (cm^2)	V_{oc} (V)	J_{sc} (mA/cm^2)	FF (%)	Test center	Description
III–V multijunctions							
Five-junction cell (bonded) 2.17/1.68/1.40/1.06/0.73 eV	38.8±1.2	1.021	4.767	9.564	85.2	NREL	Spectrolab
InGaP/GaAs/InGaAs	37.9±1.2	1.047	3.065	14.27	86.7	AIST	Sharp
GaInP/GaAs (monolithic)	31.6±1.5	0.999	2.538	14.18	87.7	NREL	Alta devices
Multijunctions with c-Si							
GaInP/GaInAs/Ge;Si (spectral split minimodule)	34.5±2.0	27.83	2.66/0.65	13.1/9.3	85.6/79.0	NREL	UNSW/Azur/Trina
GaInP/Si (mech stack)	30.5±2.0	1.005	1.45/0.69	15.3/21.5	85.1/78.2	NREL	NREL/CSEM,4-terminal
GaInP/GaAs/Si (wafer bonded)	30.2±1.1	3.963	3.046	11.9	83.0	FhG-ISE	Fraunhofer ISE
GaInP/GaAs/Si (monolithic)	19.7±0.7	3.943	2.323	10.0	84.3	FhG-ISE	Fraunhofer ISE
Perovskite/Si (monolithic)	23.6±0.6	0.990	1.651	18.09	79.0	NREL	Stanford/ASU
a-Si/nc-Si multijunctions							
a-Si/nc-Si /nc-Si (thin-film)	14.0±0.4	1.045	1.922	9.94	73.4	AIST	AIST
a-Si/nc-Si (thin-film cell)	12.7±0.4	1.000	1.342	13.45	70.2	AIST	AIST

Source: Martin A. Green: Prog. Photovoltaics: Res. Appl. 2017; 25:3–13

$$P_{max} = I_{sc} V_{oc} = I_0 V_0 (1 + \alpha \Delta T)(1 + \beta \Delta T)$$

$$= I_0 V_0 (1 + (\alpha + \beta) \Delta T + \alpha \beta \Delta T^2)$$

Ignoring the square term, we obtain

$$P_{max} = P_0 [1 + (\alpha + \beta) \Delta T] = P_0 (1 + \gamma \Delta T) \tag{3.17}$$

For example, for the M55 mono c-Si PV cell module, where $\alpha = 0.032\%/$ °C and $\beta = -0.41\%/$ °C, the theoretical maximum power temperature coefficient is $\gamma = -0.378\%/$ °C. Figure 3.21 shows the I–V curve of a certain PV cell at different temperatures. It can be seen that when the temperature changes, the voltage change is relatively large and the change in the current is relative small.

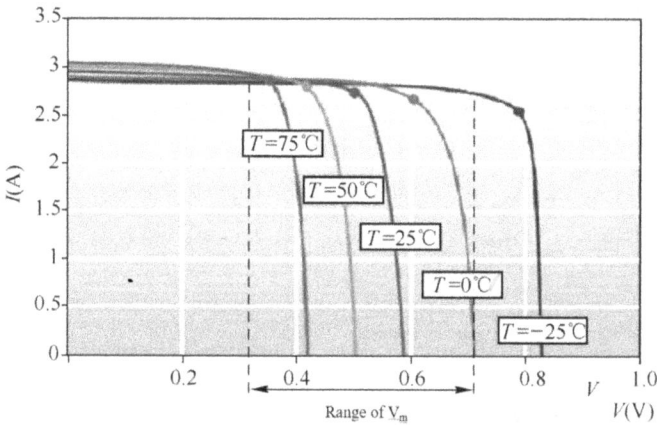

Fig. 3.21: I–V curve of a PV cell at different temperatures.

For common crystalline silicon PV cells, $\gamma = -(0.35{-}0.5)\%/$ °C. In fact, different PV cells have some differences in temperature coefficient, and a-Si PV cells have a smaller temperature coefficient than crystalline silicon cells.

In general, when the temperature rises, although the operating current of the PV cell increases, the operating voltage decreases, and the latter decrease is more drastic, so the total output power should be reduced and the PV cell should work at relatively low temperature.

3.3.3.10 Effect of solar irradiance

The open-circuit voltage V_{oc} of the PV cell is related to the magnitude of the incident light irradiance. When the irradiance is weak, the open-circuit voltage changes approximately linearly with the incident light irradiance; when the solar irradiance is strong, the open-circuit voltage changes logarithmically with the incidence light irradiance, that is, when the irradiance changes from weak to strong, the open-circuit voltage rises fast at the beginning, but the speed of rising will decrease when the solar irradiance is strong.

In the case where the irradiance of the incident light is not much larger than the STC (1,000 W/m²), the short-circuit current I_{sc} of the PV cell is proportional to the irradiance of the incident light. Figure 3.22 shows the $I–V$s of a PV cell under different irradiances. It can be seen that when the irradiance of incident light becomes timely strong, the short-circuit current of the PV cell will also be multiply increased, so the change of the incident light irradiance has a great influence on the short-circuit current of the PV cell.

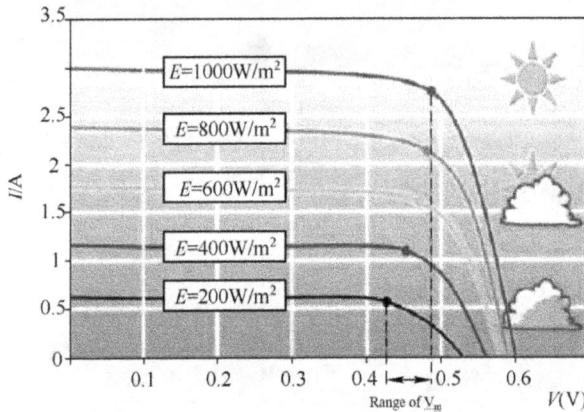

Fig. 3.22: $I–V$ curve of a certain PV cell under different irradiance.

The maximum power point of the PV cell also changes with the increase of solar irradiance. As shown in Fig. 3.22, when the solar irradiance changes from 200 to 1,000 W/m², the corresponding optimal operating voltage does not change too much, from 0.42 to 0.49 V. However, the short-circuit current changed from 0.6 to 3.0 A, which is an increase of nearly four times.

3.3.4 Factors affecting PV cell conversion efficiency

3.3.4.1 Bandgap
V_{oc} increases with the increase of E_g, I_{sc} decreases with the increase of E_g, so there is a best forbidden bandwidth to achieve the highest efficiency. As shown in Fig. 3.23, the peak efficiency is achieved within the bandgap ranging from 1.4 to 1.6 eV, and the peak efficiency increases from 26% to 29% when the solar spectrum changes from AM0 to AM1.5.

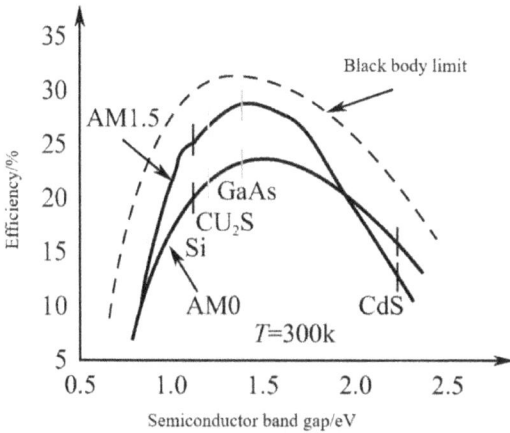

Fig. 3.23: Relationship between semiconductor bandgap and PV cell conversion efficiency.

3.3.4.2 Temperature

The temperature mainly makes a difference on V_{oc}, V_{oc} decreases with an increase in temperature, and the conversion efficiency η also decreases because of the dependence of I_0 on temperature. The I_0 equations on both sides of the p–n junction are as follows:

$$I_0 = qA\frac{Dn_i^2}{LN_D} \tag{3.18}$$

where q is the charge of one electron, D is the diffusivity of minority carriers in silicon, L is the diffusion length of minority carriers, N_D is the doping rate, and n_i is the intrinsic carrier concentration of silicon.

In the above equation, many parameters are affected by temperature, especially the intrinsic carrier concentration n_i. The intrinsic carrier concentration depends on the forbidden bandwidth (the lower the forbidden bandwidth, the higher the intrinsic carrier concentration) and the energy possessed by the carrier (the higher the carrier energy, the higher the concentration).

I_{sc} is less sensitive to temperature T. When the temperature rises, the short-circuit current I_{sc} will rise slightly, because the temperature rise reduces the bandgap of the semiconductor. When the bandgap decreases, more photons will be able to excite electron–hole pairs. However, this effect makes a little difference.

PV cell temperature sensitivity also depends on the value of the open-circuit voltage, that is, the greater the cell voltage, the less affected by the temperature.

For silicon PV cells, within a certain range, for every 1 °C increase in temperature, V_{oc} decreases by 0.4% at room temperature, and the efficiency also decreases by about the same percentage. For example, a silicon PV cell has an efficiency of 20% at 20 °C, and when the temperature rises to 120 °C, the efficiency is only 12%.

For GaAs batteries, for every 1 °C increase in temperature, V_{oc} reduces by 1.7 mV or the efficiency reduces by 0.2%.

3.3.4.3 Minority carrier lifetime

Long minority carrier lifetime will increase I_{sc} [9]. The long minority carrier lifetime will also reduce the dark current and increase the V_{oc}. In indirect bandgap semiconductor silicon, carriers usually have a lower probability of recombination than the direct bandgap, so the minority carrier lifetime is longer. In direct bandgap GaAs materials, minority carrier lifetime of more than 10 ns is very long.

The key to guarantee minority carrier lifetime of minority is to avoid the formation of recombination centers during material preparation and cell fabrication. In the fabrication process, proper processes can remove the recombination centers, thus prolonging the minority carrier lifetime. Therefore, reducing the recombination centers in silicon material and PV cell fabrication process is the key to improving the minority carrier lifetime.

3.3.4.4 Light intensity

The intensity of incident light affects the parameters of the PV cell, including short-circuit current, open-circuit voltage, FF, conversion efficiency, shunt resistance, and series resistance. The number of suns is usually used to describe the light intensity. For example, one sun is equivalent to the standard light intensity at AM1.5 air quality, that is, 1 kW/m². If the PV cell works under the light of 10 kW/m², it can be said to work under 10 suns. It is assumed that the light intensity increases by 10 times, and the J_{sc} per cell area will also increase by 10 times (removing the temperature effect), at the same time, V_{oc} also increases by (kT/q)In 10 times. The output power will increase, so the light concentration increases the conversion efficiency of the PV cell.

3.3.4.5 Doping concentration and profile distribution

Another factor that has a significant effect on V_{oc} is the doping concentration. N_D and N_A appear in the log term in the definition of V_{oc}, and their orders of magnitude is also easy to change. The higher the doping concentration, the greater the V_{oc} [10]. A phenomenon called heavy doping effect has aroused more attention in recent years. At high doping concentrations, N_D and N_A in all equations should be replaced by effective doping concentrations $(N_D)_{eff}$ and $(N_A)_{eff}$ due to the band structure deformation and changes in the electronic statistics, as shown in Fig. 3.24. Since $(N_D)_{eff}$ and $(N_A)_{eff}$ show peaks, then using a very high N_D and N_A is not significant. As the doping concentration increases, the effective doping concentration tends to saturate or even decrease, especially the minority carrier lifetime will decrease at high doping concentrations.

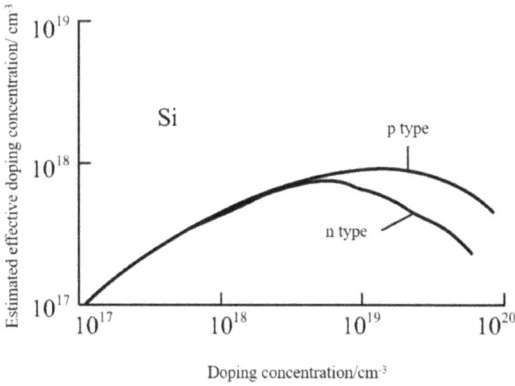

Fig. 3.24: Heavy doping effect.

The basic doping concentration is about 10^{16} cm^{-3} in silicon PV cells, and about 10^{17} cm^{-3} in direct bandgap material PV cells. In order to reduce the series resistance, the doping concentration of the front diffusion region is often higher than 10^{19} cm^{-3}; therefore, the heavy doping effect is significant in the diffusion region.

When N_D and N_A or $(N_D)_{eff}$ and $(N_A)_{eff}$ are not uniform and decrease toward the junction, an electric field is established whose direction contributes to the collection of photogenerated carriers and thus also improves I_{sc}. This nonuniform doped profile distribution is generally not available in the basic region of the cell, but it is natural in the diffusion zone.

3.3.4.6 Surface recombination rate

The low surface recombination rate helps to increase I_{sc} and improves V_{oc} due to the reduction of I_0 [11]. The aluminum back-surface field of the crystalline silicon PV cell forms a P+ layer on the back of the cell, and the electric field at the P/P+ junction prevents electrons from flowing toward the back-surface. Figure 3.25 depicts this structure.

There is an electronic barrier at the P/P+ interface. It is easy to form ohmic contact, where the electrons are also recombined. The recombination rate at the P/P+ interface can be expressed as follows:

$$S_n = \frac{N_A D_n^+}{N_A^+ L_n^+} \cot \frac{W_P^+}{L_n^+} \tag{3.19}$$

where $N_A{}^+$, $N_D{}^+$, $L_n{}^+$ are, respectively, the doping concentration, diffusion coefficient, and diffusion length in the P$^+$ region. If $W_P{}^+ = 0$, then $S_n = \infty$. If $W_P{}^+$ is comparable to $L_n{}^+$ and $N_A{}^+ \gg N_A$, then S_n can be estimated as zero; when S_n is small, I_{sc} and η will show a peak.

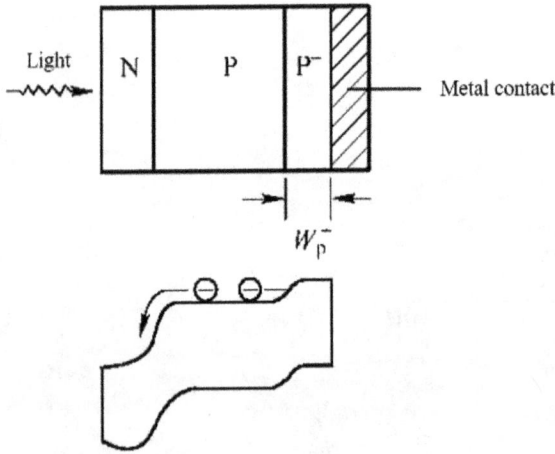

Fig. 3.25: Back-field cell.

3.3.4.7 Series and shunt resistors

In real PV cells, there must be a series resistance. The series resistance R_s is mainly composed of the bulk resistance of the semiconductor material, the contact resistance of the metal electrode and the semiconductor material, the resistance of the diffusion layer sheet, and the resistance of the metal electrode itself (as shown in Fig. 3.26), and the diffusion layer sheet resistance is the main part of the series resistance [12]. The greater the series resistance, the greater the output loss of the cell. Obviously, the series resistance can be reduced by designing the dense gate.

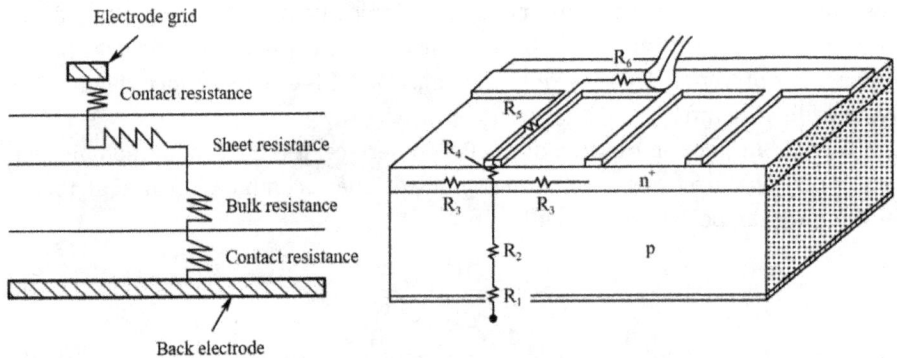

Fig. 3.26: Series resistance composition.

The shunt resistance R_{sh} is also called parallel resistance, leak resistance, or junction resistance. It is induced by the nonideality of the P–N junction and process defects and impurities in the vicinity of the junction, causing a local short circuit. Leakage current is proportional to the operating voltage.

The increase of the series resistance R_s results in that the voltage in the cell I–V curve decreases as the current increases. The decrease of shunt resistance R_{sh} results in that the current in the cell I–V curve decreases as the voltage increases. These two changes cause the cell I–V curve to deviate from the square, reducing the cell's FF, as shown in Fig. 3.27.

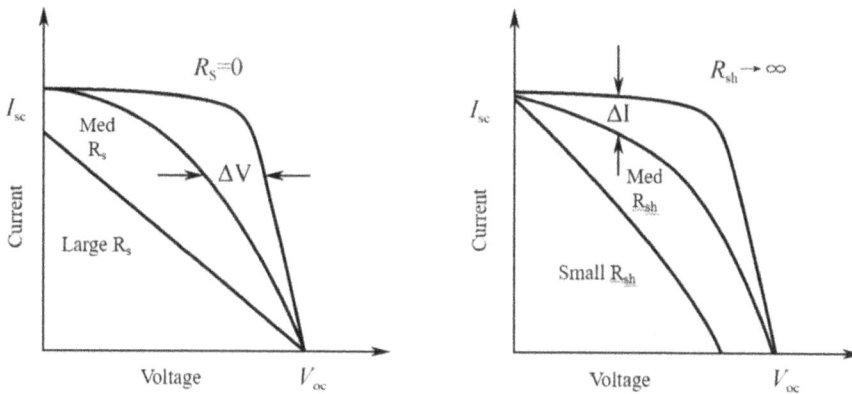

Fig. 3.27: Effect of series and shunt resistance on output characteristics of PV cells.

3.3.4.8 Absorption of light

The metal grid lines on the front of the PV cell cannot pass through the sunlight. To maximize I_{sc}, the smaller the light shielding area caused by the metal grid line is demanded. At the same time, in order to reduce R_s, the metal grid lines are generally made into dense and thin structures.

Because of the presence of sunlight reflection, not all light can enter the silicon. The reflectivity of the bare silicon surface is about 35%. The application of antireflection film can reduce the reflectivity. For a monochromatic light perpendicularly incident on the cell, theoretically a coating with a thickness of one-fourth wavelength and an index of refraction equal to $\sqrt{n}$ (n is the refractive index of silicon) can reduce the reflectivity to zero. For sunlight, multilayer coatings can achieve better results.

References

[1]   Sichou Qiu. Semiconductor surface and interface physics. Wuhan: Huazhong University of
       Science & Technology Press, 1995.
[2]   Enke Liu, Bingsheng Zhuand Jingsheng Lu, et al., The physics of semiconductors. Beijing:
       National Defense Industry Press, 2004.
[3]   Jiahua Wang, Changjian Li and Wencheng Niu. Physics of semiconductor devices. Beijing:
       Science Press, 1983.
[4]   E. S. Yang, E. K. Poon and C. M. Wu., et al. Majority carrier current characteristics in large-
       grain polycrystalline-silicon-Schottky-barrier solar cells. IEEE Transactions on Electron
       Devices, 1981, (28), 1131–1135.
[5]   M. M. Mohammad, K. C. Saraswat and T. I. Kamins. A model for conduction in polycrystalline
       silicon-part 1: Theory. IEEE Transactions on Electron Devices, 1981, 28(10), 1163–1176.
[6]   M. A. Green. Solar cells operating principles, technology and system applications[M]. Sidney:
       The University of New South Wales, 1982.
[7]   Zhongzheng Zhang, Xiaofang Cheng and Jinlong Liu. Study on load resistance output power
       of nonlinear solar cells. Acta Energiae Solaris Sinica, 2015, 36(6), 1474–1480.
[8]   M. A. Green et al. Solar cell efficiency tables (version 49) [M]. Progress in Photovoltaics:
       Research and Applications, 2017, 25, 3–13. © 2016 John Wiley & Sons, Ltd.
[9]   C Donolato. Modeling the effect of dislocations on the minority carrier diffusion length of a
       semiconductor. Journal of Applied Physics, 1998, 84, 2656–2664.
[10]   Martin A. Green. Present and future of crystalline silicon solar cells. Technical digest of the
       international PVSEC-14, Bangkok, Thailand, 2004.
[11]   S. A. Edmiston, G. Heiser and A. B. Sproul, et al. Improved modeling of grain boundary
       recombination in bulk and p-n junction regions of poly-crystalline silicon solar cells. Journal
       of Applied Physics, 1996, 80(12), 6783–6795.
[12]   S. W. Glunz. New concepts for high-efficiency silicon solar cells[C]. Technical Digest of the
       International PVSEC-14, Bangkok, Thailand, 2004.

Exercises

3.1   Outline the classification of PV cells.

3.2   What are N-type semiconductors and what are P-type semiconductors?
       How are they formed?

3.3   Briefly describe the formation of P–N junctions.

3.4   Briefly describe the working principle of silicon PV cells.

3.5   The energy of photons is hv (h is Planck's constant 6.626×10^{-34} J · s, v is the
       frequency of the light wave). Try to find the photon energy of red light with a
       wavelength of 650 nm.

3.6   The forbidden bandwidth of silicon is known to be 1.12 eV. Can infrared light
       of 1,200 nm wavelength be excited to emit electrons after being absorbed by
       silicon?

3.7   What are STCs for terrestrial PV cells?

3.8   What is open-circuit voltage and short-circuit current? Which external factors
       have a great impact on them?

3.9   The short-circuit current of a PV cell is 6.5 A. What is the short-circuit current of the PV cell when the solar irradiance is 800 W/m^2?

3.10 The open-circuit voltage of a solar module is 33.2 V, and the voltage temperature coefficient is −0.34%/ °C. When the module temperature is 50 °C, what is the open-circuit voltage of this module?

3.11 The size of a poly c-Si PV cell is 156 mm × 156 mm, V_{oc} = 625 mV, I_{sc} = 8.2 A, and FF = 79.5%. What is the conversion efficiency of the cell?

3.12 A certain size of photovoltaic module consists of mono c-Si PV cells with a conversion efficiency of 18.2%. The effective area of the module is 14,858 cm^2, and the actual measured power is 175 W. How much is the power loss of the PV module package? What is the conversion efficiency of PV module?

3.13 What are the main technical parameters for measuring PV cell performance?

3.14 What are the factors that affect the conversion efficiency of PV cells?

Chapter 4
Thin-film PV cells

4.1 Introduction

Crystalline silicon (c-Si) photovoltaic (PV) cell has been dominant in the PV industry for 40 years since the ground application started, due to its high efficiency and stability. However, conventional c-Si PV cell requires massive amount of semiconductor materials, which results in high price. Thin-film PV cell, on the other hand, reduces the amount of active material in a PV cell; therefore reduced the price and attracted interest on PV market. Thin-film PV cell has been applied to building integrated photovoltaic (BIPV) and as semitransparent, PV glazing material can be used as window glass [1].

4.1.1 Advantages and disadvantages of thin-film PV cells

Compared to c-Si PV cell, thin-film PV cell has the following advantages:

4.1.1.1 Low production cost
It can be produced at a relatively low reaction temperature of 200 °C, thus can be deposited onto glass, stainless steel plate, aluminum foil, ceramic plate, or polymer substrates, making mass production easier.

4.1.1.2 Less material
Thin-film materials have higher light absorption coefficient, thus the thickness can be as thin as 1 μm, whereas c-Si PV cell is required to be about 180 μm. Moreover, without being cut into slices like c-Si, there are fewer materials wasted.

4.1.1.3 Simple manufacturing process
It allows continuous, automated mass production. Thin-film PV cell is usually fabricated by plasma-enhanced chemical vapor deposition (PECVD) or magnetron sputtering, both can be highly automated. The manufacturing process can be performed continuously in multiple vacuum deposition chambers or done individually in one deposition chamber, making mass production possible.

https://doi.org/10.1515/9783110524833-004

4.1.1.4 Lower electricity consumption during production
Amorphous silicon (a-Si) that is used in thin-film PV cells can be produced with chemical vapor deposition (CVD), where the substrate temperature usually ranges from 200 to 300 °C, while the electrodes also have a lower discharge power density, thus largely reduced the electricity consumption compared to c-Si production.

4.1.1.5 Better high-temperature performance
Power output of PV module decreases with increasing module temperature. However, thin-film PV cells have a smaller temperature coefficient, therefore, less affected under high temperature. For example, a mono c-Si solar power plant that has theoretical power output of 1 MW only has a power output of 800 kW when the module temperature is 65 °C. However, with CdTe modules at the same temperature, the power output can be as high as 900 kW.

4.1.1.6 Better low irradiance performance
a-Si PV cell has a wider visible light spectral response range, as well as the ability to absorb scattered light, giving it a better low irradiance performance in real application. Compared to c-Si PV cell with the same nominal output power, a-Si PV cell has a higher power generation.

4.1.1.7 Suitable for BIPV
Thin-film PV cell can be made into BIPV component with different light transmittance and color, such as glass curtain wall. Moreover, it can be made on flexible stainless steel or polymer substrate, which is ideal for curved rooftop. It can also be made into foldable package for power supply of miniature instrument, computer, and portable devices in fields such as military and communication.

Thin-film PV cell also has some disadvantages:
- Low conversion efficiency: mass-produced a-Si module only has half the efficiency compared to c-Si module.
- Requires more installation area: when compared with c-Si module with the same power output, thin-film solar module requires significantly more installation space.
- Less stable: a-Si PV cell is more prone to photodegradation, accompanied by decreasing conversion efficiency under higher light intensity. This phenomenon greatly affected the wide application of thin-film PV cell.
- Higher fixed asset investment: the production process requires specific processing equipment and advanced production environment, which need higher initial investment.

4.1.2 Categories of thin-film PV cells

The different types of thin-film PV cells can be categorized by which PV material is deposited onto the substrate: silicon-based [a-Si, microcrystalline silicon (μc-Si), and polycrystalline silicon); cadmium telluride (CdTe); copper indium gallium selenide (CIS/CIGS); gallium arsenide (GaAs); dye-sensitized cell (DSC); organic PV cells (OPC); and perovskite PV cell.

Currently, CdTe, CIGS, and a-Si are three thin-film technologies often used for outdoor applications, and the remaining categories are still under development.

4.2 Amorphous silicon PV cell

4.2.1 Development history of a-Si PV cell

The a-Si thin film was obtained by R. C. Chitteck et al. through PECVD in the 1970s. In 1976, Cave and Chris Wronski from RCA Laboratories made the first a-Si thin-film PV cell with conversion efficiency of 2.4%. In 1979, Usui and Kikuchi reported that based on the original a-Si manufacturing technology, hydrogenated nanocrystalline silicon film (nc-Si:H) was made by introducing hydrogen during the production process. In 1980, Sanyo Electronics from Japan successfully made miniature calculator with a-Si PV cell. After the research development during the 1980s, the conversion efficiency and stability of a-Si PV cell made significant breakthrough [2]. In the year 1988, a-Si BIPV was put into application.

Starting from the 1990s, multijunction a-Si PV cell was put under development in order to further improve efficiency and stability problem, a-Si module with area under 1 m^2, and efficiency about 6% had become predominant [3]. In the second half of 2006, Applied Materials, the largest semiconductor supplier in the world, became interested in the development of PV industry. They made a production equipment with thin-film-transistor liquid-crystal display (TFT-LCD) that provides solution for PECVD and physical vapor deposition (PVD), and started producing thin-film PV cell with the 8.5th generation of TFT-LCD equipment. Their integrated production line is capable of producing 40 MW single-junction a-Si PV cell, and achieved 6% efficiency for 5.72 m^2 PV module. They pushed even further in 2008 and launched the production line for 65 MW a-Si/μc-Si double-junction PV cell with the same module size. The largest a-Si module, 8.5th-generation a-Si/μc-Si module produced by SunFab production line from Applied Materials, has a size of 2.2 m × 2.6 m and efficiency of 8%, as well as a stable power output of 458 W. Meanwhile, Oerlikon from Switzerland, ULVAC from Japan, and Jusung from Korea, all based on their experience in TFT-LCD industry, and provided five generations of integrated production line that produces a-Si/μc-Si module with an efficiency of 8–12% and size of 1.1 m × 1.3 m or 1.1 m × 1.4 m. Hanergy obtained the controlling stake of Apollo Solar Energy Technology by purchasing shares in September 2011. From then, Hanergy obtained silicon thin-film PV cell production line to improve from the US-based EPV, which can produce a module with a size of 0.635 m × 1.245 m and efficiency of 8%. Currently, part of the production line is still in operation.

Due to the complex properties of hydrogenated a-Si alloy, its properties are still being studied and relevant theories are being discovered. Although a-Si PV cell has made great progress, its basic theory, equipment, and process are still under development, compared to conventional c-Si PV cell.

4.2.2 Structure of a-Si PV cell

4.2.2.1 Single-junction a-Si PV cell

Silicon is a fourfold coordinated atom that is normally tetrahedrally bonded to the four neighboring silicon atoms. In a-Si, the atoms form a continuous random network. Moreover, some atoms have a dangling bond that can be passivated with hydrogen atom to form hydrogenated a-Si (a-Si:H). In device quality-level a-Si thin film, there are about 5–15% hydrogen content.

Light absorption coefficiency of common PV cell materials is shown in Fig. 4.1. It can be observed that a-Si:H material has a larger light absorption coefficiency (the solid line with circle in Fig. 4.1), higher photosensitivity, and similar distribution of

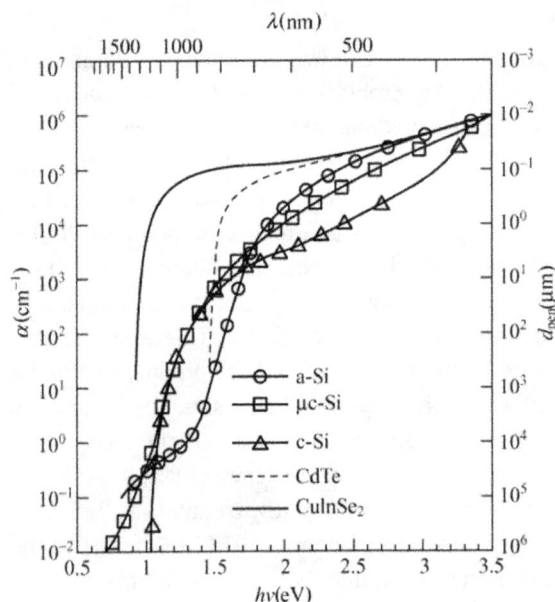

Fig. 4.1: Light absorption coefficient of common PV cell materials.

absorption peaks with sunlight, which makes it ideal for thin-film PV cell. However, there are some inevitable problems with it. First, its bandgap is about 1.7 eV. Materials with similar bandgap usually absorb sunlight within the visible light range and not sensitive to long wavelength light, which restricted the efficiency of a-Si PV cell. Second, hydrogenation is associated with light-induced degradation (LID) of the material, termed the Staebler–Wronski effect (SWE), which compromised the stability of the PV cell. People have been working on the LID problem of a-Si thin-film PV cell for a long time and made significant progress. By improving the performance of a-Si material and optimizing the light-trapping structure of the PV cell, Oerlikon has achieved over 11.0% efficiency for a-Si PV cell.

The a-Si thin-film PV cell can usually be divided into P–I–N type and N–I–P type. P–I–N type thin-film PV cell can be deposited onto substrate with high transparency such as glass or high-temperature resistance organic materials, while N–I–P type can be deposited on opaque materials such as stainless steel or organic materials with low transparency. Both the detailed structures are shown in Fig. 4.2 and the corresponding bandgap is shown in Fig. 4.3, with slanted lines inside bandgap representing band tail state. Due to the presence of band tail state in a-Si thin film, there could be some invalid photo-induced carriers. P region collects holes that have lower mobility ratio, and the distance between the photo-induced holes and the P region can be shortened by making P region facing the light. Thus, for a-Si PV cell, P region is usually the window layer of the PV cell, that is, the incident layer.

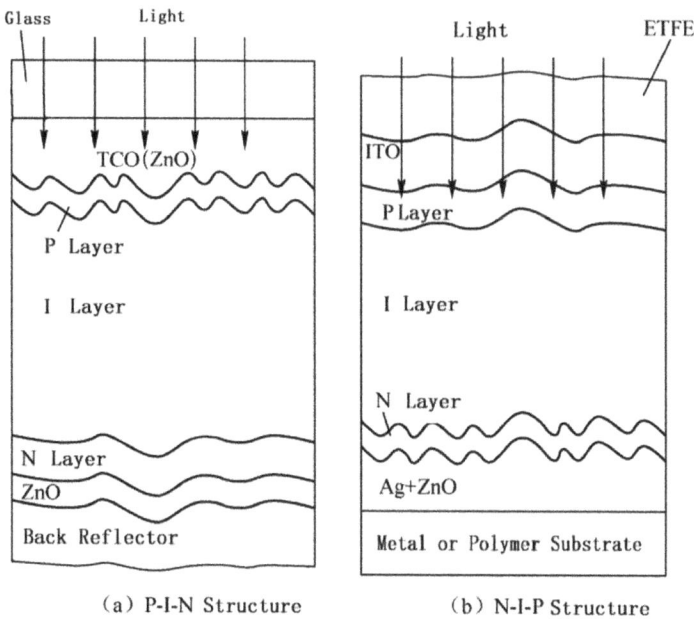

(a) P-I-N Structure (b) N-I-P Structure

Fig. 4.2: P–I–N- and N–I–P-type PV cell.

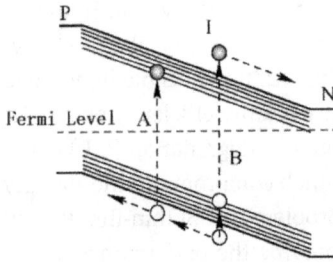

Fig. 4.3: Bandgap of P–I–N-type a-Si PV cell.

4.2.2.2 Multijunction PV cell

Multijunction PV cell utilizes the difference in photonic bandgap of materials through spectral splitting, which allows the absorbance of a broader range of wavelengths, improving the PV cell's sunlight to electrical energy conversion efficiency compared to conventional single-junction PV cell [4]. Meanwhile, each junction has a smaller thickness and improves the PV cell stability. Theoretically, open-circuit voltage of multijunction PV cell is the sum of open-circuit voltages of each subcell, while the short-circuit current is the smallest one of all short-circuit current of all subcells. Fill factor (FF) is determined by FFs and currents of all subcells. Thus, matching design is crucial in obtaining high efficiency for multijunction PV cell. On the junction of two subcells, N region of the top PV cell is connected to the P region of the bottom PV cell, which forms a P–N junction in the opposite direction where electrons move through the region by tunnel effects. To enhance this tunnel effect, in real application, microcrystalline P region and microcrystalline N region are used to increase the mobility of the carrier.

4.2.2.2.1 a-Si/a-Si double-junction structure

During the manufacturing process, different bandgaps can be achieved by adjusting the deposition parameter in the intrinsic layer of top PV cell and bottom PV cell. Since the adjustment range for a-Si bandgap is relatively small, intrinsic of the bottom PV cell is much thicker than that of the top PV cell in order to maintain sufficient current for the bottom PV cell. The thicknesses are usually 100 and 300 nm for top and bottom PV cells, respectively, and bandgaps are 1.8 and 1.7 eV, respectively. The corresponding PV cell structure and quantum efficiency (QE) are shown in Fig. 4.4, and the bandgap is shown in Fig. 4.5.

4.2.2.2.2 a-Si/a-SiGe double-junction structure

a-Si germanium (a-SiGe) is an ideal intrinsic material to improve the long wave response of the bottom PV cell. Bandgap of a-Si thin film can be lowered by adding germanium, and the bandgap can be adjusted by changing the ratio of germane to silane in plasma. The best Ge/Si ratio is 15–20% for a-SiGe double-junction PV cell,

(a) Structure
(b) QE Curve

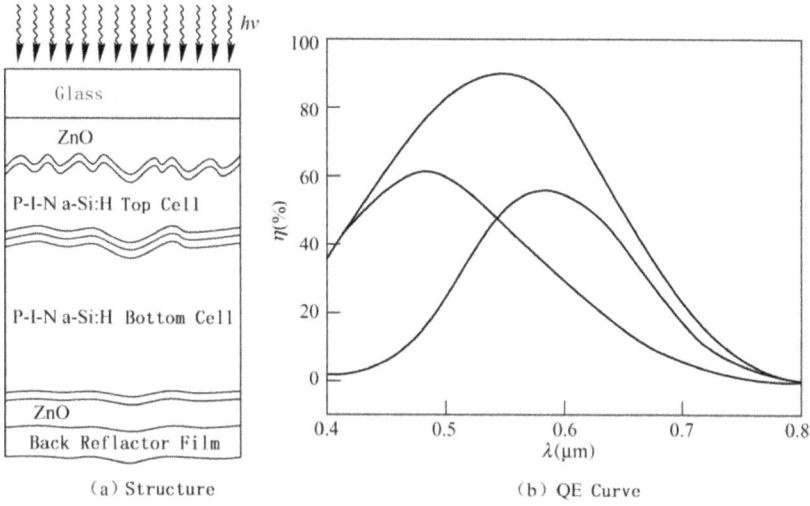

Fig. 4.4: a-Si/a-Si PV cell structure and QE curve.

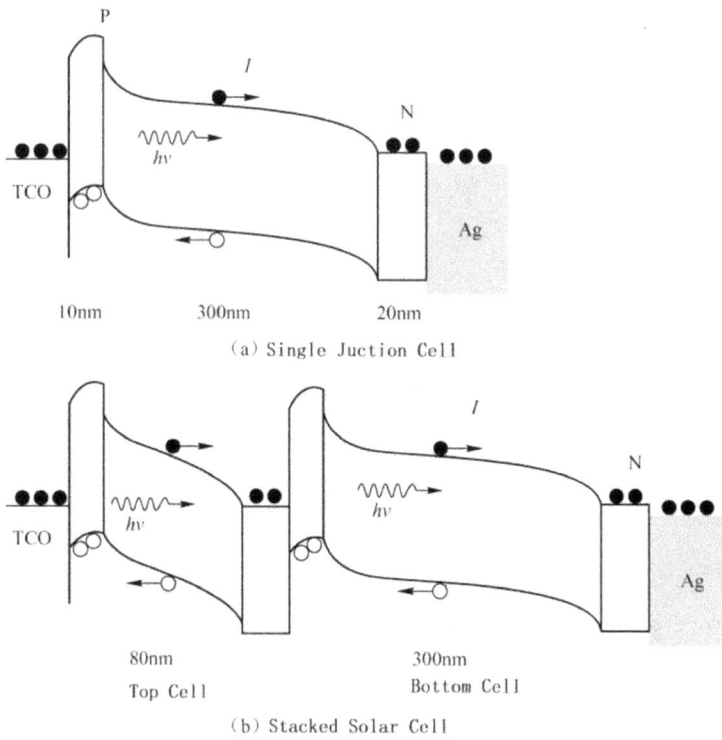

(a) Single Juction Cell

80nm
300nm

Top Cell
Bottom Cell

(b) Stacked Solar Cell

Fig. 4.5: Bandgap of a-Si single- and multijunction PV cells.

while the corresponding bandgap is about 1.6 eV. The structure of such PV cell is shown in Fig. 4.6.

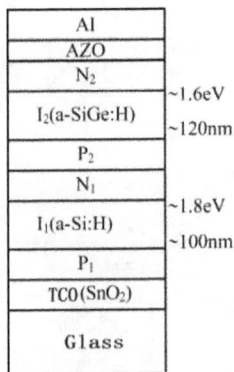

| Al |
| AZO |
| N_2 |
| I_2(a-SiGe:H) |
| P_2 |
| N_1 |
| I_1(a-Si:H) |
| P_1 |
| TCO(SnO_2) |
| Glass |

~1.6eV
~120nm

~1.8eV
~100nm

Fig. 4.6: Structure of a-Si/a-SiGe double-junction PV cell.

4.2.2.2.3 a-Si/µc-Si double-junction structure [5]

The bandgap of µc-Si is about 1.1 eV. Compared to a-SiGe, µc-Si has both better long wave response and stability. The structure of a-Si/µc-Si double-junction PV cell and corresponding QE curve is shown in Fig. 4.7.

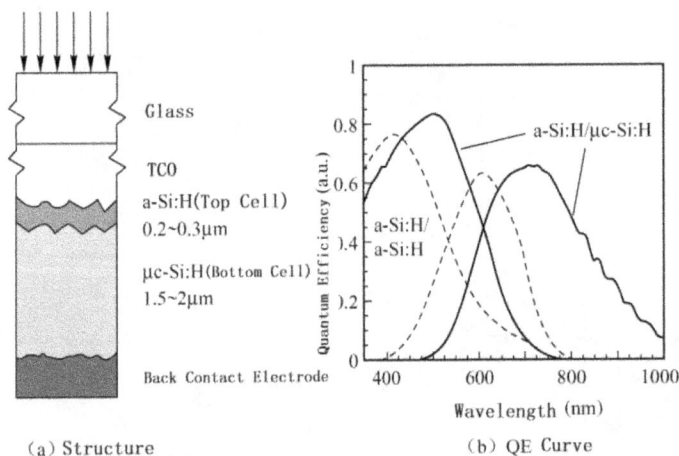

Glass

TCO

a-Si:H(Top Cell)
0.2~0.3µm

µc-Si:H(Bottom Cell)
1.5~2µm

Back Contact Electrode

(a) Structure

a-Si:H/µc-Si:H

a-Si:H/
a-Si:H

Quantum Efficiency (a.u.)

Wavelength (nm)

(b) QE Curve

Fig. 4.7: a-Si/µc-Si PV cell structure and QE curve.

Compared to a-Si PV cell, µc-Si PV cell has a much better current. Thus, a ZnO-reflective middle layer is used between junctions to reduce the thickness of a-Si PV cell (illustrated in Fig. 4.8). This method can enhance current matching as well as the QE of PV cell. By optimizing the reflectivity of the reflective middle layer, Kaneka

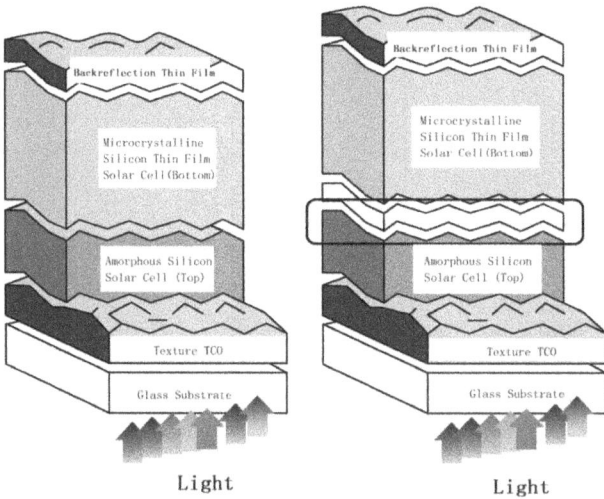

Light   Light

(a)Solar Cell Structure with Intermediate Reflectance

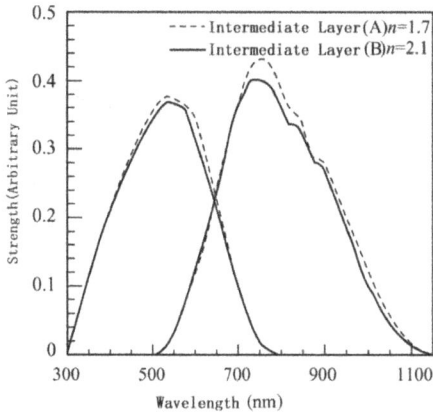

(b) Performance Comparison of Solar Cells with Different Refractive Index

Fig. 4.8: Performance of PV cells with different reflective middle layers.

Corporation of Japan obtained a-Si/μc-Si PV cell with efficiency of 14.7% (J_{SC} = 14.4 mA/cm^2, V_{OC} = 1.41 V, FF = 72.8%), and the module efficiency of 13.2%.

In a-Si/μc-Si PV cell, μc-Si thin film has to be over 1 μm thick. However, the deposition speed for μc-Si with current manufacturing process is relatively slow. Sanyo developed a localized plasma confinement (LPC-CVD) process, as shown in Fig. 4.9, which allows rapid manufacture of high-performance μc-Si thin film. It achieved the deposition speed of 4.1 nm/s by using pyramid nozzles and pumping holes, enabling stable plasma generation under very high pressure (1,000–2,000 Pa). Chemicals such as SiH$_2$ are evacuated out of the system during deposition, which improved the quality

Rapid Exhaust

ConicalNozzle

Gas Supply

Silicon

Plasma

Transparent conductive oxide film

Glass Substrate

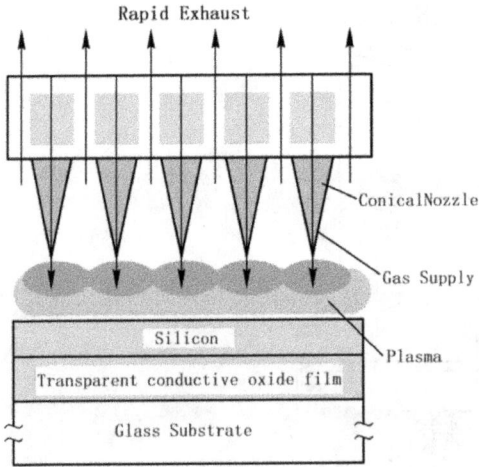

Fig. 4.9: Illustration of LPC-CVD process.

of product. With this technique, Sanyo achieved initial efficiency of 11.1% ($V_{OC} =$ 161.7 V, $I_{SC} = 1.46$ A, FF = 72.4%, $P_m = 171$ W) on 1.1 m × 1.4 m module, and stabilized the efficiency of 10%, while the deposition speed for μc-Si is over 2.4 nm/s.

a-Si/μc-Si double-junction PV cell is the dominant structure of high-efficiency thin-film PV cell. First, PV cell with this structure has higher conversion efficiency than the conventional a-Si PV cell. Second, it has better stability. Last but not least, production cost is lowered by using germane (GeH$_4$) during the manufacturing process.

4.2.2.2.4 a-Si/a-SiGe/a-SiGe triple-junction structure

Sunlight can be more effectively utilized through triple-junction PV cell. Its spectral response covers the spectral region of 300–950 nm, and the FF is also higher than that of a single-junction PV cell. Current of the top cell is usually the smallest one in all three PV cells in a triple-junction PV cell, thereby limiting the short-circuit current of the triple-junction PV cell and increasing the FF. The triple-junction structure can effectively improve the efficiency and stability of the PV cell, whose structure is shown in Fig. 4.10.

4.2.2.2.5 a-Si/a-SiGe/μc-Si triple-junction structure

The spectral response of the PV cell can be extended to 1,100 nm with a-Si/a-SiGe/μc-Si triple-junction structure, while it can also achieve the highest efficiency among all multijunction thin-film PV cells. Compared to the a-Si/a-Si/a-SiGe triple-junction PV cell, this PV cell has higher short-circuit current and FF, as well as improved stability.

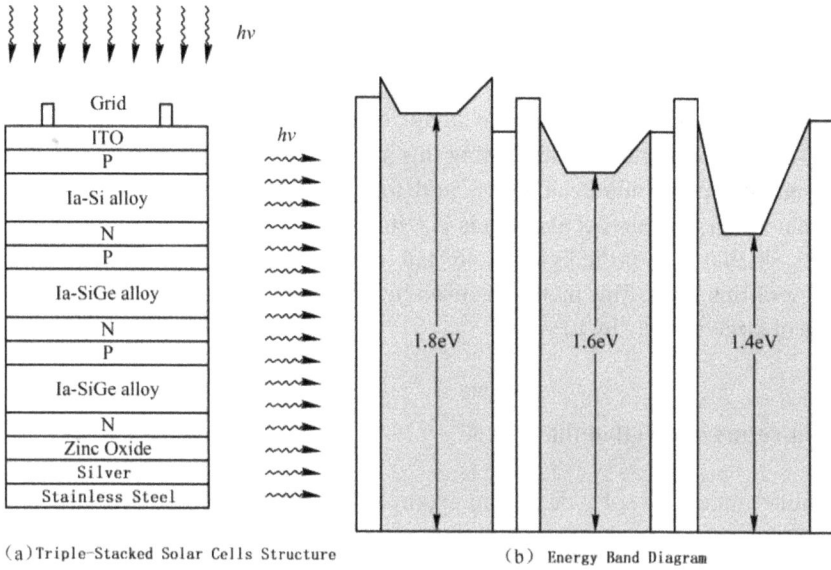

(a) Triple-Stacked Solar Cells Structure   (b) Energy Band Diagram

Fig. 4.10: a-Si/a-Si/a-SiGe triple-junction PV cell structure.

The US Uni-solar used this structure to obtain 16.3% thin-film PV cell, which is the highest efficiency in the world (as shown in Fig. 4.11). This was achieved by adding an N-region layer between a-SiGe/μc-Si PV cells, which constitutes the P–N tunnel junction. The N-region is a highly conductive nc-SiO$_x$:H layer. While it is acting as the N-region, this material also acts as a light-reflecting layer to increase the light absorption of intermediate a-SiGe, thus improving the current matching of junctions.

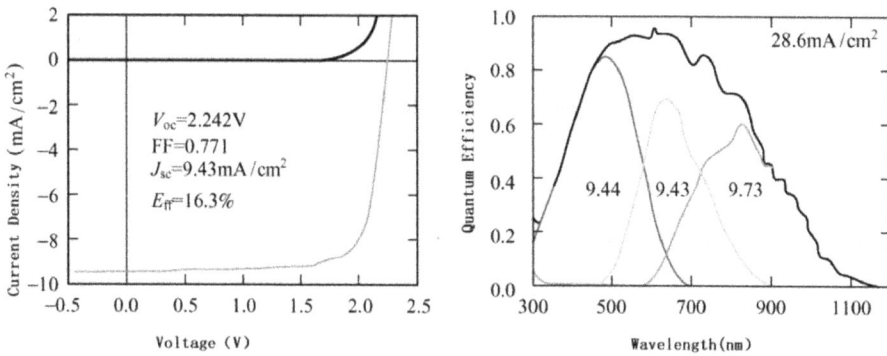

Fig. 4.11: $I–V$ and QE curves of Uni-solar triple-junction PV cell.

4.2.2.2.6 a-Si/nc-Si/μc-Si triple-junction structure

The use of a-Si/nc-Si/μc-Si triple-junction structure can effectively improve the stability of the PV cell. With this structure, Uni-solar obtained PV cell with an initial conversion efficiency of 14.1%, and the stabilized efficiency after 1,000 h of irradiance is 13.3%. The difficulty in fabricating this structure mainly focuses on the current mismatch among middle, bottom, and top layers. In order to improve the current matching, a semireflective film is inserted between the middle and the bottom layers, so that part of the light is reflected into the middle layer, and increased the current of this layer. This method is used by Sharp and increased the stabilized efficiency of large module by 10%.

4.2.3 Fabrication of a-Si thin-film PV cell

With double-junction a-Si PV cell as an example, the main process flow is shown in Fig. 4.12.

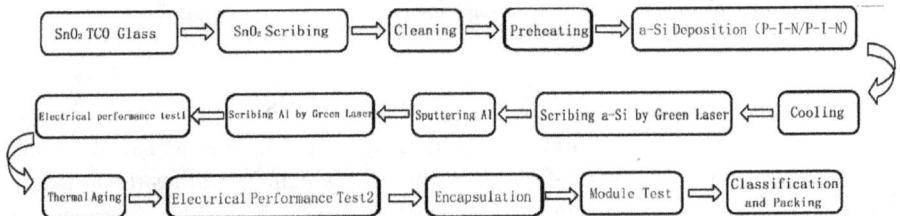

Fig. 4.12: Process flow of double-junction a-Si PV cell.

4.2.4 Commercialization of a-Si thin-film PV cell

In recent years, silicon-based thin-film PV cells, represented by a-Si, have seen slow progress in PV technology and their efficiency has been stagnated at around 10%. Many manufacturers based on these technologies have already withdrawn from the PV industry.

Hanergy, which specializes in thin-film PV cell, adopted several advanced PECVD technologies from Apollo Solar Energy Technology (original from CHRONAR or EPV) in several plants in China in 2016. a-Si thin-film module produced by them has a specification of 1,245 mm × 635 mm and nominal power of 60–65 W. Tokyo Electron (formerly Oerlikon) applied PECVD to produce nearly 100 MW of a-Si/a-SiGe/a-SiGe triple-junction PV modules and a-Si/nc-Si double-junction glass–glass modules, and specifications for these are 1,300 mm × 1,100 mm with nominal power of 110–130 W. Their product is almost the last one standing among the first industrialized thin-film PV cells.

4.3 Cadmium telluride PV cell

Although the conversion efficiency of CdTe PV cell is lower than that of the c-Si PV cell (its efficiency is still higher than that of a-Si PV cell), it has considerably low price and stable performance. Although CdTe PV cell has been studied since very early age, people were concerned about the toxicity of cadmium and did not put CdTe PV cell into large-scale application. It was later proved that as long as it is properly handled, no safety issues will arise in either the production process or during the application lifetime. With the promotion of the First Solar Company in the USA, it has become the fastest growing thin-film PV cell in recent years. At present, CdTe PV cell occupies the largest thin-film production and is widely used around the world.

In 1963, Cusano announced the first heterojunction CdTe thin-film PV cell with a structure of N-CdTe/P-Cu_{2-x}Te with an efficiency of 7%, but the P–N junction of this structure was poorly matched. In 1969, Adirovich first deposited CdS and CdTe thin films on transparent conductive glass, and developed the now commonly used CdTe PV cell structure. In 1972, Bonnet et al. reported on a PV cell with a conversion efficiency of 5–6%, using CdS_xTe_{1-x} film with graded bandgap as the absorber layer. In 1991, Chu et al. reported on an N-CdS/P-CdTe PV cell with a conversion efficiency of 13.4% [6]. In 2001, Wu et al. reported on an N-CdS/P-CdTe PV cell with an efficiency of 16.5%. In July 2011, First Solar announced that it had obtained PV cell with efficiency of 17.3%, which broke the record for CdTe PV cell. Through years of R&D investment, First Solar announced a new record in 2016 that has a conversion efficiency of 21.0%.

4.3.1 Characteristics of CdTe materials and cells

From the physical properties of CdTe, its main characteristics can be summarized as follows:

(1) CdTe is a II–VI compound semiconductor. It is a direct bandgap material with a visible light absorption coefficient of 104 cm^{-1} or more. Only 1 μm of CdTe is needed to absorb more than 99% of visible light (wavelength < 826 nm), that is only 1/100 the thickness mono c-Si. Therefore, CdTe can be manufactured with a small amount of absorbing layer material, which significantly reduced cost and energy consumption during the production.

(2) Being the most important physical property of CdTe, it has a bandgap width of 1.5 eV. After calculating the correlation between the conversion efficiency and the bandwidth of PV cell, the ideal data result confirmed that CdTe matches the terrestrial solar spectrum very well. GaAs PV cell, which has a very close bandgap with CdTe, has already achieved a conversion efficiency of 25%. It is believed that the conversion efficiency of CdTe PV cells still has room to be improved.

(3) The CdTe material is a binary compound. The Cd–Te bond energy is as high as 5.7 eV, and the cadmium element is extremely stable in nature. Therefore, it is chemically stable at room temperature and has a melting point of 321 °C. The temperature will not exceed 100 °C when the module is being used, and thus CdTe will not decompose and diffuse during regular operation and is insoluble in water, which makes CdTe very safe while used.

(4) The phase diagram of $Cd_{1-x}Te_x$ alloy is simple. When the temperature is lower than 320 °C, only solid CdTe (Cd:Te = 1:1) is present; and excess elemental Cd and Te are present when both Cd and Te are present. Without the formation of alloy with other ratio, the production process has a relatively low prerequisite, and leads to semiconductor thin film that is not sensitive to the environment and history of the preparation process. Therefore, the uniformity and yield of the product are high, and it is ideal for mass production.

(5) When the temperature is higher than 400 °C in a vacuum environment, the solid CdTe will sublime and form steam directly from the solid surface. However, when temperature is lower than 400 °C, or when the ambient pressure is higher, the sublimation rapidly weakens. This feature is not only favorable to vacuum rapid film preparation, such as closed-space sublimation (CSS) and vapor transport deposition (VTD), but also guarantees the safety during production of CdTe – once the equipment's vacuum or high-temperature environment is not present, CdTe vapor will quickly condense into solid particles or lumps and not diffuse into air to harm the human body.

(6) With respect to the PV cell itself, CdTe has a small temperature coefficient, and maintains a great performance under low irradiance level. As the energy gap of semiconductors changes with temperature, the efficiency of PV cells will be reduced. Therefore, compared with c-Si PV cells with the same nominal power in the same light environment, CdTe can generate on average 5–10% more energy throughout the year.

In CdTe thin-film PV cell, CdS is used as a window layer and CdTe the absorption layer. The energy band matching between the two layers and the electrodes is an important factor deciding PV cell conversion efficiency and the stability. However, cadmium and tellurium have strong toxicity and will become pollutants if not handled properly. These are the main obstacles to its development.

4.3.2 Structure of CdTe PV cell

CdTe thin-film PV cell can adopt a variety of structures such as homojunction or heterojunction. Currently, the N-CdS/P-CdTe heterojunction structure is most commonly used in the world. In this structure, N-CdS and P-CdTe heterocrystals mismatch and bandgap mismatch are small, thus PV cells with better performance can be obtained. Figure 4.13 shows the structure of high-efficiency CdTe thin-film PV cell.

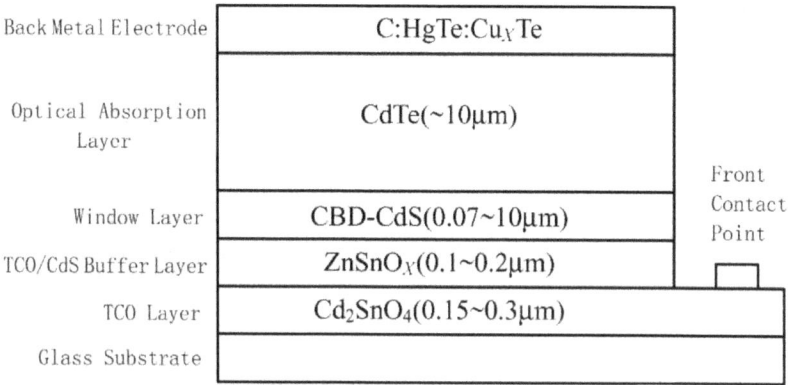

Back Metal Electrode	C:HgTe:Cu$_X$Te	
Optical Absorption Layer	CdTe(~10μm)	
		Front Contact Point
Window Layer	CBD-CdS(0.07~10μm)	
TCO/CdS Buffer Layer	ZnSnO$_X$(0.1~0.2μm)	
TCO Layer	Cd$_2$SnO$_4$(0.15~0.3μm)	
Glass Substrate		

Fig. 4.13: Structure of high-efficiency CdTe thin-film PV cell.

The efficiency of CdTe PV cell can be improved by the transparent conductive oxide (TCO) front electrode. Cd_2SnO_4 is a TCO material with high light transmittance and low electric resistance, which can effectively improve the light transmittance and electrode contact of the cell, thereby improving the short-circuit current and the FF of the cell. However, its cost is higher than the traditional SnO_2 conductive glass, which is currently widely used in the industry due to its low cost. ZTO ($ZnSnO_X$) is a buffer layer for CTO/CdS, which is a highly transparent and high-resistance material. Compared to the more traditional choice of buffer layer, $ZnSnO_4$, ZTO, and $ZnSnO_4$ have similar bandgap width ($\approx$3.6 eV) and resistivity (1–10 Ωcm after annealing), and the light transmittance of ZTO can be further improved after annealing. Using ZTO as a buffer layer, the localization of CdTe/TCO decreases when the CdS layer is thin. This is because ZTO has a larger bandgap and conductivity and matches CdS, and ZTO can also better prevent short circuit inside the cell after etching. Except from TCOs, intrinsic ZnO and the like can also be used for the TCO/CdS buffer layer.

After long-term experiments and theoretical studies, it has been found that CdS is the best heterojunction material to be used with CdTe [7]. Common preparation methods for CdS include chemical bath deposition (CBD), sputtering, and high vacuum evaporation (HVE). As the window layer, the thickness of CdS in CdTe cell needs to be controlled at about 100 nm, a thickness where the material cannot be crystallized well. Depositing $CdCl_2$ or vacuum heat treatment on the surface of the deposited CdS layer can recrystallize CdS and improve its contact with CdTe, which reduces the lattice mismatch.

4.3.3 Manufacturing of CdTe thin-film PV cells

4.3.3.1 Manufacture process of CdTe thin-film PV cell

Industrial CdTe thin-film PV cells are generally prepared on glass substrates. The process flow is shown in Fig. 4.14, which mainly focuses on the preparation of TCO (or purchasing SnO_2 conductive glass directly), deposition of TCO buffer layer, CdS, CdTe, metal buffer layer and metal back electrodes, and so on. In addition, the process includes laser scribing and laminating. During the CdTe manufacturing process (refer to Fig. 4.14), steps of "CdS" and "CdTe" require special equipment, and all other steps can use the common equipment with the silicon thin-film process. Due to space limitations, only the preparation of CdTe films is briefly described here.

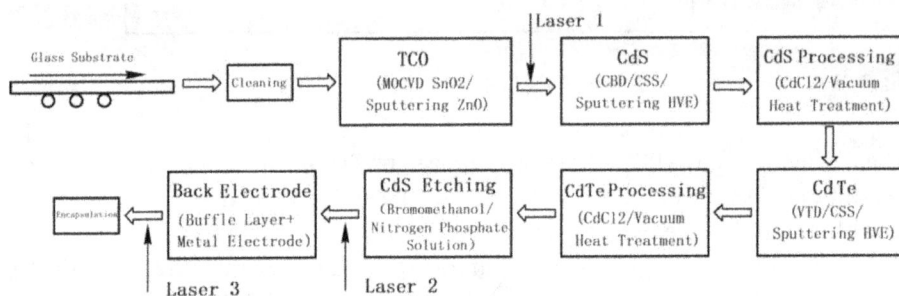

Fig. 4.14: Manufacturing process of CdTe thin-film PV cell.

4.3.3.2 CdTe thin-film preparation process

Commonly used preparation methods for N-CdS/P-CdTe thin films include CSS, VTD, sputtering, HVE, and electrodeposition. The methods for producing CdTe thin films that are proven to have commercialization value include CSS, VTD, and sputtering. Commonly used preparation methods for CdS include CBD, sputtering, CSS, and HVE. Figure 4.15 shows the commonly used preparation process of several CdTe thin films currently in the industry.

4.3.3.2.1 Vapor transport deposition

The core technology of First Solar is the VTD process. The principle is to load semiconductor powder through a preheated inert gas into a vacuum chamber and fully vaporize in a tumbling evaporation chamber to become saturated gas, and then to spray onto a colder glass substrate through the opening of the evaporation chamber, creating a supersaturated gas that condenses into a thin film. The advantages of this process are as follows:

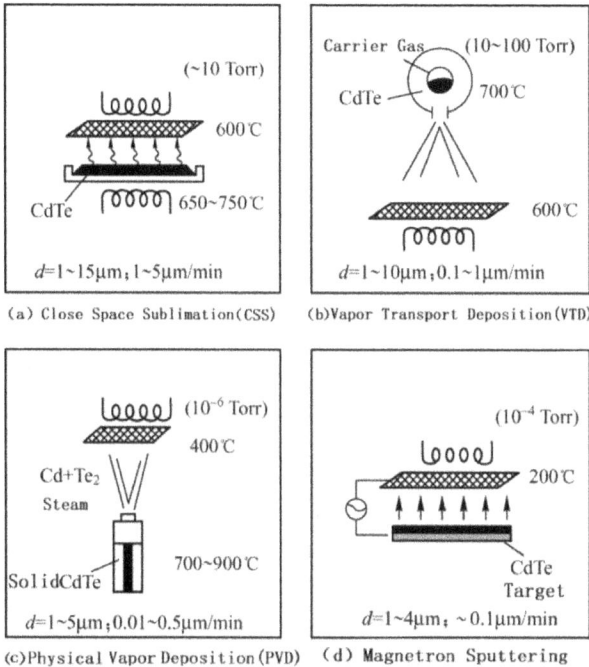

Fig. 4.15: Commonly used reparation processes of CdTe PV cells.

- No need to open the vacuum chamber to add or replace raw materials. The carrier gas is fed from the vacuum chamber during production, and lowering the time and cost for production and maintenance.
- The rapid deposition rate can not only meet the requirements of rapid production but also save semiconductor materials. The utilization rate of raw materials has reached nearly 90%.
- Easy to achieve uniform growth over a large area and high yields.

The disadvantage is that this technique only applies to materials that have large variations in saturated vapor pressure with changing temperature, and small variation in chemical composition and structure with changing temperature.

The patent for this technology is maintained confidential by First Solar. Some companies are also conducting independent research and development for this technique. For example, some engineers who left First Solar in Toledo, Ohio, USA, have successfully developed atmospheric pressure VTD technique based on the VTD technique, which saves equipment cost of the vacuum device. Some companies have also developed a technique for simultaneously coating VTDs with two parallel glass plates placed vertically. With the same equipment, production capacity of the same equipment can be doubled. The production of this company is expanding. The improved

VTD techniques of two companies are expected to be lower in cost than that of First Solar.

4.3.3.2.2 Closed-space sublimation

Compared to VTD, CSS can achieve faster thin-film deposition. The re-evaporation of Cd and Te from a CdTe substrate limits the rate and utilization of CdTe deposition on substrates above 400 °C. This can be suppressed by deposition at higher pressures ($\approx$1 Torr), but the migration of the material from the source to the substrate is controlled by diffusion; thus, the source and substrate must be very close. In CSS, CdTe source material is contained in a boat. The source boat and substrate cover act like radiant heater that transfers heat to the CdTe source and substrate, while the insulating sheet between the boat and the substrate acts as a heat insulator, and the temperature gradient between the boat and the substrate can be maintained during deposition.

The disadvantage of this technology is that it is usually not easy to control the thickness of CdTe thin film. Thickness over 10 nm can occur from time to time, which is much thicker than the required 1 nm. Large particle (5–10 nm) used made it hard to prepare ultrathin devices. The preparation process of high-efficiency PV cell requires the use of acid etching process to obtain a Te-rich surface layer and a Cu_xTe transition back electrode layer, which reduces the utilization rate of semiconductor materials and increases process complexity.

To achieve uniformity in a large area, there is a certain process requirement at each fill of the source boat for the smoothness of the source surface and the uniformity of heating over a large area. Due to the fast consumption of raw materials, the distance between the source and the substrate is very short, and it is necessary to frequently open vacuum equipment to replace or add raw materials, which increases the time and cost of maintenance.

The technical data of CSS has been disclosed. National Renewable Energy Laboratory (NREL) used this method to produce the most efficient small-area PV cell. An improved method of this technique is currently successfully commercialized by Abound Solar, USA. In 2009, they successfully produced competitive solar modules. Chinese manufacturers such as Advanced Solar Power have also achieved mass production with this method.

4.3.3.2.3 Magnetron sputtering

This technique forms deposition of CdTe thin films by RF magnetron sputtering onto compound targets. The transport of Cd and Te is achieved by bombardment of the CdTe target with Ar^+ ions and subsequent diffusion to the substrate and agglomeration. This technique has the lowest temperature in all CdTe production techniques that are currently being commercialized. Normally, the deposition is carried out at a substrate temperature of less than 300 °C. and a pressure of about 10^{-4} Torr.

Compared with VTD and CSS, magnetron sputtering technology has the advantages of simple process (no chemical corrosion), easy access to equipment, smooth film, small particles, better uniformity, easy control of deposition rate, and so on. It is applicable to ultrathin CdTe (≤1 nm), which makes it suitable for the application of BIPV. The disadvantage is that the deposition speed is relatively slow and does not apply to the production of thicker CdTe films.

There are also techniques such as PVD and HVE. Table 4.1 lists the latest efficiency records for CdTe PV cells and modules.

Tab. 4.1: Latest efficiency records of CdTe PV cells and modules.

Institution	Area (cm^2)	VOC (V)	JSC/ISC (mA/cm^2)/(A)	FF (%)	Efficiency (%)	Testing facility	Testing time	Type
First Solar	1.0623 (ap)	0.8759	30.25	79.4	21.0 ± 0.4	NEWPORT	Aug 2014	CdTe PV cell with glass substrate
First Solar	0.4798 (da)	0.8872	31.69	78.5	22.1 ± 0.5	NEWPORT	Nov 2015	CdTe PV cell with glass substrate
First Solar	7038.8 (ap)	110.6	1.533 (ISC/A)	74.2	18.6 ± 0.6	NREL	Apr 2015	CdTe module

Note: ap, aperture area; da, designated illumination area; J_{SC}, short-circuit current density.

4.3.4 Commercialization of CdTe thin-film PV modules

At present, First Solar stands out in all companies across the world that produces CdTe thin-film PV modules. Its predecessor, Solar Cell Inc., was established in 1986. It actively developed CdTe thin-film PV cells with the support of research institutions. After its acquisition in 1999, it was renamed to First Solar, whose production plants are located in Perrysburg, USA, and Kulim, Malaysia. In 2005, the production capacity was only 25 MW/year; in 2006, it began to expand rapidly after initial public offering (IPO). It produced 1.11 GW of CdTe thin-film solar modules in 2009, and became one of the top ten PV companies in the world. In 2010, it increased by 26% to reach 1.4 GW. The output in 2015 was 2.52 GW. At present, its global PV market share has reached to about 5%, and its products have been used in many large-scale PV power plants around the world. In the second quarter of 2016, its average conversion efficiency was 16.2%, and the highest conversion efficiency was 18.6%, which was equivalent to the conversion efficiency of poly c-Si PV modules.

After Professor Wu Xuanzhi, who created the CdTe PV cell record in the USA, returned to China, he founded Advanced Solar Power (Hangzhou) Inc. in May 2008.

The main technical parameters of the CdTe modules produced by Advanced Solar Power and First Solar are shown in Tab. 4.2.

Tab. 4.2: Main technical parameters of First Solar and Advanced Solar Power.

Institution	Country	Capacity in 2015	Module size	Highest efficiency	Average production efficiency	Technique
First Solar	USA	2.8 GW	0.6 m × 1.2 m	21.0%	16.2%	VTD
Advanced Solar Power	China	40 MW	0.6 m × 1.2 m	—	12.5%	CSS

4.4 Copper indium gallium selenide thin-film PV cell

The conversion efficiency of thin-film PV cell is lower than that of c-Si PV cell. With the significant decrease in the production cost in recent years, the development of thin-film PV cell has been greatly affected. At present, the highest efficiency among commercialized thin-film PV technologies is CIGS, which has stable performance and can be made into flexible PV cells. It has great potential in areas such as BIPV and portable power supply. Some people in the industry believe that CIGS is the most promising thin-film PV cells in the future.

The development of CIGS PV cells originated from Bell Laboratories, USA, in 1974. Wagner et al. first developed single-crystal $CuInSe_2$(CIS)/CdS heterojunction PV cell, and the efficiency of which increased to 12% in 1975. From 1983 to 1984, Boeing used ternary co-evaporation to produce CIS polycrystalline thin-film PV cell with a conversion efficiency of more than 10%, which attracted attention for film-type CIS PV cells. In 1987, ARCO made significant progress in this field. They obtained CIS thin-film PV cell with a conversion efficiency of 14.1% by sputtering Cu and In prelayers, followed by the H_2Se selenization process. Later, after being acquired, ARCO was renamed to Shell Solar, and they spent 10 years to launch the first commercial CIS module in 1998. In 1989, Boeing introduced Ga element to prepare a CIGS thin-film PV cell, which significantly increased the open-circuit voltage. In 1994, the US NREL adopted a three-step co-evaporation process. The efficiency of CIGS thin film made with this method has been leading the industry, and in 2008, conversion efficiency of CIGS thin-film PV cell reached 19.9%. This record was refreshed in 2010 by the Center for Solar Energy and Hydrogen Research Baden-Württemberg (ZSW), Germany, with an efficiency of 20.3%. In May 2016, ZSW announced that they achieved a 22.6% conversion efficiency for CIGS PV cell on glass substrate, which created a new record. In order to further improve the efficiency, 11 research teams from eight European countries including Germany, Switzerland,

France, Italy, Belgium, and Luxembourg formed a research alliance in 2015, and announced the implementation of the "Sharc25" program with the aim of increasing the efficiency of CIGS thin-film PV cell to 25%.

4.4.1 Characteristics of CIGS PV cell

CIGS PV cell has the following features:
(1) CIGS is a direct bandgap semiconductor material and is most suitable for thin-film formation. Its light absorption coefficient is extremely high, and the thickness of the film can be reduced to about 2 μm, which can greatly reduce the consumption of raw materials. At the same time, the preparation methods of the thin-film materials used in PV cells are mainly sputtering and CBD, which can achieve large-area uniformity, and the cost of CIGS PV cell is further lowered.
(2) The addition of Ga in $CuInSe_2$ can make the bandgap of semiconductor to vary from 1.04 to 1.67 eV, which is ideal for adjusting and optimizing the forbidden bandwidth. If the content of Ga is varying across the film thickness direction and forms a gradient bandgap, back-surface field will be generated, more current output will be obtained, and the bandgap near the P–N junction will be increased to form a V-shaped bandgap distribution. The ability to achieve such bandgap trimming is the greatest advantage of CIGS-based PV cell compared to Si-based PV cell.
(3) CIGS can form high-quality crystals with few defects and large crystal grains on glass substrate, and this crystal grain size cannot be achieved by other polycrystalline films.
(4) CIGS has the highest light absorption coefficient in known semiconductor materials, up to 10^5 cm^{-1}.
(5) CIGS is a semiconductor material with no LID effect or SWE. Illumination may even increase its conversion efficiency. Experimental results show that CIGS PV cell has a longer lifetime than mono c-Si PV cell.
(6) For Si-based semiconductors, alkali metal such as Na must be avoided as much as possible [8]. However, in the CIGS system, trace amounts of Na could increase conversion efficiency and yield. Therefore, Na doping is also an important factor for using soda-lime glass as a substrate for CIGS, in addition to low cost and similar expansion coefficients.

4.4.2 Structure of a CIGS thin-film PV cell

The improvement in performance of CIGS thin-film PV cell is largely attributed to the optimization of PV cell structure. Figure 4.16a–c shows the structure of the early, mid-term (1985), and current CIGS PV cells, respectively. As the research progressed,

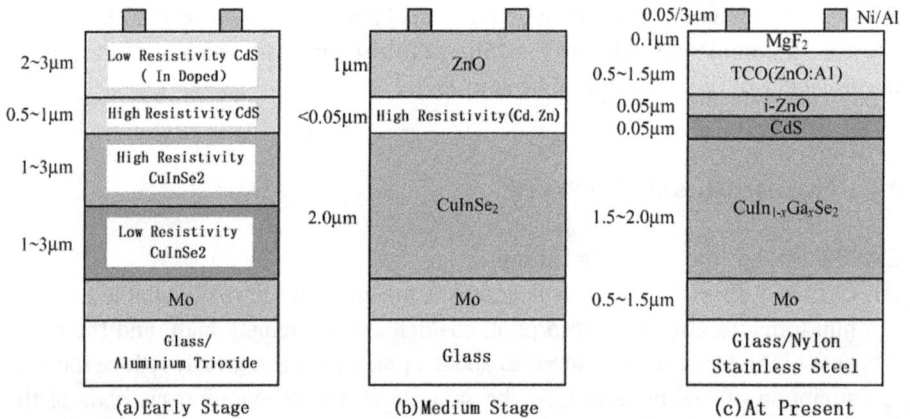

Fig. 4.16: Illustration of structures of CIGS thin-film PV cell.

the absorption layer, buffer layer, and window layer of the PV cell are all different than the original structure. The absorption layer is the key part of a PV cell. The doping of Ga increases the bandgap width of the absorption layer and improves the open-circuit voltage. Meanwhile, the gradient bandgap can be formed by varying Ga content, thereby optimizing the bandgap matching with other layers, and helps the transmission of photogenerated carriers. Therefore, $CuIn_{1-x}Ga_xSe$ has become the prevalent material of the absorption layer, and has a thickness of about 1.5–2.0 µm. High-resistance CdS buffer layer will increase the series resistance of the PV cell; thus, its thickness is generally controlled at about 0.05 µm. The window layer is generally composed of a thin layer of intrinsic ZnO and a thick layer of TCO.

To further improve PV cell performance, several types have been developed in recent years: CIGSS PV cell, whose components are CuInGaSSe; CZTSS PV cell, whose components are $Cu_2ZnSnS_4Se_4$; and CZTS PV cell whose components are Cu_2ZnSnS_4.

4.4.3 Fabrication of CIGS thin-film PV cell

4.4.3.1 Manufacturing process of CIGS thin-film PV cell
In industry, CIGS thin-film PV cell is generally prepared on a glass substrate or a flexible substrate. The general process flow is shown in Fig. 4.17. Many processes are similar to other thin-film PV cells and will not be described here in detail. The following content only discusses the preparation of CIGS thin films.

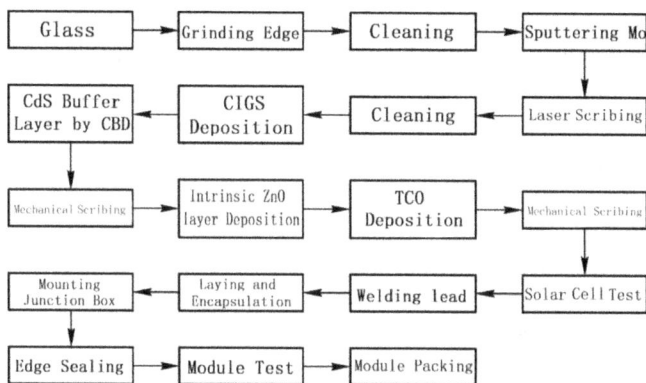

Fig. 4.17: Manufacturing process of CIGS thin-film PV cell.

4.4.3.2 Preparation of CIGS film

There are many methods for preparing CIGS thin films, which can be divided into two major categories: vacuum deposition and non-vacuum deposition, as shown in Tab. 4.3 [9]. Based on current research and commercialization status, the multi-component co-evaporation method and the selenization method after sputtering are mostly used ones.

Tab. 4.3: Main methods for CIGS thin-film deposition.

	Prefabricated layer deposition (low substrate temperature)		Selenization/annealing/ recrystallization (high substrate temperature 450–600 °C)	
Vacuum	Method	Material	Preprocess	Selenization
	Sputtering	Cu, In, Ga sequentially deposited layers, with or without Se	None	Selenization in Se/S, H_2Se/S gas
	Evaporation			
Nonvacuum	Electrodeposition		Selectively purified by H_2 annealing	Recrystallization under Se/S, H_2Se/S gas
	Spraying	Cu, In, Ga, Se/S compounds		
	Printing	Cu, In, Ga, O compounds	Adhesive elimination, reduce to alloy with H_2	Selenization in Se/S H_2Se/S gas

4.4.3.2.1 Multicomponent co-distillation

Currently, the most efficient CIGS thin-film PV cell is manufactured on a labora-tory-scale co-evaporation process [10]. Figure 4.18 shows a schematic diagram of a CIGS film prepared by co-evaporation. Cu, In, Ga, and Se evaporation sources provide four elements for the film formation. Atomic absorption spectroscopy and electron impact emission spectroscopy are used to monitor parameters such as film composition and the evaporation rate of source in real time, and to precisely control the growth of the film. The substrate temperature during the deposition of high-efficiency CIGS PV cells is generally higher than 530 °C, and the temperature of each evaporation source must be individually adjusted to control the evapora-tion rate of the elements, thereby controlling the stoichiometric ratio of deposited $CuIn_{1-x}Ga_xSe_2$ thin film. In general, target temperatures for Cu is 1,300–1,400 °C, for In is 1,000–1,100 °C, for Ga is 1,150–1,250 °C, and for Se is 300–350 °C.

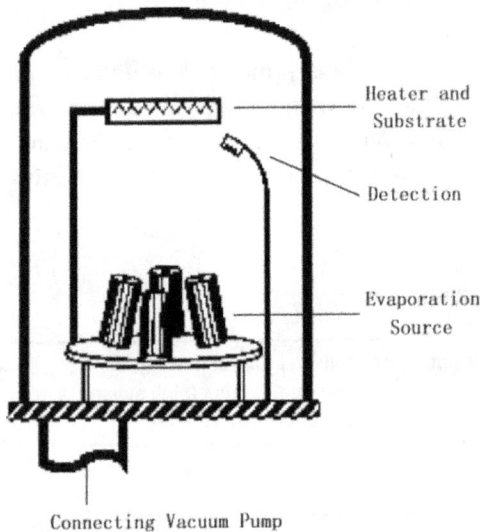

Fig. 4.18: Schematic diagram of producing CIGS thin film with multicomponent co-evaporation.

Since the adhesion coefficient of Cu, In, and Ga on the substrate is relatively high, the composition of the thin film and the growth rate can be controlled by the atomic fluxes of Cu, In, and Ga. The relative ratio of In and Ga defines the width of the bandgap. Se has a high vapor pressure and a low adhesion coefficient, so the vola-tilized Se flux must be greater than the total amount of Cu, In, and Ga, and exces-sive Se will evaporate from the surface of the film. If the amount of Se is insufficient, In and Ga may form In_2Se and Ga_2Se. By studying the growth kinetics of CIGS thin films, we found that the variation in Cu evaporation rate strongly influ-ences the growth mechanism of thin films. According to the evaporation process of

Cu, the co-evaporation process can be divided into one-step method, two-step method, and three-step method. Since the diffusion rate of Cu in the film is sufficiently fast, no matter which process is adopted, the film thickness is substantially uniformly distributed.

The one-step method is to keep the flow rates of the four evaporation sources unchanged during the deposition process, as shown in Fig. 4.19a. This process is relatively simple to control and is suitable for commercial production, but the grain size in thin film is small and a gradient bandgap cannot be formed.

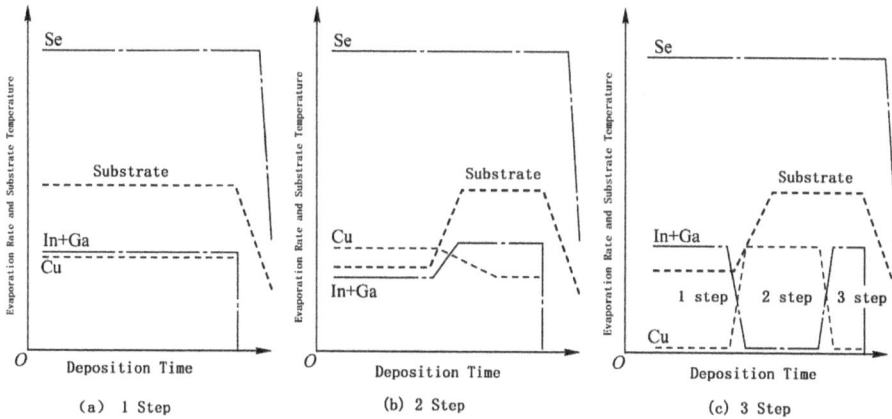

Fig. 4.19: Manufacture process of CIGS thin film with multicomponent co-evaporation.

The two-step process is also known as Boeing double-layer process. The variation in substrate temperature and evaporation source flow of this process is shown in Fig. 4.19b. First, a Cu-rich CIGS thin film is deposited at a substrate temperature of 400–450 °C, with a small grain size and a low resistivity. The second layer is a Cu-depleted CIGS film deposited at a high substrate temperature (550 °C). This layer has large grain size and high resistivity. During the process, the surface of the CIGS film is covered with Cu_xSe. Above 523 °C, Cu_xSe exists in liquid phase, which increases the mobility of the constituent atoms, and finally obtains a film with a large grain size.

The three-step process is shown in Fig. 4.19c. In the first step, substrate temperature is 250–300 °C, and 90% of In, Ga, and Se elements formed a $(In_{0.7}Ga_{0.3})_2Se_3$ prefabricated layer using co-evaporation method with a Se/(In + Ga) flow ratio greater than 3. In the second step, Cu and Se are evaporated at a substrate temperature of 550–580 °C until the film is slightly rich in Cu. In the third step, while maintaining the substrate temperature at the second step, the remaining 10% In, Ga, and Se are co-evaporated to form an In-rich thin layer on the surface of the thin film, and finally a $CuIn_{0.7}Ga_{0.3}Se_2$ thin film with almost perfect stoichiometric ratio is obtained. The film prepared by this process has a smooth surface, compact crystal grains, large grain size, and a Ga dual-gradient bandgap. This method can eventually result in a CIGS PV cell with higher conversion efficiency.

4.4.3.2.2 Sputtering and selenization

The process of this method is first sputter Cu, In, and Ga on the Mo electrode to form a prefabricated layer, and then reacted with H_2Se or an atmosphere containing Se to obtain a thin film satisfying the stoichiometric ratio. The process does not require much on equipment. Naturally it becomes the first choice for commercial production. However, the content and distribution of Ga in the selenization process are not easy to control; thus, it is difficult to form a double-gradient structure. Therefore, a one-step vulcanization process is sometimes introduced to the post-selenization process. Part of the doped S atoms replace the Se atoms, forming a wide bandgap of $Cu(In,Ga)S_2$ on the surface of the film, which can reduce the interface recombination of the film and increase the open-circuit voltage of the cell.

The technical difficulties of this process mainly lie in the selenization process. In the selenization process, Se can be easily lost due to the formation of In_2Se and Ga_2Se, and the composition in different positions of the film is mismatched, which leads to nonuniformity of the film. Therefore, control of the selenization process is particularly important. As shown in Fig. 4.20, the entire process is divided into a low-temperature stage, a rapid temperature increase stage, and a high-temperature stage. The low-temperature stage can effectively prevent the formation of dense CIGS film on the surface layer that can affect the internal selenization, so that the

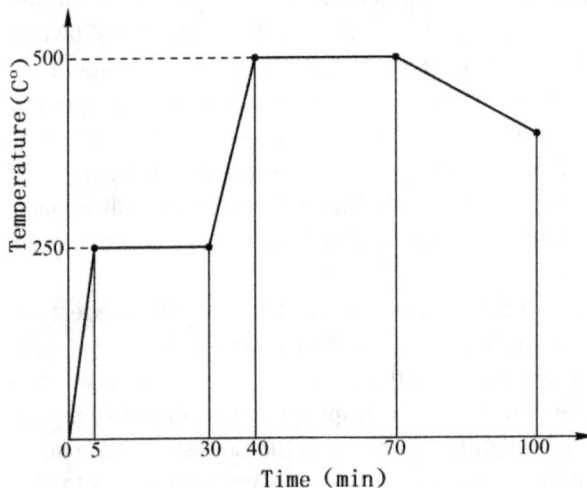

Fig. 4.20: Selenization process.

Se and prefabricated layers react fully. The rapid temperature rise is to avoid volatilization of In_2Se and Ga_2Se. The high-temperature stage can help with the full growth of CIGS grains.

4.4.4 Commercialization of CIGS thin-film PV cell

Due to the high technical difficulty of CIGS thin-film PV cell, the atomic ratio of multiple components in the PV cell is hard to control, and the repeatability problem is a bottleneck that limits its development. Therefore, the commercialization of CIGS thin-film PV cell is developing slowly. However, with the advancement in technology and theory, there have been some great progresses in recent years. In April 2014, Solibro of Hanergy Group was confirmed by test authority that it created a record of 21.0% conversion efficiency cell. In May 2016, ZSW announced that it achieved the highest efficiency of 22.6% with CIGS cell on glass substrate, setting a new record.

MiaSolé, USA, which was acquired by Hanergy, maintained the highest record of efficiency of commercial CIGS module. After they set a record of 14.3% in September 2010, it was announced in December that it passed test in NREL and the module efficiency reached 15.7%. Since 2015, Hanergy manufacturing plant located at Heyuan, Guangdong, adopted the roll-to-roll production method to achieve continuous production of the absorber layer. The CIGS is sputtered on a flexible stainless steel substrate, which makes process control easier and reduces cost.

Solar Frontier, Japan, has maintained a 17.5% efficiency record for Cd-free module since June 2014. In March 2016, they announced that its Cd-free CIGS thin-film PV cell has created a new record of 22.0% conversion efficiency. Table 4.4 shows the efficiency records of the CIGS PV cells from various institutions, and Tab. 4.5 shows the efficiency records of the CIGS submodules of three manufacturers.

In 2010, the global production capacity of CIGS thin-film PV cell was 712 MW. In 2011, it expanded to about 2 GW, within which Solar Frontier expanded its production to 1 GW. In 2016, the global production capacity was maintained at 1.3 GW.

The major manufacturers of CIGS thin-film PV cell are mainly located in the USA, Germany, and Japan. The main technical route adopted is the selenization after sputtering and co-evaporation. The technical routes adopted by CIGS manufacturers are shown in Tab. 4.6.

The conversion efficiency of CIGS thin-film PV cell is the highest among all commercialization thin-film PV cells, and the average conversion efficiency of the modules sold in the market is gradually increasing. The main parameters of the world commercial CIGS modules in 2016 are shown in Tab. 4.7.

China started late in the development of CIGS PV cell production. Currently, Hanergy, China Building Materials (CNBM), joint venture company of Shenhua, Shanghai Electric, and MANZ are engaged in the commercialization of CIGS thin-

Tab. 4.4: Efficiency records of CIGS PV cells from various institutions.

Institution	Area (cm^2)	V_{OC}(V)	J_{SC} (mA/cm^2)	FF (%)	Efficiency (%)	Test authority	Test time	Type of cell
Solibro	0.9927 (ap)	0.757	35.70	77.6	21.0 ± 0.6	FhG-ISE	Apr. 2014	CIGS cell (glass substrate)
ZSW	0.4092 (da)	0.7411	37.78	80.6	22.6 ± 0.5	FhG-ISE	May 2016	CIGS cell (glass substrate)
Solar Frontier	0.512 (da)	0.7170	39.45	77.8	22.0 ± 0.5	FhG-ISE	Mar. 2016	CIGSS cell (Cd-free, glass substrate)
IMRA Europe	1.115 (da)	0.5073	31.95	60.2	9.8 ± 0.2	NEWPORT	Apr. 2016	CZTSS cell
UNSW	1.067 (da)	0.6585	20.43	56.7	7.6 ± 0.1	NREL	Apr. 2016	CZTS cell
IBM Solution	0.4209 (ap)	0.5134	35.21	69.8	12.6 ± 0.3	NEWPORT	Jul. 2013	CZTSS thin-film cell
UNSW	0.2379 (da)	0.6732	21.25	66.3	9.5 ± 0.2	NREL	Sep. 2016	CZTS thin-film cell (glass substrate)

Note: ap, aperture area; da, designated illumination area; J_{SC}, short-circuit current density.

Tab. 4.5: Efficiency records of CIGS submodules from three manufacturers.

Manufacturer	Area (cm^2)	V_{OC}(V)	J_{SC} (mA/cm^2)	FF (%)	Efficiency (%)	Test authority	Test time	Type
Hanergy/Solibro	15.892 (da)	0.701	35.29	75.6	18.7 ± 0.6	FhG-ISE	Sep. 2013	CIGS (4 cells)
Solar Frontier	808 (da)	47.6	0.408 A	72.8	17.5 ± 0.5	AIST	Jun. 2014	CIGS (Cd-free, 70 cells)
Hanergy/MiaSolé	9703 (ap)	28.24	7.254 A	72.5	15.7 ± 0.5	NREL	Nov. 2010	CIGS minimodule

Note: ap, aperture area; da, designated illumination area; J_{SC}, short-circuit current density.

film PV cell. Hanergy has already sold products at its Solibro plant in Germany. CNBM adopted Avancis' sputtering + post-selenization technology to build a 1.5 GW plant in Jiangyin, Jiangsu Province. Its scale will be comparable to that of Solar Frontier, Japan. The joint venture between Shenhua, Shanghai Electric, and MANZ will build a 306 MW plant in Chongqing, using technologies from MANZ, Germany.

Tab. 4.6: Technical routes adopted by CIGS manufacturers in 2016.

Company	Country of production	Substrate	Area (cm²)	Nameplate efficiency (highest efficiency) (%)	Technical route	Capacity (MW)	Status in build
Shanghai Electric MANZ (Wurth Solar)	Germany	Glass	1,200 × 600	13–16 (16)	Co-evaporation	30	306 MW factory in build
Hanergy/Solibro	Germany	Glass	1,190 × 789.5	14.2–16.0	Co-evaporation	145	>5 factories in build (>1,500 MW)
CNBM/Avancis	Germany	Glass	622	13 (17.9)	Sputtering + post-selenization	120	4 factories in build (15 00 MW)
Hanergy/Global Solar	USA	Stainless steel	8,390 (3,822)	10.5 (13)	Co-evaporation	4.2	Built
Solar Frontier	Japan	Glass	1,257 × 977	12.2–13.8 (16.3)	Sputtering + post-selenization	1,100	
Hanergy MiaSolé	USA	Stainless steel	10,000	15.7	Sputtering + post-selenization	150	>5 factories in build (>15 00 MW)
BOSCH SOLAR (Johanna)	Germany	Glass	15,000	—	Sputtering + post-selenization + post-vulcanization	30	Built

Tab. 4.7: Key parameters of world commercial CIGS modules in 2016.

Ranking	Manufacturer	Module efficiency (%)	Power (W)	Size (mm^2)	Comment
1	Hanergy/Miasole (USA)	11.5–13.5	200–230	1,710 × 999	Stainless steel substrate
2	Hanergy/Solibro (Q-Cells) (Germany)	11.0–13.0	100–120	1,190 × 789.5	Glass substrate
3	Solar Frontier (Japan)	12.6–13.8	150–170	1,257 × 977	Glass substrate
4	CNBM/Avancis (Germany)	12.3–13.8	130–145	1,587 × 664	Glass substrate
5	Hanergy/Global Solar (USA)	9.7–10.5	275–300	5,745 × 495	Polyester flexible substrate
6	Shanghai Electric/ MANZ (Germany)	12.5–14.0	90–105	1,200 × 600	Glass substrate
7	BOSCH (Germany)	11.6–13.5	130	1,417 × 791	Glass substrate

CIGS PV cell efficiency is the highest among commercialized thin-film PV cells and is equivalent to that of poly c-Si modules. It is stable in performance, can be mass produced, and has a low cost and uniform color. It can be used not only for various types of PV power plants, but also as flexible modules on stainless steel substrates. Moreover, it can be used by architects for BIPV, also for portable power and special market applications, with good market development potentials.

4.5 Perovskite PV cells

Perovskite-based PV cell has attracted attention in the PV industry due to its low cost of raw materials and manufacturing process, along with the rapid increase in conversion efficiency [11]. Some even think that perovskite PV cell will replace silicon-based PV cells.

4.5.1 Development history of perovskite PV cell

Perovskite material is a material with the same crystal structure as calcium titanate ($CaTiO_3$), which was discovered by Gustav Rose in 1839 and later named by Russian Mineralogist L. A. Perovski. Perovskite structure has a ABX_3 form in general, and typical perovskite crystals have a special cubic structure, as shown in Fig. 4.21. In

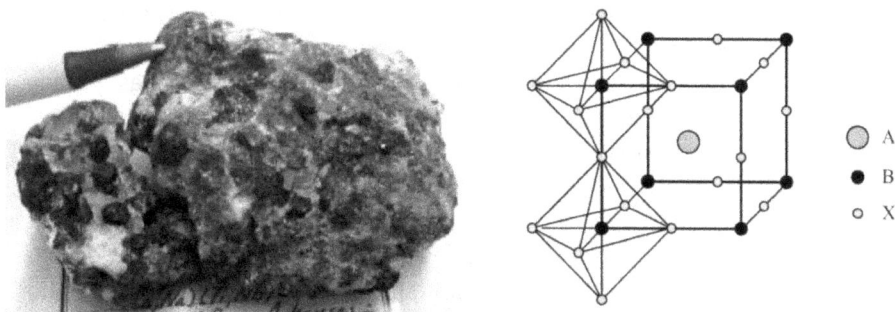

Fig. 4.21: Perovskite material and its crystal structure.

the cubic structure of perovskite crystals, element A is a bulky cation that resides in the center of the cube; element B is a smaller cation that resides in the eight vertices of the cube; element X is an anion and resides in 12 midpoints of the edges of the cube. If the crystal structure of a material is consistent with this structure, then such material can be called a perovskite material.

In the 1980s, organic–inorganic complex perovskite materials began to appear. The structural characteristics of these materials are that cation A in ABX_3 is a small organic molecule, while B and X are inorganic ions. After the introduction of organic small molecules, such perovskite materials can be dissolved in regular solvents. This peculiar crystal structure gives it many unique physical and chemical properties, such as light absorption and electrocatalysis, and it is widely used in chemistry and physics. The perovskite family now contains hundreds of substances, ranging from conductors, semiconductors, and insulators, many of which are artificially synthesized, thus bringing much convenience to the application of materials. Typical organic–inorganic complex perovskites include lead methylamine iodide ($CH_3NH_3PbI_3$) and lead bromide methylamine ($CH_3NH_3PbBr_3$), which are semiconductors and have great light absorption properties. In 2009, Professor Gong Junli from Yokohama University in Toyama, Japan, applied lead methylamine iodide and lead bromide methylation to dye-sensitized PV cell, and achieved a photoelectric conversion efficiency of up to 3.8%, starting a new era for novel perovskite PV cell.

Perovskite PV cell is low in material cost and manufacturing process, and is flexible. The bandgap of it can be adjusted by changing the composition of the raw materials. Perovskite layers with different bandgap widths can be stacked together to form a multijunction perovskite PV cell, making it possible to have a higher efficiency than Si-based PV cells.

Since then, the perovskite PV cell's structural design and supporting materials have continued to develop, and its efficiency has increased to 22.1% in just 7 years. Such a rapid technological leap has never happened in PV field. Taking the dominant poly c-Si cell as an example, in 1985, the laboratory efficiency of poly c-Si PV cell was about 15%, and it increased to 20.4% in 2004. It took 20 years to increase the efficiency by only 5%, and then, there has been barely any progress from 2004 to 2015, the efficiency increased to 20.6% in 11 years. Perovskite PV technology has sprung up in a very short period of time and has quickly surpassed poly c-Si technology. One of the top ten scientific breakthroughs in 2013 was perovskite PV cell, and the efficiency increased rapidly (from 2.2% in 2006 to 20.1% in 2014). Perovskite PV cell has become five times more efficient in 7 years, and the efficiency has doubled in the past 2 years.

In 2013, Snaith announced a perovskite PV cell without titanium dioxide nanoparticles, simplifying the PV cell structure and improving its efficiency to more than 15%. At the Materials Research Society conference in April 2014, Yang Yang, a materials scientist at the University of California Los Angeles, revealed that the efficiency of perovskite PV cell in his lab reached 19.3%. In April 2016, Hong Kong Polytechnic University made new breakthroughs. Perovskite-type silicon PV cell has created the world record with conversion efficiency of 25.5%. Many people predict that its efficiency will reach new heights in the coming years.

4.5.2 Features of perovskite PV cell

$CH_3NH_3PbI_3$-type perovskite material is a direct bandgap material, which means that the perovskite has strong light absorption. c-Si is an indirect bandgap material; thus, the silicon wafer must be 150 μm or thicker to achieve saturation absorption of incident light. The perovskite only needs 0.2 μm to achieve saturation absorption, which is nearly a thousand times smaller than the thickness of silicon. Therefore, perovskite PV cell consumes much less active materials than c-Si PV cell.

Perovskite materials have high carrier mobility. The carrier mobility reflects the speed of positive and negative charges generated under light irradiation, and higher mobility means that generated charges can move to the electrode faster.

The carrier mobility of the perovskite material is almost in equilibrium, that is, the mobility of electrons and holes in the perovskite material is almost the same. As a comparison, the carrier mobility of c-Si is unbalanced, and its electron mobility is far greater than the hole mobility. As a result, the current output will saturate when the incident light intensity is high enough. This limits the photoelectric conversion efficiency of silicon PV cell under higher light intensity.

The carrier recombination in perovskite crystals is almost completely radiation type. This is an extremely important advantage of perovskite materials. When the electrons and holes in the perovskite recombine, a new photon is released, which is reabsorbed by the nearby perovskite crystals. Therefore, perovskites have a very high utilization of incident photons, and the amount of heat generated under illumination is low, whereas the carrier recombination in c-Si is almost entirely a nonradiation type, when the electrons and holes in c-Si recombine, the energy they carry will be converted to heat and cannot be reused for electric power generation. Therefore, the theoretical limit of photoelectric conversion efficiency of perovskites is significantly higher than that of silicon material PV cells. At present, the maximum efficiency of mono c-Si PV cell is 25.6%. This record has been maintained for many years and it is unlikely that there will be a major breakthrough in the near future. The perovskite's radiation-type recombination nature makes it possible to achieve the same efficiency as GaAs PV cell, even exceeding 29%.

Perovskite materials are soluble so that the perovskite materials can be formulated into a solution and applied to a glass substrate like a paint. For high-efficiency PV cells, the solubility of perovskites is an unprecedented advantage, and perovskite is the only soluble material with efficiency higher than 20%. A few years ago, Nanosolar attempted coating method to produce CIGS PV cell. However, the CIGS materials are insoluble and CIGS particles were merely dispersed in liquids. Therefore, such coating method does not promote crystal growth. The truly soluble perovskite material, formed by coating and precipitated from the solution, is a spontaneous crystallization process, which has a high possibility of making a high-performance PV cell.

4.5.3 Commercialization of perovskite PV technology and its problems

Since 2013, with the rapid development of perovskite PV technology, it has become a hot topic in PV research. Many universities and research institutions around the world are engaged in the research and development of perovskite PV technology. In China, large companies that have started the development of perovskite technology include Huaneng Group, Trina Solar, and Shenhua Group. Small or start-up companies include Weihua Solar and Qingdao Grene. According to the public data available, Weihua Solar is currently the only company in China that has established a large-scale perovskite module pilot production line. An experimental line of perovskite PV modules with a size of 45 cm × 65 cm has been established. The coating and printing processes are used throughout. The laboratory efficiency of their product is 21.5% and the module efficiency is 12.7%. The large gap between laboratory efficiency and module efficiency shows that there are still many problems that need to be solved toward the mass production. However, compared to the resources that are invested by a small enterprise like Weihua Solar, this is already a remarkable progress. Table 4.8 shows the efficiency records of perovskite PV cells and minimodules.

Scientists discovered in the latest research that the conversion efficiency of a perovskite-structured OPC can be as high as 50%, which is twice the conversion efficiency of PV cells on current market and can significantly reduce the cost of PV cells. Related research was published in the latest issue of *Nature*.

Tab. 4.8: Efficiency records of perovskite PV cells and minimodules.

Institution	Area (cm^2)	V_{OC} (V)	J_{SC} (mA/cm^2)	FF (%)	Efficiency (%)	Test authority	Test time	Type
KRICT/ UNIST	0.9917 (da)	1.104	24.67i	72.3	19.7 ± 0.6	Newport	Mar. 2016	Cell
SJTU/ NIMS	36.13 (da)	0.836	20.20	71.5	12.1 ± 0.6	AIST	Sep. 2016	Minimodule (10 cells)
KRICT/ UNIST	0.0946 (ap)	1.105	24.97	80.3	22.1 ± 0.7	Newport	Mar. 2016	Cell
Stanford/ ASU	0.990 (ap)	1.651	18.09	79.0	23.6 ± 0.6	NREL	Aug. 2016	Perovskite/Si multijunction cell

Note: ap, aperture area; da, designated illumination area; J_{SC}, short-circuit current density.

Of course, perovskite PV cells have a long way to go to replace silicon PV cells and need to overcome many technical and nontechnical difficulties:

- The PV cell is not very stable. Perovskite materials have poor endurance to air and water. Currently used perovskite materials can be decomposed by air and dissolved in water and organic solvents, resulting in short device lifetime.
- Reproducibility of PV cell efficiency is poor. Although the efficiency of many perovskite PV cells reported so far is above 15%, the reproducibility is poor, and the performance of a group of PV cells prepared under the same conditions shows a large statistical deviation in efficiency. It can be seen that the manufacturing process is not mature enough.
- The PV cell materials are toxic. The light-absorbing material in the current high-efficiency perovskite PV cell generally contains lead, and it will bring environmental problems if used in large-scale application. Lead-free perovskite materials with high photoelectric conversion efficiency need to be developed.
- Urgent need for commercial equipment development. Because it is hard for large-area thin films to maintain uniformity, the currently reported high-efficiency perovskite PV cells have a working area of only about 0.1 cm^2, and there is still a long way to go from prototype to practical application. Therefore, it is necessary to develop commercial equipment that is able to produce large-area perovskite PV cell with stable performance.
- It needs to be tested with a long period of practical application. So far, perovskite PV modules have not yet been commercialized in mass production, and there is no practical application in scale. To become a mature and reliable solar power source, long-term practical tests are required.

In summary, perovskite PV cell has great potential, and it is expected to achieve the same efficiency as GaAs PV cell and a lower manufacturing cost than a poly c-Si PV cell. Recent technical advances have shown that perovskite PV technology does not have a fundamental problem for commercialization; therefore, perovskite PV cell technology may be just around the corner.

4.6 Dye-sensitized cell

DSC is a novel chemical PV cell based on redox reaction. Its mechanism is similar to the photosynthesis process of plants. It consists of nanoporous semiconductor film, dye photosensitizer, redox electrolyte, counterelectrode, and the conductive substrate.

Figure 4.22 shows a schematic diagram of the structure and mechanism of a DSC. The upper and lower ends of the DSC are glass substrates coated with a transparent conductive film, and in the middle, a porous conductive film made of nanosized titanium dioxide (TiO_2) that adsorbed photosensitizing dye and was filled with redox electrolyte solution. When the dye molecules absorb sunlight, the electrons are excited from the original ground state to an excited state, undergoing an oxidation reaction with titanium dioxide, and these electrons are injected into the conduction band of the nanoporous semiconductor. The electrons quickly run to the surface and are collected by the electrodes and passed to the external circuit. Electrons returning from the other end of the electrode are trapped by ions in the electrolyte. The dye in the

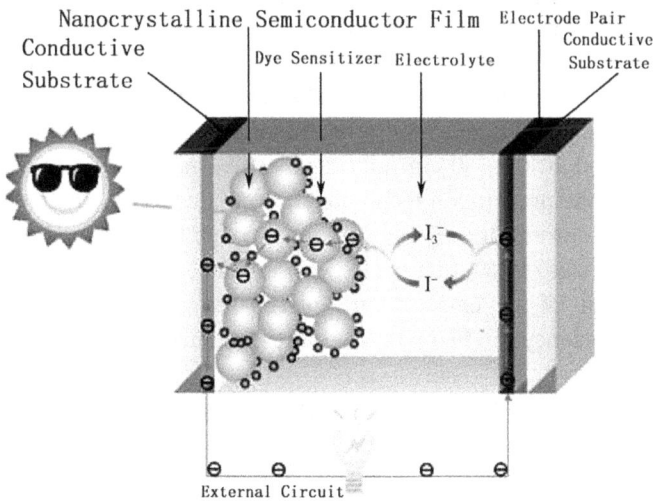

Fig. 4.22: Illustration of structure and mechanism of DSC.

oxidized state is reduced by the electrolyte and returned to the oxidized dye molecule. The oxidized electrolyte then undergoes electron reduction at the positive electrode and the electrolyte restored to the ground state, which completes the electron transport cycle. As long as there is sunlight and is connected to an external circuit, this cycle can be continued.

In 1991, a team led by Professor M. Grätzel from Swiss Federal Institute of Technology in Lausanne (École Polytechnique Fédérale de Lausanne, EPFL) used nanoporous membranes instead of the previous flat electrodes to achieve a breakthrough of 7.1% (AM1.5) in conversion efficiency on the DSC [12]. Later, the USA, Japan, and other countries invested a large amount of funding into research, making DSC one of the most active research areas in the PV industry. In addition to the low cost, high efficiency, and the potential market in the future, another main reason that attracts investors is the low capital investment compared to the capital investment of several hundred million dollars for other commercialized thin-film PV cell technologies.

Japan is a world leader in theoretical and applied research in DSC. Earlier, Sharp Corporation and Arakawa et al. reported the photoelectric conversion efficiency of DSC modules of 6.3% (26.5 cm^2) and 8.4% (100 cm^2), respectively. Han et al. developed a W-type DSC module with an efficiency of 8.2% in 2009. Its area is 50 mm x 53 mm and the active area is as high as 85%. Fujikura Corporation uses Ni as the gate electrode in a DSC PV cell with an area of 10 cm × 10 cm, and the efficiency of the entire module reached 5.1% (effective area is 68.9 cm^2). In 2005, a large-area high-performance plastic DSC production line was jointly developed by Peccell Corporation, Fujimori Industries, Ltd., and Showa Denko, and was successfully tested. Screen printing method was used to achieve low-cost continuous production, and the large-area DSC module manufactured had a length of 2.1 m, a width of 0.8 m, a thickness of 0.5 mm, and a weight density of 800 g/m^2. It is the world's largest and lightest DSC module. This module has output voltage as high as 100 V, even under indoor lights. Miyasaka et al. from Yokohama University, Japan, developed and manufactured 30 cm × 30 cm fully flexible DSC module, based on low-temperature TiO$_2$ electrode preparation technology. It contains 10 cells with an output voltage of 7.2 V and current of 0.25–0.3 A. Yongseok Jun et al. from Korea studied the influence of the size of TiO$_2$ film on DSC performance, and produced a DSC module with an area of 10 cm × 10 cm. Its efficiency reached 6.3%. Photoelectric conversion efficiency can be up to 6.6% after the TiO$_2$ film was added with a scattering layer.

In recent years, research of DSC has attracted great attention from people, and has made great progress in commercializations. In October 2006, G24i Power Limited was established in South Wales, UK. It mainly used the joint technology of Konarka and Professor M. Grätzel. In October 2009, G24i began providing commercial DSC modules for Mascotte Industrial Associates, Hong Kong.

After more than 20 years of development, DSC PV cell has made great progress in its technology and commercialization. However, its development still faces some bottlenecks. First, the traditional DSC can only absorb visible light at wavelengths less than about 650 nm, while other wavelength of light is barely used, and it is urgent to develop DSC with full-spectrum absorption characteristics. Second, most DSC anodes use TiO_2 nanocrystalline thin films, which have severely reduced the electron transportation and electrolyte penetration, due to large grain boundary and narrow channel space. The structure of the anode thin film needs to be improved and a thin-film preparation technology suitable for large area production was developed. Finally, the large-area DSC preparation process is immature and the PV cell is not stable, and it is necessary to develop an efficient, low-cost process that is suitable for large-area PV cell production, such as all-solid-state cells and flexible cells. Therefore, to design new types of long exciton life dyes and electrolytes, to improve cell efficiency and stability, and to develop all-solid-state and flexible devices are the necessary tasks for DSC to be used in application. It is expected that significant progress will be made in the near future. Table 4.9 shows the efficiency records of dye-sensitized PV cells and minimodules.

Tab. 4.9: Efficiency records of dye-sensitized PV cells and modules.

Institution	Area (cm^2)	V_{OC} (V)	J_{SC} (mA/cm^2)	FF (%)	Efficiency (%)	Test authority	Test time	Type
Sharp	1.005 (da)	0.744	22.47	71.2	11.9 ± 0.4	AIST	Sep. 2012	Cell
Sharp	26.55 (da)	0.754	20.19	69.9	10.7 ± 0.4	AIST	Feb. 2015	Minimodule (7 cells)
Sharp	398.8 (da)	0.697	18.42	68.7	8.8 ± 0.3	AIST	Sep. 2012	Submodule (26 cells)

Note: ap, aperture area; da, designated illumination area; J_{SC}, short-circuit current density.

The main advantages of DSC are abundant raw materials, low cost, and relatively simple process, and has great advantages in mass production. Meanwhile, all raw materials and production processes are nontoxic and nonpolluting, and some materials can be fully recycled, which is important for protecting the environment. However, dye-sensitized PV cell also need to improve efficiency and stability; hence, it takes time to achieve large-scale commercialization.

4.7 Organic semiconductor PV cell

A P–N junction PV cell can be made by doping with an organic material with semi-conductor properties, such as poly(p-phenyl vinyl) and polyaniline. Ion doping can also turn some plastic films into semiconductors. Although this type of PV cell has a low cost, its photoelectric conversion efficiency is far below satisfaction and its ability to resist photoaging is not ideal. Currently, it is still in research stage and has not yet come into practical application. Toshiba Corporation, Japan, is currently holding its leading position in organic semiconductor PV cells. Its latest R&D results are shown in Tab. 4.10.

Tab. 4.10: Efficiency records for organic semiconductor PV cells and minimodules.

Institution	Area (cm^2)	V_{OC} (V)	J_{SC} (mA/cm^2)	FF (%)	Efficiency (%)	Test authority	Test time	Type
Toshiba	0.992 (da)	0.780	19.30	74.2	11.2 ± 0.3	AIST	Oct. 2015	Cell
Toshiba	26.14 (da)	0.806	16.47	73.2	9.7 ± 0.3	AIST	Feb. 2015	Minimodule (8 cells)

Note: da, designated illumination area; J_{SC}, short-circuit current density.

4.8 Market and development prospects of thin-film PV cell

With the excellent performance of CdTe and CIGS thin-film PV cells, the compound annual growth rate of thin-film PV cells from 2001 to 2016 was 46%, which is 40% higher than that of c-Si PV cells. In 2001, the global market for thin-film PV cells was only 14 MW, accounting for only 2.8% of the total PV market. In 2005, the market scale exceeded 100 MW for the first time, and was 6% of the market, and in 2007 it was 10%. By 2009, it had reached 2.141 GW, made up of ~25% of the market, and it has doubled to reach more than 4 GW in 2016, accounting to about 7% of the total production of PV modules. In 2009, the share of thin-film PV cell in the global PV market grew steadily. Many institutions at that time, such as PV Market Outlook 2010 by EPIA, 2010 Solar Technologies Market Report by NREL, PV Status Report 2011 by European Commission Joint Research Center (JRC), Thin Film 2010: Market from Outlook to 2015 by GTM Research, were all very or overoptimistic about the future of thin-film PV cells. Due to the sharp drop in the price of c-Si PV cells, there has been no significant improvement in the efficiency of silicon-based thin-film PV cells in the past 5 years, and a large number of silicon-based thin-film PV cell factories have been shutdown, with growth only seen in CIGS and CdTe areas. Therefore, market share of thin-film PV module did not catch up with the speed with c-Si PV modules. Among the top 10 global thin-film PV cell manufacturers in 2010, only First Solar of the USA, Solar Frontier of Japan, and Solibro and MiaSolé acquired by Hanergy are still in business.

Compared to the slump in silicon-based thin-film PV cells, other types of thin-film PV cells are still developing. In recent years, the efficiency record of compound thin-film PV cells has been continuously broken: the laboratory efficiencies of CIGS cells and CdTe cells exceeded 22%, and the efficiency of the CdTe modules was over 16.4%, which is comparable to that of poly c-Si modules. The market application scale is also expanding. In recent years, the practical application research of perovskite PV cells has also developed rapidly. If these three types of thin-film PV cells could have further breakthroughs in technology, with their low cost, low-temperature coefficients, good weak light response, and integration with buildings, thin-film PV cells still have a promising future in terms of application.

References

[1]   A. Shah, Thin film silicon solar cells [M], Chapter 2–5©2010, First edition, EPFL Press, 2010.

[2]   A. V. Shah, et al.. Thin-film silicon solar cell technology. [J] Progress in Photovoltaics: Research and Applications, 2004, (12), 113–142.

[3]   R. H. Bossert, et al. Thin-film solar cells: technology evaluation and perspectives, Netherland Agency for Energy and the Environment, 2000; (Nov.) report number DV 1. 1.170.

[4]   W. Shinohara, et al. Recent progress in thin-film silicon photovoltaic technologies[C]. 5th World Conference on Photovoltaic Energy Conversion, 6–10 September 2010, Valencia, Spain.

[5]   Michio Kondo and Akihisa Matsuda. An approach to device grade amorphous and microcrystalline thin films fabricated at higher deposition rates. [J] Current Opinion in Solid State and Materials Science, 2002, (6), 445–453.

[6]   X. Wu, et al. 16.5% efficient CdS/CdTe polycrystalline thin-film solar cell[C]. 17th European Photovoltaic Solar Energy Conference, 22–26 October 2001, Munich, Germany.

[7]   I Visoly-Fisher, et al. How polycrystalline devices can outperform single-crystal ones: thin film CdTe/CdS solar cells. [J] Advanced Materials, 2004, 16(11), 879–883.

[8]   A. Rockett. The effect of Na in polycrystalline and epitaxial single-crystal $CuIn_{1-x}Ga_xSe_2$. [J] Thin Solid Films, 2005, 480–481, 2–7.

[9]   T Negami, et al. Production technology for CIGS thin film solar cells. [J] Thin Solid Films, 2002, 403–404, 197–203.

[10] Liangqi Ouyang, Daming Zhuang, Yu Zhan, et al.. Preparation of 17.5% efficiency CIGS battery by magnetron sputtering four-dimensional target method. Acta Energiae Solaris Sinica, 2016, 37(11), 2994–2998.

[11] Na Yi, Progress in perovskite solar cell technology. Solar Power, May, 2016.

[12] B O' Regan and M. Grätzel. A low-cost high-efficiency solar cell based on dye-sensitized colloidal TiO_2 films. Nature, 1991, 353, 737–739.

Exercises

4.1 Compared to c-Si PV cells, what are the advantages and disadvantages of thin-film PV cell?

4.2 Briefly summarize the types of thin-film PV cells that are currently in application and under development.

4.3 What are the main structures of multijunction a-Si PV cell?

4.4 What are the main difficulties while putting a-Si PV cells into large-scale application?

4.5 Briefly summarize the advantages and disadvantages of CdTe thin-film PV cell.

4.6 What are the main methods of commercialized CdTe thin-film production?

4.7 Briefly summarize the advantages and disadvantages of CIGS thin-film PV cell.

4.8 What are the main methods of commercialized CIGS thin-film production?

4.9 What is the mechanism of dye-sensitized PV cell?

4.10 Briefly describe the features and development potential of perovskite PV cell.

4.11 Explain the development prospects of thin-film PV cells. What are the advantages and disadvantages of large-scale applications?

Chapter 5
Concentrator PV and solar tracker

Power generation of PV cell is directly related to solar irradiance. In a certain range, the greater the irradiance, the more power can be generated by the PV cell. Therefore, the use of concentrating, tracking, and other methods is an effective way to increase PV power generation [1].

The application of concentrated solar power is divided into concentrator photovoltaic (CPV) power and concentrated solar power (CSP) which uses solar thermal process to generate electricity. This chapter focuses on CPV.

5.1 CPV power generation

CPV technology uses optical devices to concentrate direct sunlight onto PV cells, increasing the irradiance on PV cells, which can increase power generation [2].

When describing the concentration level of the system, the concentration ratio is often used. The concentration ratio refers to the ratio of the concentrated energy density per unit area to the incident energy density when the optical system is used to collect the radiant energy. For example, when the concentration ratio is 1,000, it means that the surface of the PV cell is exposed to light that is 1,000 times stronger than normal sunlight. It is commonly referred to as "1000X." A 1000X CPV means that, compared to a regular PV cell, only one thousandth of an area of CPV cell is needed to achieve the same power generation as a nonconcentrating system. In addition, the commonly used geometric concentration ratio refers to the ratio of the geometric light receiving area of the optical device used to gather solar energy in the geometric area of the CPV cell. However, due to factors such as aberrations and chromatic aberrations in the optical system, reflection, absorption and scattering losses of sunlight through the concentrator, and the light intensity on the surface of the cell is not uniform, when the geometric concentration ratio is 1000X, the actual average light intensity is smaller than 1,000 times the normal light intensity.

As a CPV system needs to be equipped with light concentrating and tracking devices, it is relatively complex. If it is applied to small-scale home and commercial PV systems, it will not have a cost advantage and will bring about more maintenance problems. CPV systems are more suitable for areas with sufficient energy, preferably large-scale PV power plant in areas with an average direct irradiance (DNI) greater than 5.5 to 6 $kWh/m^2/d$ (or 2,000 $kWh/m^2/y$), and an installed capacity of 1–1,000 MW.

https://doi.org/10.1515/9783110524833-005

5.1.1 Advantages and disadvantages of CPV

5.1.1.1 Advantages
Compared to crystalline silicon and thin-film cell power generation, CPV power generation has many advantages.

(1) High power generation efficiency
The highest efficiency of high-concentration PV cell now exceeds 44%. Even low-concentration PV cells are much more efficient than nonconcentrating PV cells. At present, the efficiency of crystalline silicon PV cell is about 23%, and the efficiency of thin-film PV cell is about 13%. CPV cell maintains the highest photoelectric conversion efficiency record in PV technology.

(2) Less land occupation
With the same amount of power generation, the CPV system covers only about half area of the crystalline silicon PV cell power generation system. For example, Concentrix's CPV system requires only 6 to 8 acres of land per MW. Moreover, the land can also be comprehensively used, such as grazing animals or planting crops within the power plant, as shown in Fig. 5.1.

Fig. 5.1: Comprehensive utilization of CPV power plant.

(3) Convenient to install on site

Due to the high degree of integration, installation on the site is very convenient and the complete set of power station systems can be completed in a very short time from the end of the approval process to the installation.

(4) Can be used comprehensively

For high-concentration ratio CPV power plants, hot water produced by cooling can be used in other ways in addition to power supply.

5.1.1.2 Disadvantages

(1) The demand for light resources is high, and it is necessary to build power plants in areas with high direct radiation.
(2) CPV systems (especially high-concentration systems) do not absorb scattered sunlight. If direct sunlight diverges slightly from the cell, there would be a sharp drop in power generation. Therefore, high-precision solar trackers are often required.
(3) CPV cells will increase the temperature during operation, so heat dissipation measures are generally required.
(4) As the real practical application time of CPV systems is not long and the scale is small, further practical tests are needed.

5.1.2 CPV system components

Compared with conventional PV systems, CPV systems have similar balance of system (BOS), only the front array is different. The regular flat plate PV cell array consists of PV cell modules, a holder and a base, a connection cable, a combiner box, which are relatively simple. In addition to these, a CPV system requires a number of other components, which are described below.

5.1.2.1 CPV cells

Different from regular PV system, in the CPV system, the PV cell operates under high-intensity sunlight and high temperature conditions, and the current is much larger than that of regular cell, so there are special requirements for the cell [3].

According to different degree of concentration, CPV systems generally use special single-crystalline silicon cell or III-V multijunction PV cell. There are also thin-film PV cells used occasionally.

(i) Monocrystalline silicon PV cells

Since monocrystalline silicon PV cells are stable in performance and relatively inexpensive, high conversion efficiency monocrystalline silicon PV cells are generally used in low-concentration systems to avoid additional cost for manufacturer special PV cells. Under the light-concentrating condition, there are high requirements for the performance of the PV cell, and measures should be taken to reduce the series resistance of the cell and the loss of the tunnel junction.

At the same time, the grid lines of the concentrating cells are relatively dense, and typical grid lines account for approximately 10% of the cell area, in order to meet the needs of high current densities.

In addition, PV cells are prone to aging because they are often exposed to intense light, so they should be designed and manufactured specifically.

(ii) III-V Multijunction PV cell

The III-V PV cells are widely used in medium- and high-concentration systems, and the so-called III-V PV cells are compounded semiconductor PV cells made of group III and V elements in the periodic table. Compared to silicon-based materials, multijunction PV cells based on III-V semiconductors have extremely high photoelectric conversion efficiency and are about twice as high as crystalline silicon PV cells. For traditional silicone cells, III-V semiconductors have a much higher temperature resistance than silicon, a low temperature power coefficient, and marvelous high photoelectric conversion efficiency at high irradiance, and therefore can be applied to high-concentration technology, which means that it produces as much power and only requires a smaller PV cell chip. A unique aspect of the multijunction technique is that different materials can be selected and combined so that their absorption spectra and the solar spectrum are broadened. At present, the most widely used are germanium, indium gallium arsenide (or gallium arsenide), and gallium indium phosphorus triple junction cell. In this multijunction PV cell, not only the lattice constants of the three materials are substantially matched, but also each semiconductor material has a different forbidden bandwidth, so that it is possible to absorb the solar spectrum of different wavelength bands, and thus the broad spectrum sunlight can be fully absorbed. The corresponding spectrum of III-V PV cells such as gallium, arsenic indium gallium, gallium indium phosphorus, and so on as described before is 300 to 1,750 nm, and can fully absorb the solar radiation energy. Figure 5.2 shows the corresponding graph of the quantum efficiency of a typical triple-junction III-V PV cell. The conversion efficiency records of various CPV cells are shown in Tab. 5.1.

In order to pursue higher conversion efficiency, some research institutes have been developing four-junction or even five-junction PV cells. Undoubtedly, PV cells with four junctions or more will have more uniform spectral response and a wider spectral range, and will be more effective for improving the conversion efficiency of PV cells. NREL has been studying multijunction cells with a conversion efficiency of 50%.

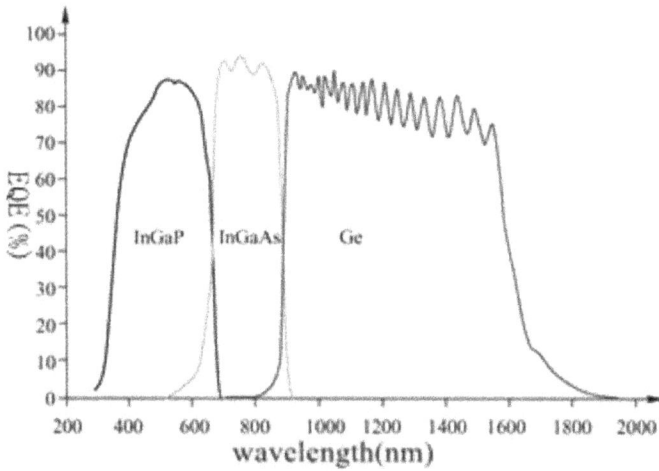

Fig. 5.2: QE curves of III-V PV cells.

Tab. 5.1: Terrestrial concentrator cell and module efficiencies.

Under ASTM G-173-03 direct beam AM1.5 spectrum at a cell temperature of 25 °C.					
Classification	Efficiency (%)	Area (cm²)	Intensity (suns)	Test center (date)	Description
Single cell					
GaAs	29.3 ± 0.7	0.09359 (da)	49.9	NREL (10/16)	LG electronics
Si	27.6 ± 1.2	1.00 (da)	92	FhG-ISE (11/04)	Amonix back contact
CIGS (thin film)	23.3 ± 1.2	0.09902 (ap)	15	NREL (3/14)	NREL
Multijunction cell					
GaInP/GaAs; GaInAsP/GaInAs	46.0 ± 2.2	0.0520 (da)	508	AIST (10/14)	Soitec/CEA/FhG-ISE 4 j bonded
GaInP/GaAs/ GaInAs/GaInAs	45.7 ± 2.3	0.09709 (da)	234	NREL (9/14)	NREL, 4 j monolithic
InGaP/GaAs/ InGaAs	44.4 ± 2.6	0.1652 (da)	302	FhG-ISE (4/13)	Sharp, 3 j inverted metamorphic
GaInP/GaInAs	34.2 ± 1.7	0.05361 (da)	460	FhG-ISE (4/16)	Fraunhofer ISE 2 j
Minimodule					
GaInP/GaAs; GaInAsP/GaInAs	43.4 ± 2.4	18.2 (ap)	340	FhG-ISE (7/15)	Fraunhofer ISE 4 j (lens/cell)

Tab. 5.1 (continued)

Classification	Efficiency (%)	Area (cm^2)	Intensity (suns)	Test center (date)	Description
Under ASTM G-173-03 direct beam AM1.5 spectrum at a cell temperature of 25 °C.					
Submodule					
GaInP/GaInAs/Ge; Si	40.6 ± 2.0	287 (ap)	365	NREL (4/16)	UNSW 4 j split spectrum
Module					
Si	20.5 ± 0.8	1875 (ap)	79	Sandia (4/89)	Sandia/UNSW/ENTECH (12 cells)
Triple-junction	35.9 ± 1.8	1092 (ap)	N/A	NREL (8/13)	Amonix
Four-junction	38.9 ± 2.5	812.3 (ap)	333	FhG-ISE (4/15)	Soitec
Notable exceptions					
Si (large area)	21.7 ± 0.7	20.0 (da)	11	Sandia (9/90)	UNSW laser grooved
Luminescent minimodule	7.1 ± 0.2	25 (ap)	2.5	ESTI (9/08)	ECN Petten, GaAs cells

Source: Martin A. Green: Prog. Photovolt: Res. Appl. 2017; 25:3–13

5.1.2.2 Concentrator

There are many types of concentrators, and there are many classification methods [4].

(1) By shape

(i) Spot-focus concentrator: The solar radiation forms a focal point (or focal spot) on the surface of a PV cell.

(ii) Line-focus concentrator: The solar radiation forms a focal line (or focal band) on the surface of a PV cell.

(2) By imaging attributes

(i) Imaging concentrator

According to the optical principle, when the light is focused on an extremely small area by the light-concentrating optical system, the image of the object can be clearly presented at the convergence of the light. In 1979, Welford and Winston proposed that the goal of PV concentrators is not to reproduce the exact image of the sun, but to maximize the collection of energy, and imaging concentrators are not ideal PV concentrators.

(ii) Nonimaging concentrator

The ultimate goal of nonimaging concentrator design is to obtain the maximum intensity of light per unit area, which is essentially an optical "funnel" that requires a large area of incident light to be refracted or reflected and then concentrated to a small area, to achieve the purpose of concentrating energy. Sunlight passes through the concentrator and can reach or exceed the brightness of the sun. Imaging optics usually cannot achieve the ideal concentrating level. In a report published in 2002 by O'Gallagher et al., according to theoretical and practical analysis, in a nonimaging concentrator, when the solar incident angle is 0° to 42.2°, the solar concentrator is able to collect the total solar energy. Therefore, nonimaging optics applied to solar concentrators can not only obtain a high-concentration ratio, but also a larger acceptance angle and a smaller volume. They are ideal as a nontracking static concentrator application.

(3) By concentrating methods

Concentrators can be classified into three types: reflective concentrators, refractive concentrators, and light-guide solar optic (LSO). These are described in detail below.

(i) Reflective concentrator

Reflective concentrators collect sunlight onto the PV cell by reflecting [5]. Due to the different reflection methods, it can be divided into the following two types.

– Slot-shaped plane concentrator

The slot-shaped plane concentrators are usually made of a plane mirror with a slot shape, and the parallel light is reflected by the slot-shaped plane mirror and concentrated on the PV cell at the bottom, as shown in Fig. 5.3. It can increase the solar radiation intensity projected on the surface of the PV cell, and the light collection ratio can range from 1.5 to 2.5. The angle of the mirror depends on the angle of inclination and latitude and the design of the assembly, which is usually fixed.

– Parabolic concentrators

The parallel light after reflected by the parabolic concentrator can be concentrated onto a focal point, as shown in Fig. 5.4. If a PV cell is placed at the focal point, incident sunlight can be collected on the PV cell, increasing the intensity of radiation projected onto the surface of the PV cell. Although the production of parabolic mirrors is more complicated than that of flat mirrors, the effect of concentrating light is much better. Therefore, many parabolic concentrators are used in low-concentration power generation systems.

In order to further increase the concentration ratio, some concentrators also use a method of secondary parabolic focusing. When parallel sunlight is incident

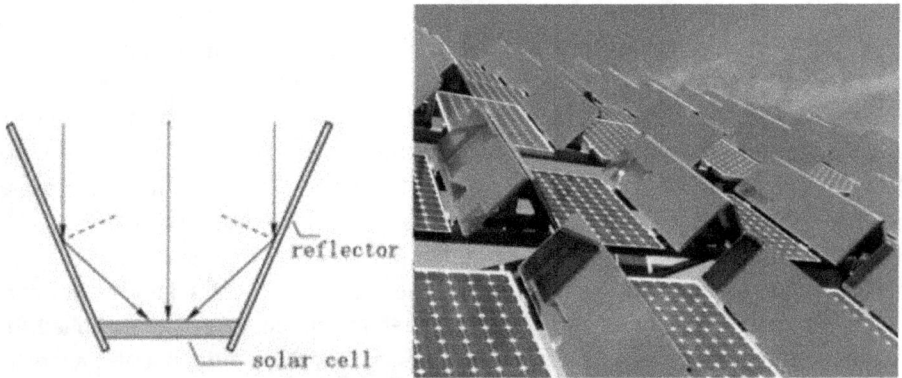

Fig. 5.3: Slot-shaped plane concentrator. (left) Schematic diagram; (right) Real application.

Fig. 5.4: Parabolic concentrator. (left) Schematic diagram; (right) Real application.

on the first relatively large 1# parabolic mirror, the sunlight is focused on the focal point of a second, smaller 2# parabolic mirror, and then reflects the sunlight to the PV cell through the second reflector. After two reflections, the solar radiation intensity can be further increased. Figure 5.5 shows the principle of secondary parabolic concentrator.

Welford and Winston later improved the parabolic concentrator and developed a compound parabolic concentrator (CPC). Two-dimensional CPC consists of multiple parabola lines, which can further increase the effect of concentration.

In addition to parabolic trough concentrator, parabolic concentrators can also be dish type, as shown in Fig. 5.6. In this way, a large parabola can be divided into numerous small reflection areas, so that the curved surface represented by each reflection area is extremely smooth and is almost a plane, which greatly reduces the processing difficulty and cost of each reflection area.

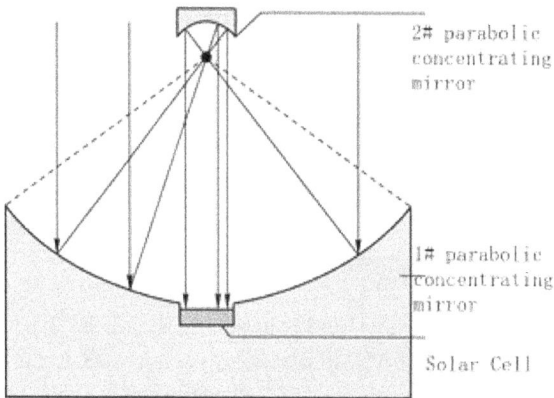

Fig. 5.5: Illustration of secondary parabolic concentrator.

Fig. 5.6: Parabolic dish concentrator.

There are also other types such as hyperboloid concentrators.

(ii) Refractive concentrator

The refractive concentrator collects solar rays on the PV cell by means of refraction so as to enhance the solar irradiation intensity.

The refractive concentrator can be a conventional continuous lens or a Fresnel type lens [6]. Fresnel lenses have the following advantages.

- The Fresnel lens can be made thin and light when the aperture is large.
- Using a Fresnel lens as a concentrator allows a larger aperture than a conventional lens, that is, the Fresnel lens can have a very low Fresnel number.

– The material used to make the Fresnel lens can be plastic glass, which is not only cheaper than glass, but also convenient for mass production.

It is well known that an ordinary spherical convex lens can collect light, but a concentrator device generally used for a PV cell is relatively large, and if a conventional spherical convex lens is used, its thickness is going to be really large. In order to reduce the thickness and mass, and to save material, Fresnel lens is usually used, which is made by using the principle that light refracts at the interfaces of different media, and has the same function as a regular lens. The Fresnel lens used in application is obtained by continuously dividing and connecting a convex lens, and is generally formed by injection molding a plexiglass or molding a sheet of acrylic plastic or polyolefin material. The surface of the lens is a glossy surface; the other side is composed of a series of concentric annular sections with different angles. The cross section has a zigzag shape. Its shape is designed by using the principle of light refraction and based requirements on relative sensitivity and receiving angle, so as to meet the requirements of short focal length and large aperture [7]. As shown in Fig. 5.7. The Fresnel lens is also the main component of the CPV system. On one hand, it concentrates sunlight, and on the other hand, it also protects the inner components of PV module. It is part of the PV module housing. Based on cost and outdoor reliability considerations, most high-concentration PV (HCPV) now use transmissive concentrators.

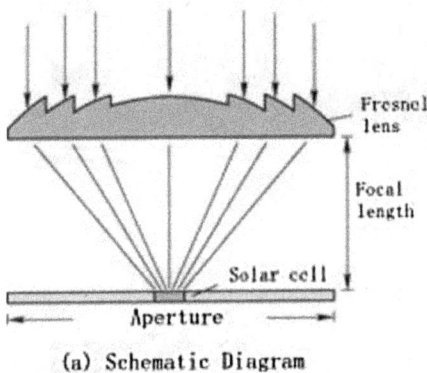

Fig. 5.7: Fresnel lens concentrator.

A high-quality Fresnel lens must have a smooth surface, clear texture, light weight, high light transmittance, and not prone to degradation. Its thickness is generally about 1 mm. There are a variety of process manufacturing technologies, such as injection molding and hot pressing and continuous roll pressing of PMMA, and thermoforming silicone on glass to obtain a Silicone on glass (SOG) Fresnel lens. They require a more complex manufacturing process. Parameters such as transmittance, spot uniformity, focal length, aberration, process uniformity, UV resistance, sand-scraping resistance, and so on are all important factors for evaluating the performance of Fresnel lenses.

The Fresnel lens has two types of focusing, spot and line concentration, and its corresponding tracking system type can be two-dimensional tracking and one-dimensional tracking, respectively. According to different applications, different concentrating methods and tracking formats can be selected.

(iii) Light-guide solar optic (LSO)

The LSO generally adopts a two-stage optical system. The first-stage system is a single planar light-insertion molded polymeric slab that is transmissive to sunlight. The second-stage system is a wedge-shaped plate. The light guided by first-level system multiple-reflects inside the second-level system's wedge to reach the end of the second-level system. The schematic diagram is shown in Fig. 5.8.

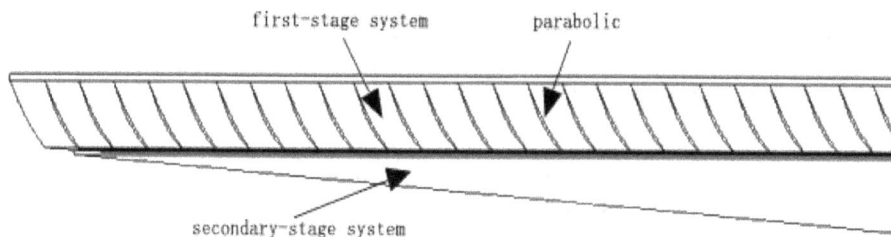

Fig. 5.8: Schematic illustration of light-guide solar optic.

Since the slab light-guide concentrator uses multiple internal reflections to fold the optical path, the overall thickness of the slab waveguide concentrator will be greatly reduced compared to the aforementioned reflective and refractive concentrators. Under the premise of maintaining a high-concentration ratio, it is still possible to achieve a lower system thickness, even without frames. Figure 5.9 is a real product of Sun Simba from Morgan Solar.

Light-guide concentrators can be processed using glass or optical grade plastics, because this type of product has a delicate design and requires complex processing before commercialization of this technology. The price of slab light-guide concentrators are relatively high.

5.1.2.3 Solar tracker

With the increase of the concentration ratio, the angle range that the CPV system can receive light will become smaller. In order to ensure that the sunlight can always

Fig. 5.9: Sun Simba from Morgan Solar.

reach the PV cell accurately, system with concentration ratio greater than 10, in order to ensure the concentrating effect, should use the tracking system. Especially for high-concentration systems, as long as the Sun slightly deviates from the cell, its power generation will drop sharply. The larger the concentration ratio, the higher the tracking accuracy should be. When the concentration ratio is 1,000, the tracking accuracy error is required to be less than $\pm 0.3°$ or even $\pm 0.1°$ (the tracking error is shown in Fig. 5.10. The high-concentration system must be equipped with a high-precision tracking device. The solar tracker has become one of the key components of HCPV systems. According to statistics, most of the time failure of HCPV systems are related to the failure of the solar tracker.

The classification and structure of the solar trackers are given in section 5.2.1.

5.1.2.4 Cooling components

Under the concentrated solar irradiance condition, the temperature of the CPV module will rise. Because the temperature coefficient of CPV cell power output is negative, the CPV cell power will decrease when the temperature increases. In order to reduce the efficiency loss caused by the high module temperature, it is necessary to consider the cooling problem, and appropriate measures should be taken to keep the temperature of the CPV cell within a certain range. Usually, the upper limit of the cell temperature is determined by reaching 80% of the conversion efficiency

Fig. 5.10: Tracking error of solar tracker.

under normal temperature. For crystalline silicon PV cells, the upper limit is about 100 °C. For III-V CPV cells, the upper temperature limit can be slightly higher.

If the temperature of the CPV cell is high, cooling methods must be taken. The cooling modes are divided into active cooling and passive cooling. Active cooling is accomplished through active components (usually through cooling water); passive cooling is accomplished by relying on air convection and heat radiation without any active working components. Figure 5.11 shows two types of cooling systems for CPV cells. On the left is a passive cooling system with a radiator, and on the right is an active cooling system with a cooling fluid.

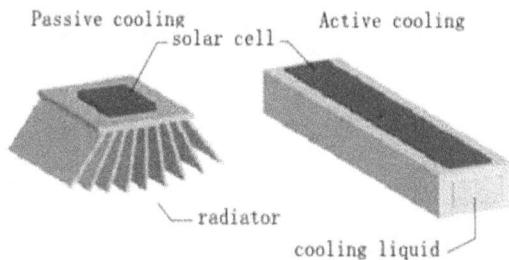

Fig. 5.11: Two types of cooling systems for CPV cell.

Due to high energy density of CPV technology, after the photoelectric conversion of high-efficiency PV cells, there is still a lot of heat to be emitted. When designing cooling system of CPV, it is necessary to consider critical parameters, heat capacity and thermal resistance, and calculate the amount of heat that needs to be treated (with direct light of 1,000 W/m^2). Extreme conditions also need to be considered, for example, there is no grid-connected power generation in the PV system, the operating temperature of the cell must still meet the requirements of the cell and its packaging components.

Because active cooling has many limitations in terms of cost and maintenance, many concentrators currently use passive cooling. To improve the cooling efficiency, the heat-conducting layer needs to be minimized to reduce the thermal contact resistance. The commonly used passive cooling structure is shown in Fig. 5.12, so that the heat passes through the following dielectric layers in sequence: cell chip → packaging materials → copper clad ceramic substrate → thermally conductive adhesive → radiator → surface heat exchange or heat radiation to the outside.

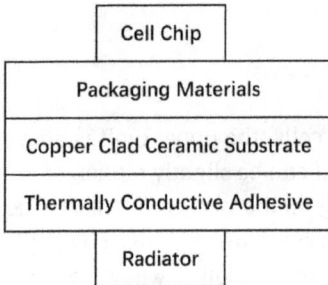

Fig. 5.12: Typical concentrator heat transfer structure.

In some occasions, only using passive cooling may not be effective enough. Water cooling and other methods are then required. Water cooling requires special configuration of water supply and circulation equipment. If water is needed, piping, valves, and pumps are needed, which requires higher investment. However, it can also be used comprehensively in some cases. For example, in the 25 kW concentrating system built at Dallas Airport in the USAin 1982, in addition to generating power, active cooling is used to pass water to cool the PV cells, and the resulting hot water can be used for nearby hotels.

5.1.3 CPV systems

Driven by the world oil crisis, in 1976, the US government prepared a budget of $1.25 million to support research on CPV technology, and thereafter the budget increased year by year, and reached $6.2 million by 1981. With the support of government projects, Sandia National Laboratories in the USA developed a 1 kW array in 1976 with a PV conversion efficiency of 12.7%. This array was later called Sandia1, as shown in Fig. 5.13. This prototype used a point-focusing Fresnel lens, a concentrating ratio of 50X, cold water cooling, dual-axis tracking, a concentrating crystalline silicon PV cell, and an analog closed-loop tracking control system, of which Fresnel lens, dual-axis tracking, and analog closed-loop tracking control system are still widely used in today's CPV systems. At the end of the 1970s, a prototype developed by Ramón according to this concept used a SOG point-focusing

Fig. 5.13: Sandia 1 CPV system.

Fresnel lens with a concentrating ratio of 40X, crystalline silicon PV cell with of 5 cm diameter, and the heat sink was passively cooled with radiators, as shown in Fig. 5.14. This prototype was tested and assessed. Soon after, SpectroLab also developed a 10 kW CPV system with a conversion efficiency of 10.9% and a concentrating ratio of 25X. Later, in Germany, Italy, and Spain, various replicas ranging from 500 W to 1 kW appeared. They made improvements in some components but have not been commercialized due to high costs.

In the 1980s, after some long-term exploration and development, some institutes began to implement small-scale real-world application demonstrations and established a number of small-scale CPV power plant, as shown in Tab. 5.2.

CPV systems can usually be classified into low-concentration PV (LCPV), medium-concentration PV (MCPV), and high-concentration PV (HCPV) according to concentration ratio. However, the specific classification standards for low, medium, and high are not unified, and it is also changing with the development of technology. Originally, system with concentrating ratio of 2 to 10 would be considered LCPV, 10 to 100 is MCPV, and more than 100 is HCPV. While GTM Research classification method is: 2 to 8 – LCPV, 10 to 150 – MCPV, and more than 200 is HCPV. Now, it is generally believed that when the concentrating ratio is 2 to 100, it is LCPV, 100 to 300 is MCPV, and over 300 or even over 1,000 is HCPV.

Fig. 5.14: Ramón CPV system.

5.1.3.1 Low-concentration PV (LCPV)

LCPV system is designed on the premise that the hour angle is not tracked. This type of concentrating system mostly uses crystalline silicon PV cells as a power generation chip. Concentrator usually use slot or planar reflection type. Several reflectors are placed around the cell to increase the amount of sunlight received on the surface of the cell. Reflective concentrators have lower concentrating ratios. If they are equipped with a simple tracking device, the effect of concentrating will be increased. For nondemanding and error-insensitive applications, single-axis trackers can be used, that is, only tracking in the east-west direction. Due to the low-concentrating ratio, it is generally not necessary to provide a dedicated cooling system.

Many manufacturing companies have conducted long-term research and development of LCPV systems. In order to compete with ordinary flat-panel fixed PV systems, many different technical routes have been adopted, and most of them use reflective concentrating, such as SunPower's system has concentrating ratio of 7X, and is equipped with a tracker; Skyline High Gain Solar, Inc. uses a 10X compound parabolic concentrator with a tracker; Abengoa Solar NT uses 1.5X and 2.2X reflecting mirrors with a tracker, with monocrystalline silicon chip; JX Crystals uses concentrating ratio of 3X; Megawatt Solar uses concentrating ratio of 20X. Solaria uses a refractive concentrator with concentrating ratio of 2X with a tracker.

SunPower began developing CPV systems in 1985. The C-7 type low-concentration system with tracker, as shown in Fig. 5.15, has power of 14.7 kW each module, and the

Tab. 5.2: Early CPV systems.

System type	Concentrating device	Concentrating ratio	Cell type	Cooling	Tracking	Application	Capacity	Year
Point focus on single cell	Lens	$50 < x < 500$	Si	Passive	Dual axis	Panel Ramón Areces – UPM Madrid	1 kW	1978
						PCA – Ansaldo Italy	1 kW	1978
						Soleras village in Saudi Arabia	350 kW	1981
						Sky Harbor Airport in Phoenix	250 kW	1982
						POCA Alpha Solarco array	10 kW	1989
						Amonix IHPVC system in Nevada Power Company	18 kW	1995
						Amonix CPV California Polytechnic University Pomona	15 kW	1998
Point focus on jointed cell	Dish parabola	$x < 500$	Ga		Dual axis	Solar farm in Broken Hill (Australia)	1 MW	In progress

(continued)

Tab. 5.2 (continued)

System type	Concentrating device	Concentrating ratio	Cell type	Cooling	Tracking	Application	Capacity	Year
Line focusing	Lens	15 < x < 60	Si	Passive	Dual axis	Entech-3 M/Austin system	300 kW	1990
						Entech-20 kW PVUSA system	20 kW	1991
						Entech-CSW solar park	100 kW	1995
						Entech-TU electric energy park	100 kW	1995
						PVI – Clean Air Now, Los Angeles, CA	30 kW	1996
						PVI, Sacramento municipal utility district	30 kW	1996
						PVI Arizona public service	3 kW	1996
				Active		E-systems-PV/thermal system for Dallas airport	25 kW	1982
	Mirror			Passive	Single axis	EUCLIDES system	450 kW	1999
				Passive		PV/thermal acurex-system	60 kW	1981
				Active		BDM corp. office block, Albuquerque	50 kW	1982

Source: Crating/BOOK1/31/12/02

system voltage is 1,000 V and consists of 108 pieces of 136 W cells, and has module efficiency of 20.1%. It uses parabolic concentrator with concentrating ratio of 7X, the material is hot dip galvanized steel and stainless steel glass mirror, using maintenance-free bearings, facing north-south horizontal single-axis tracking, with tracking angle of −75 °~+75 °. It can resist wind speed up to 40 m/s. The company claims that with its C-7 tracker, its cost of electricity generation can be reduced by 20% compared to other competing technologies.

Fig. 5.15: SunPower C-7 LCPV.

At the Flower Port in Shanghai, China, JX Crystals, USA designed and installed the 125 kW LCPV power generation system produced by SunPower. After the completion of the second phase, the capacity reached 330 kW. In 2013, SunPower built a 1 MW low-concentration power plant on Arizona State University Polytechnic Campus, and a 7 MW concentrating power plant in Arizona in 2014.

Sevilla, Spain built a 1.2 MW low-concentration power plant with 154 dual-axis trackers, as shown in Fig. 5.16, each with 36 CPV modules, and a reflective concentrator with concentrating ratio of 1.5X to 2.2X. The conversion efficiency is 12%, the power station covers an area of 295 000 m^2, and the annual power generation is 2.1 GWh.

5.1.3.2 Medium-concentration photovoltaic (MCPV)

Point-focus or line-focus concentrators can be used in MCPV. When the point-focus concentrator is used, its properties are the same as those of the high-concentration

Fig. 5.16: LCPV in Sevilla, Spain.

concentrator, and the effect of biaxial tracking is ideal. With a line-focus type concentrator, it is best to place the focal line in the east-west direction. The MCPV technology is not widely used in the market yet.

Skyline Solar of the USA adopted a parabolic reflective CPV with a concentration ratio of 14X and crystalline silicon PV cell. Figure 5.17 is the Skyline Solar 14X concentrating array. By the end of 2011, production capacity reached 100 MW. In 2009, 24 kW was installed and in 2010, 83 kW was installed. The 500 kW concentrating power plant in Durango, Mexico has been completed.

The CPV power plant constructed by Solaria in 2011 in Pontinia, Italy, has a capacity of 585 kW. In 2012, a 1.1 MW solar power plant was built in California. In March of the same year, a concentrating power plant was built in Puglia, Italy, using single-axis tracking with a capacity of 2 MW. In December of the very same year, a 4.1 MW concentrating power plant was constructed in New Mexico, as shown in Fig. 5.18, and also with a single-axis tracking system. In 2012 and 2013, two 1 MW and 2 MW concentrating power plants were built in Sardinia, Italy.

Fig. 5.17: Skyline Solar 14X concentrating array.

Fig. 5.18: Solaria, 4.1 MW concentrating power plant.

5.1.3.3 HCPV system

HCPV system is usually composed of three parts: a concentrator, a tracker, and a balance component [8]. Among them, the balance components are basically the same as the conventional crystalline silicon and thin film solar power systems. The concentrator consists of a concentrating cell, an optical system, a cooling system, and a module frame. In the range of HCPV, point-focusing nonimaging concentrators are mainly used. With this type of concentrator, even if the incident angle of the Sun changes only by 0.5°, the irradiance on the PV cell will be reduced by half. Therefore, it is necessary to provide a sophisticated solar tracking device. Passive cooling is usually used because it does not require the use of cooling water and is particularly suitable for use in hot, dry areas. It is less affected by the effect of low conversion efficiency under high temperature compared to other technologies. With the continuous decrease in the price of cell chips, high-concentration systems have great advantages in terms of efficiency and cost. HCPV has the largest share in the global CPV market and will become the main development direction of concentrating technology.

HCPV system has a relatively high technical threshold and cross many industrial areas. It covers semiconductor materials and process manufacturing, semiconductor packaging, optical design and manufacturing, automation control, mechanical design and manufacturing, and metal processing. Products of the HCPV industry include multijunction cell epitaxial materials, photoelectric conversion chips, light receiver components, concentrators, and dual-axis trackers.

In the report published by GTM Research, *Concentrating Photovoltaics 2011: Technology, Costs and Markets*, they point out that the installation capacity of CPV systems will be 5 MW in 2010 and will increase to 1,000 MW in 2015. At present, the CPV market is dominated by three system integrators. The CPV system capacity that has been completed or is being installed or has been contracted includes: Concentrix Solar, 181 MW (all in the USA); Amonix, 127 MW (115 MW in the US, 12 MW in Spain); SolFocus, 39 MW (US 32 MW, Portugal 5 MW, Greece 2 MW), others 13 MW (including 3 MW by SolarSystems in Australia).

Founded in 2005 in Freiburg, Germany, Concentrix Solar is a company separated from the Fraunhofer Institute for Solar Energy Systems and specializes in the development and production of CPV technology. In 2007, Concentrix Solar's CPV technology was awarded the German Economic Innovation Award. In September 2008, a fully automated 25 MW production line was built. In December 2009, it was acquired by Soitec Group, France, so the name Concentrix-Soitec is now commonly used. Concentrix Solar's CPV technology uses Fresnel lenses to collect sunlight with concentrating ratio of 500X and use III-V triple junction PV cells (GaInP/GaInAs/Ge). In order to ensure that the Sun is concentrated on the CPV modules, a dual-axis solar tracking system is adopted, and the system efficiency of the Concentrix concentrating technology is 27%.

In 2010, Concentrix Solar built a 1.37 MW CPV power plant in Questa, New Mexico, as shown in Fig. 5.19, which was the largest in the world. The company's technology is also being promoted in countries such as Germany, Spain, Italy, South Africa, and Egypt.

Fig. 5.19: Concentrix CPV power plant.

In October 2011, Soitec introduced its fifth-generation CPV system designed specifically for grid-scale power plants. Soitec's "Concentrix" technology includes a 28 kW array design with trackers over 100 m² in area. The system conversion efficiency was increased to 30% to reduce the cost of levelized cost of electricity (LCOE).

Amonix, USA, designed the first 20 kW high-concentration system in 1994. Since then, they have successively developed a 6 generations of the system, which improved the efficiency and performance continuously. The latest generation of system is the Amonix 7700 CPV system, as shown in Fig. 5.20, which is currently the world's largest pedestal-mounted solar system. The tracker has a width of 70 feet and a height of 50 feet. It has 7560 Fresnel lenses and concentrates 500 times of sunlight onto multijunction gallium arsenide PV cells. Each can generate 60 kW of electricity. The Amonix 7700 CPV System is said to use land more rationally than other solar technologies: only 5 acres are required to install a rated capacity of 1 MW, while other solar technologies require 10 acres. Some Amonix systems (such as the system in Pomona, California) have been operating safely for more than 12 years.

From 2006 to 2008, Amonix CPV solar energy equipment was installed in Navarra, Spain, for a total of 3 phases, with a total capacity of 7.8 MW. It has been the largest concentration power plant in the world. From 2008 to 2009, Amonix

Fig. 5.20: Amonix 7700 array.

installed the seventh 300 kW concentrating system in Las Vegas, using III-V multi-junction cells, with an efficiency of 25% and an energy efficiency of 2,500 kWh/kW. In October 2011, Amonix finished the construction of the largest concentrating power plant in North America in Hatch, New Mexico, with a capacity of 5 MW. It consists of 84 dual-axis trackers. The power of the modules installed on each tracker is 60 kW, and III-V multijunction concentrating cells are used. With dual-axis tracking, system efficiency can reach 29%. The electricity generated can be used by 1,300 families. In 2011, Amonix installed a total of 35 MW concentrating power systems in southwestern US. In May 2012, a 30 MW CPV power plant with a capacity of 30 MW was built in the St. Louis Valley, Alamosa, Colorado, USA. This power plant covers an area of 225 acres and is composed of 500 Amonix 7700 dual-axis trackers. It is boosted to 115 kV and integrated into the grid. It has been put into normal operation.

SolFocus installed a 1 MW concentrating power plant at Victor Valley College in California, USA. It consists of 122 SF-1100 arrays, each with a capacity of 8.4 kW and covers an area of 6 acres. The annual power generation is 2.5 million kWh, which can meet about 30% of the college's electricity consumption. In 2012–2013, nine concentrating power plants with capacities of 1.0 to 1.6 MW were built in Mexico, Italy, and the USA.

Suncore, China completed a 3 MW CPV solar energy demonstration project in Golmud, Qinghai in 2010. It uses a 500-times lens and a dual-axis tracking system with an average conversion efficiency of 25%. On November 2012, phase 1 HCPV power station with a capacity of 57.96 MW was built in Golmud, where the altitude is 9,000 feet above sea level, and 2300 CPV trackers were used. Each tracker has 56 modules with a power of 450 W, 100 sets of 500 kW inverter are used. The second phase with a total capacity of 79.83 MW, was completed in 2013, using 3168 CPV trackers, each tracker has 56 modules, each module has a power of 450 W, and using 120 sets of 120 kW inverter. The total capacity of the two phases reached 136.79 MW, becoming the largest HCPV power plant in the world, as shown in Fig. 5.21.

Fig. 5.21: 140 MW HCPV power plant in Golmud, Qinghai, China.

The world's second largest CPV power plant is located in Touwsrivier, as shown in Fig. 5.22, 150 km northeast of Cape Town, South Africa, with a capacity of 44.19 MW, and an area of 212 hectares. Hundred million US dollars was invested into this power plant. Using 1500 Soitec CX-S530-II CPV trackers, each tracker has 12 subarrays. Each subarray has 12 modules totaling 2,455 W, 60 inverters were deployed. The whole project was completed in December 2014. The generated electricity can supply 23,000 households.

The world's third-largest CPV power plant is located in Alamosa, Colorado, USA, as shown in Fig. 5.23. It was completed in March 2012 and has a capacity of 35.28 MW, covering an area of 225 acres. There are 504 Amonix 7700 CPV trackers. The tracker is 70 feet wide and 50 feet high. There are 7 subarrays and each subarray has a power of 10 kW. There are 504 inverters with an AC capacity of 70 kW and the annual power generation capacity is 76 GWh.

Fig. 5.22: 44.19 MW CPV power plant in South Africa.

Fig. 5.23: 35.28 MW CPV power plant in Alamosa, Colorado, USA.

5.1.4 Current situation of CPV

According to the report published by Fraunhofer ISE on November 17, 2016, the current technical situation of CPV power generation is shown in Tab. 5.3 [9].

Tab. 5.3: Current situation of CPV technology.

Type	Efficiency (lab record)	Efficiency (commercial product)
PV cell	46.0 % (ISE, Soitec, CEA)	38%–43%
Minimodule	43.4% (ISE)	
Module	38.9% (Soitec)	27%–33%
System (AC)		25%–29%

Source: Fraunhofer ISE Progress in PV

From 2006 to 2015, a total of 54 CPV power plants with a capacity of more than 1 MW were installed in 12 countries, including China, the USA, France, Italy, and Spain. By 2016, the cumulative installed capacity of global CPV plants was 360 MW, with concentrating ratios above 400, and instantaneous efficiency of the commercial HCPV system is 42%. The IEA hopes that the efficiency can be increased to 50% by the middle of the 2020s. In December 2014, the lab efficiency of cell with four or more junctions has reached 46%. Under outdoor operating conditions, the CPV module efficiency has exceeded 33%. Certified by Fraunhofer ISE, Soitec PV cell has an efficiency record of 46%, and the record for minimodule efficiency is 43.4%. In 2015, under the concentrating standard test conditions, the Soitec CPV module efficiency has reached 38.9%, and the commercial application of CPV module efficiency was over 30%.

According to industry surveys and literature, the price of 10 MW CPV PV plants in 2013 was between €1,400/kW and €2,200/kW (including installation), from which the cost of electricity for CPV plants was calculated to be €0.10/kWh to €0.15/kWh. Even considering that there are many uncertainties in the market development, due to the technical progress of CPV, if the installation volume continues to increase, by 2030, the price of CPV PV system, including installation, is expected to be between €700/kW and €1,100/kW, then electricity cost of a CPV power plant can be reduced to between €0.045/kWh and €0.075/kWh.

However, after the financial crisis in 2008, especially after 2014, the cost of crystalline silicon PV cells fell drastically, making the cost advantages of CPVs no longer exist, and the development of CPV industry was greatly affected. In 2015, only Soitec installed six concentrating power plants with capacities from 1.1 to 5.8 MW in France, China, and the USA. Since then, some large-scale CPV manufacturing companies, such as Suncore, Soitec, Amonix, Solfocus, and so on have ceased production of

CPV, and it has entered a downgrade. There are many uncertainties whether CPV industry will rise again. If HCPV can make further breakthroughs in technology, it may also regain a place in the field of CPV power generation.

5.2 Solar tracker system

The Sun moves from east to west every day, and the altitude and azimuth angles are constantly changing. Meanwhile, during the year, the solar declination also changes back and forth between −23.45° and +23.45°. Of course, changes in the east-west direction are the main ones. In the horizontal coordinate system, the azimuth angle of the Sun changes by 180° almost every day, and the change of the Sun declination angle during a year is 46.90°. If the array can face the Sun at any time, more solar radiation energy can be received, thereby increasing the PV system's power generation. This requires the installation of a solar tracker.

Solar trackers are mechanical devices used to direct PV modules to the Sun or direct sunlight to PV cells. Previously, it was mainly used to meet the needs of CPV power generation systems, especially for HCPV and CSP systems, for which the tracker is an essential equipment. In recent years, it has been applied in the general PV cell arrays.

5.2.1 Classification of trackers

Solar trackers can be divided into PV tracker and CPV tracker, each of which can be classified based on the number of tracking axes, drive architecture and transmission method, and supporting structure.

5.2.1.1 By application

(1) Nonconcentrating tracker
The nonconcentrating module tracker is a device that minimizes the incident angle between the light and the PV array. The PV module can receive direct light and scattered light at various angles. This means that using a nonconcentrating tracker allows the array to generate electricity efficiently without facing the sun. The function of the nonconcentrating tracker system is to increase the amount of power generated by direct light, and at the same time increase the power generating time compared to fixed PV system, thereby increasing the power generation. In PV systems, the energy generated by a direct sunlight decreases in cosine relation with the angle between the incident light and the array. The tracker with an accuracy of ±5 can use more

than 99.6% of the light in the direct light for energy conversion. Therefore, very high precision trackers are generally not required for nonconcentrating systems.

(2) Concentrating tracker

The concentrating tracker is used to ensure the light path of the CPV system. The tracker directs the concentrating module to the Sun or concentrates the sunlight onto the PV receiver. Direct solar radiation instead of scattered light is the main source of energy for CPV modules. The special light path design allows direct light to focus on the modules. If the concentration is not accurately maintained, the power output will drop significantly. If the CPV module concentration is one-dimensional, a single-axis tracker is required; if the CPV module concentration is two-dimensional, a dual-axis tracker is required. In a concentrating module, the need for tracking accuracy is generally related to the half-angle that the module can receive. If the solar pointing error is less than the half-angle that the module can receive, the module power can generally output more than 90% of the rated power.

5.2.1.2 By number of axes

PV trackers can be classified into single-axis trackers and dual-axis trackers based on the number and orientation of the axes of rotation.

(1) Single-axis tracker

The single-axis tracker has one degree of freedom in the rotation axis, and there are several different implementations, including a horizontal single-axis tracker, as shown in Fig. 5.24, a vertical single-axis tracker, as shown in Fig. 5.25, and an tilted single-axis tracker, as shown in Fig. 5.26 [10]. The axis of the horizontal single-axis tracker is horizontal to the ground, and the axis of the vertical single-axis tracker is vertical to the ground. During the day, the tracker moves from east to west, and all single-axis trackers between horizontal and vertical are tilted single-axis trackers.

Fig. 5.24: Horizontal single-axis tracker.

Fig. 5.25: Vertical single-axis tracker.

Fig. 5.26: Tilted single-axis solar tracker.

The angle of inclination of the tilted single-axis tracker is often limited by the need to reduce the wind profile and the need to reduce the height of one head lift. The polar-aligned tilted single-axis tracker is a special tilted single-axis tracker in which the tilt angle is equal to the latitude of the installation site so that the axis of rotation of the tracker is aligned with the axis of rotation of the Earth.

The axis of the single-axis tracker is usually aligned with the meridian, and it is also possible to align any ground orientation based on a more advanced tracking algorithm. When simulating the system, the orientation of the modules with respect to the axis of rotation is important. The module surfaces of the horizontal and tilted single-axis trackers are generally parallel to the axis of rotation. The modules' trajectory while tracking the Sun sweeps over a portion of the cylinder or cylinder with respect to the axis of rotation. The module surface of a vertical single-axis tracker generally forms an angle with the axis of rotation. When the module tracks the sun, the sweep path forms a conical surface symmetrical to the axis of rotation.

(2) Dual-axis tracker

The dual-axis tracker has two degrees of freedom for rotation, and the two axes are usually perpendicular to each other [11]. The axis fixed to the ground is called a main axis, and the one fixed to the main axis may be called a second axis. There are several implementations of the dual-axis tracker, which are categorized by the principal axis relative to the direction of the ground. The two general approaches are as follows.

(i) Tilted dual-axis tracker

The main axis of the tilted dual-axis tracker, as shown in Fig. 5.27, is parallel to the ground, and the second axis is generally perpendicular to the main axis. The axis of the tilted tracker is generally aligned with the east-west latitude line or the north-south meridian. The polar-aligned dual-axis tracker is one of them.

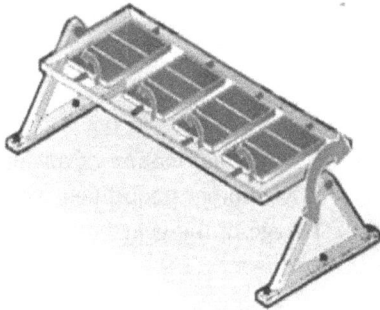

Fig. 5.27: Tilted dual-axis tracker.

(ii) Azimuth-altitude dual-axis tracker

The main axis of the azimuth-altitude dual-axis tracker, as shown in Fig. 5.28, is perpendicular to the ground, and the second axis is generally perpendicular to the main axis.

Fig. 5.28: Azimuth-altitude dual-axis tracker.

Dual-axis trackers have two common drive and control structures: decentralized drives and linked drives. There are multiple concrete implementations. In a decentralized drive structure, each tracker and spindle are independently driven and controlled; in a linked drive structure, one drive system drives multiple axes simultaneously. This allows multiple identical axes or multiple trackers to be arranged in an array in one tracker.

With the advance of technology, the tracking system has become more and more widely used in recent years. The selection of the tracking system is recommended to meet the following requirements: (1) Horizontal single-axis tracking systems should be installed at low latitudes; (2) Tilted single-axis and tilted vertical single-axis tracking system and dual-axis tracking system are suitable for installation in middle and high latitudes; (3) It is not recommended to use a passive control tracking system in areas that are susceptible to sensor contamination [12].

5.2.1.3 By driving power

There are roughly three types of power drivers for solar trackers.

(1) Electric drive

The electric drive system converts electric energy into rotational motion of an AC motor, a DC brush motor, or a DC brushless motor. The motor is equipped with gearbox deceleration to achieve high torque. The last stage of the gear box transmits linear or rotational motion to push the track of the tracker.

(2) Hydraulic drive

The hydraulic drive system uses a hydraulic pump to generate hydraulic pressure. Hydraulic pressure goes through valves, various pipes to hydraulic motors and hydraulic cylinders. Hydraulic motors and hydraulic cylinders will straight or rotary transmit motion to the tracker according to a pre-designed mechanical motion.

(3) Passive drive

The passive system uses hydraulic differential pressure to drive the tracker axis. Differential pressure is obtained from different thermal gradients created by different shades, driving the tracker to move to balance the pressure difference.

5.2.1.4 By control type

(1) Passive control

Passive solar trackers often rely on the forces of the environment to create fluid density changes that provide internal forces that are used to track the sun. Zomeworks, one of the earliest tracker manufacturers, used this technology.

(2) Active control

Active solar trackers use externally supplied power to drive circuits and actuators (motors, hydraulics, etc.) to allow modules to track the sun, both open-loop and closed-loop.

(i) Open-loop control

Open-loop control is a tracking method that does not use a sensor that directly senses the position of the sun, but uses a mathematical calculation of the Sun position (based on time, date, location, etc. within a day) to determine the direction and inclination of the tracker, and thus to drive the tracker. Open-loop control does not mean that the actuator itself does not provide feedback control. The actuator may be a servo motor with an encoder. It may be a PID controller and the like. Open-loop control refers to feedback without actual tracking error in the control algorithm.

(ii) Closed-loop control

Closed-loop control is the use of certain kind of feedback (such as the change of the power of the optical solar position sensor, or the power output of the module) to determine how to drive the drive system and the active tracking of the module position. It is a mixture of sun position calculation method (open-loop calendar code) and closed-loop active tracking of sun position sensor data.

In the early stage, there was timing control. A quartz crystal was used as the vibration source to drive the stepping mechanism. Each step was driven once every 4 mins. Each time the vertical axis rotates by 1°, 360° can be finished in a day, and single-axis intermittent active tracking is performed.

5.2.2 Application of tracking system

The tracking system was previously mainly used with CPV and CSP. In recent years, since the tracking system can increase the power generation of fixed PV arrays by 15% to 25%, more and more ground-mounted PV systems are beginning to use tracking methods. According to statistics, the installed capacity of tracked PV power plants installed in the world in 2015 was 5 GW, and the average annual installed capacity of trackers in 2013–2016 increased by 83%.

The suppliers of global tracker systems are relatively concentrated. In 2015, four vendors: NEXTracker, Array Technologies, First Solar, and SunPower supplied tracker systems that accounted for 72% of the world's total. The tracker market is dominated by single-axis tracker. The percentage of dual-axis trackers used in high-concentration systems is less than 4%.

The world's largest single tracking power plant is the Solar Star power plant in Rosamond, California, which is currently the world's third largest PV power plant with a capacity of 579 MW. It was started in 2013 and completed in June 2015. It has a total of 1.72 million PV modules and uses SunPower's single-axis tracking technology, as shown in Fig. 5.29, which increases power generation by up to 25%. In the 3-year installation process, 650 jobs were provided, and there were 15 full-time operation and maintenance personnel jobs.

Fig. 5.29: Single-axis tracking system of Solar Star power plant, USA.

The 50 MW dual-axis tracking PV power plant in the desert area of AL-Jag bob in Libya covers an area of 2.44 km^2. It adopts dual-axis tracking to increase the power generation by 40% compared to a fixed array. The annual power generation is 128.5 GWh, and the average module efficiency is 16.6%. CO_2 emissions is reduced by 85,527 t a year, making system energy repayment time 4 years. It is divided into 50 subarrays, each subarray has 125 trackers, each tracker is 8 kW, the total capacity is 1 MW. There are 5,000 modules in total covering an area of 290,180 m^2.

The largest tracking PV power station in China is the Yellow River water solar–hybrid power station in Qinghai. The total capacity of the project is 2 GW, of which 60.5 MW uses horizontal single-axis tracking, as shown in Fig. 5.30, and a total of 1,320 tracking systems are used. The product was developed by Arctech Solar, Jiangsu.

According to statistics, as of 2016, the installed capacity of tracking PV power plants installed worldwide is approximately 14.8 GW, 1.5 GW is being installed, and 2.9 GW has already been signed. GTM's research report *The Global PV Tracker Landscape 2016* predicts that the installed capacity of PV tracking systems will increase by 19% in 2017, and will increase by 21% in 2017–2021, and will reach 37.7 GW in 2021, accounting for about half of the installed capacity of PV power stations on the ground, as shown in Fig. 5.31.

The global tracker market will have a value of $1 billion in 2015 and will increase to $4.9 billion by 2021.

Fig. 5.30: 60.5 MW power plant with single-axis tracking system, Qinghai, China.

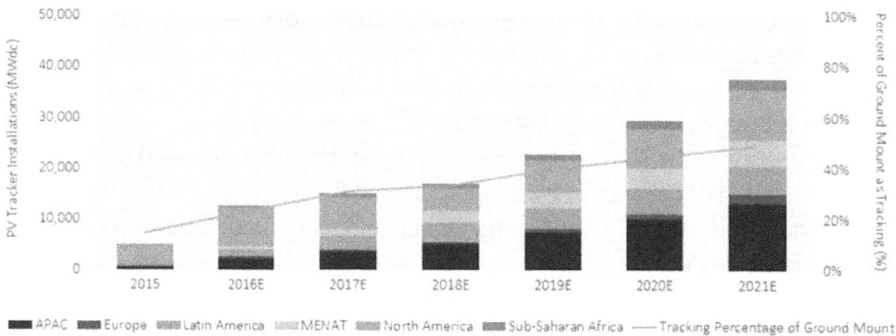

Fig. 5.31: PV tracker market prediction.
Source: GTM Research report.

In addition, there is a more special tracking system – central tower, which is characterized by a large area of PV cells arranged adjacent to each other on the top of the central tower, control scattered plane mirror to reflect the Sun and concentrates to the top of the tower. This concentrated-reception concentrating PV system has been built in Australia with a capacity of 154 MW PV power plant, as shown in Fig. 5.32. The construction cost of the power station is $95 million, which is funded by the Australian government. The cost of power generation is expected to be 10 cents/kWh.

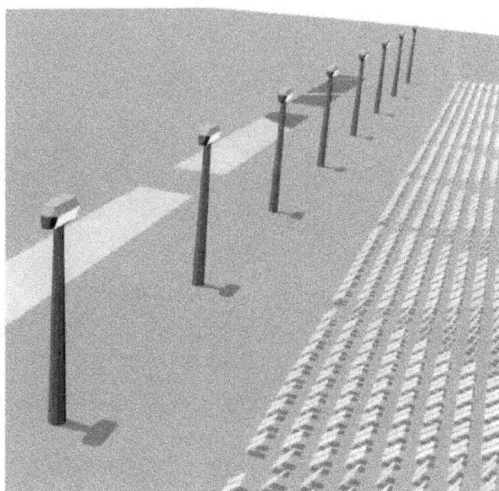

Fig. 5.32: Central tower CPV system.

In summary, due to the sharp price drop of crystalline silicon PV cells in recent years, CPV power generation has been greatly affected, and the development of the industry has entered a low point. However, because of its outstanding advantages, there is still a considerable market demand in some regions where the direct solar radiation is high, and with the continual development and improvement of the concentrating system, if a further breakthrough in technology is achieved, CPV power generation still has a chance in PV market.

The tracker is originally configured as a component of a CPV power generation system, which can significantly increase the amount of power generated by the PV system, and has a relatively high cost performance. In recent years, the use of trackers in fixed flat array power generation system has achieved good results, and it has been rapidly promoted, and the proportion of tracker installed on ground PV systems has increased significantly. It can be expected that with the advancement of technology, the performance of solar tracking systems will continue to increase, and the scope and scale of use will also continue to expand, playing a more and more important role in the field of PV power generation, and making greater contributions to the realization of cheaper prices for PV power generation.

References

[1]   Sarah Kurtz. Opportunities and challenges for development of a mature concentrating photovoltaic power industry. Prepared under Task No. PVA7.4401 NREL/TP-5200-43208 Revised June 2011.

[2]   F. Muhammad-Sukki. et al., Solar concentrators. International Journal of Applied Sciences (IJAS), 2010, 1(1). October. http://www.scribd.com/doc/42622141/International-Journal-of-Applied-Sciences-IJAS-olume-1-Issue-1.

[3]   Yiping Wang. et al., Research status of concentrated photovoltaic cells and systems. Acta Energiae Solaris Sinica, 2011, 32(03), 433–438.

[4]   F Muhammad-Sukki. et al., Solar concentrators. International Journal of Applied Sciences, 2010, 1(1).

[5]   I. S. Hermenean. et al., Modelling and optimization of a concentrating PV-mirror system. International Conference on Renewable Energies and Power Quality(ICREPQ'10) Granada (Spain), 2010.

[6]   Arthur Davis. Fresnel lens solar concentrator derivations and simulations. Proc. of SPIE Vol. 8129, 81290J: 1–15, 2011, SPIE. http://www.reflexite.com/tl_files/EnergyUSA/pa pers/Fresnel-lens-solar-concentrator-derivations-and-simulations_Davis_SPIE-2011_8129-17. pdf.

[7]   R. Leutz and A. Suzuki. Nonimaging fresnel lenses- Design and performance of solar concentrators. ISBN 3-540-41841-5. Springer- Verlag Berlin Heidelberg New York, September 6, 2001.

[8]   Rongwen Yu. The latest development of high-concentration photovoltaic technology. Solar Power, May, 2015.

[9]   Simon P Philipps. et al., Current status of concentrator photovoltaic (CPV) technology. Fraunhofer ISE | NREL CPV Report 1.2 TP-6A20-63916 3 /26, February 2016.

[10]   S. Ibrahim, D. Mehmet and Ç. Ilhami. Application of one-axis sun tracking system. Energy Conversion and Management, 2009, 11(50), 2709–2718.

[11]   I. Visa, D. Diaconescu, et al., On the incidence angle optimization of the dual-axis solar tracker. 11th International Research/Expert Conference TMT Hamammet, Tunisia, 2007: 1111–1114.

[12]   Y Aldali and F Ahwide. Evaluation of a 50 MW two-axis tracking photovoltaic power plant for AL-jagbob, Libya: energetic, economic, and environmental impact analysis. World academy of science, engineering and technology. International Journal of Environmental, Chemical, Ecological Geological and Geophysical Engineering, 2013, 7(12), 811–815.

Exercises

5.1   Summarize the advantages and disadvantages of CPV systems.

5.2   Summarize the main components of CPV arrays and their functions.

5.3   What are the types of CPV cell materials currently used in the market?

5.4   Briefly describe the working principle of refractive concentrators and reflective concentrators.

5.5   Which CPV systems need to have a solar tracker?

5.6   What are the main components of the solar tracker and the tracking control methods?

5.7 Why cooling is important to CPV system, which ways can I use?

5.8 What are the types of CPV systems? What kinds of PV cells and concentrators are used respectively?

5.9 The diameter of a Fresnel lens used in a concentrating system is 300 mm, and the output power is 120 W under standard test conditions. Please determine the conversion efficiency of the concentrating system.

5.10 Summarize the role of the solar tracker. Under what circumstances is it better to use a tracker?

Chapter 6
PV module manufacturing

The efficiency and production cost of PV modules vastly depend on the manufacturing process. The manufacturing process has made astonishing advances in recent years, making it possible for the large-scale application of PV power generation. The main topic in this chapter is the manufacturing process of crystalline silicon (c-Si) PV cells and PV modules.

There are a series of complex production procedure from raw silicon ore to c-Si module, which is illustrated in Fig. 6.1: polycrystalline silicon (poly c-Si) ore → silicon ingot (rod) → silicon wafer → cell → module.

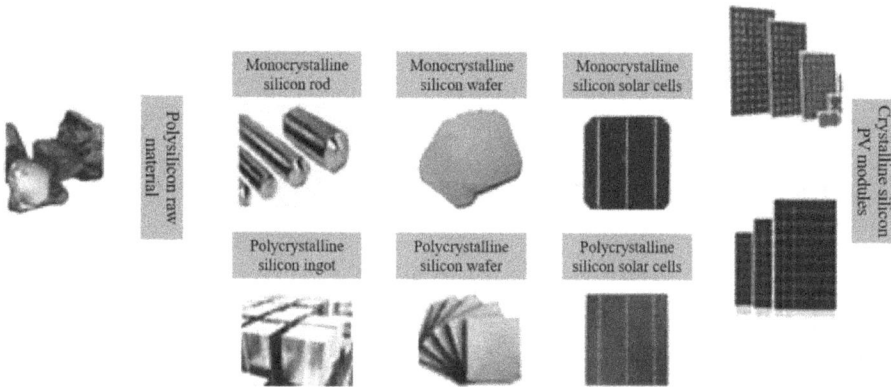

Fig. 6.1: Illustration of c-Si PV module manufacturing process.

6.1 Silicon material manufacturing

Silicon material is the most essential material in PV industry, while poly c-Si is an important intermediate product in silicon product industry chain. According to the purity level, poly c-Si can be divided into the following categories: metallurgical-grade silicon (MG-Si), solar grade polycrystalline silicon (SG-Si) and electronic grade polycrystalline silicon (EG-Si). High-purity poly c-Si usually referred by the industry, which has "seven nines" purity (99.99999%), also called 7 N, is the most commonly used raw material for silicon PV cell.

https://doi.org/10.1515/9783110524833-006

6.1.1 MG-Si manufacturing

MG-Si is a silicon of about 98.5% purity. About 10% of the world consumption of metallurgical purity silicon is refined to produce high-purity poly c-Si, which would be used later in PV industry [1]. MG-Si can be obtained by reducing quartzite or sand with highly pure coke. The reduction is carried out in an electric arc furnace, the reaction is follows:

$$SiO_2 + 2C \rightarrow Si + 2CO$$

There are a series of complex chemical reactions inside the reactors, and at the bottom of the furnace, the temperature is over 2,000 °C. The main reaction is as follows:

$$SiC + SiO_2 \rightarrow Si + SiO + CO$$

In the middle part of the furnace, the temperature is usually about 1,500–1,700 °C, and the main reaction is

$$SiO + C \rightarrow SiC + CO$$

At the top of the furnace where the temperature is lower than 1,500 °C, the reverse reaction dominates:

$$SiC + CO \rightarrow SiO_2 + C$$

Iron can be used as a catalyzer in the reaction to prevent SiC from producing. The result product liquid silicon is precipitated at the bottom of the electric arc furnace and will be drained from the furnace periodically. Extracted liquid silicon will be blown with oxygen or a mixture of oxygen and fluorine to further purify the product. The product will then be poured into a sink, and solidifies to be MG-Si. MG-Si produced in this way contains a fair amount of metal impurities such as iron, copper, zinc, and nickel. The process is shown in Fig. 6.2.

6.1.2 High-purity poly c-Si manufacturing

The most effective purification method for high-purity polysilicon is to use silicon halides such as silane (SiH_4) and chlorosilane (SiH_xCl_y) reducing high-purity silicon halides to high-purity silicon. In the process of chemical purification, many intermediate products and by-products will be produced, so the recycle and reuse of by-products is an important step of polysilicon production. The improvement of the traditional Siemens method (modified Siemens method) can achieve 100% recycle and reuse of by-products. This is an important measure in recent years to decrease the polysilicon production cost and to reduce the pollution and energy consumption.

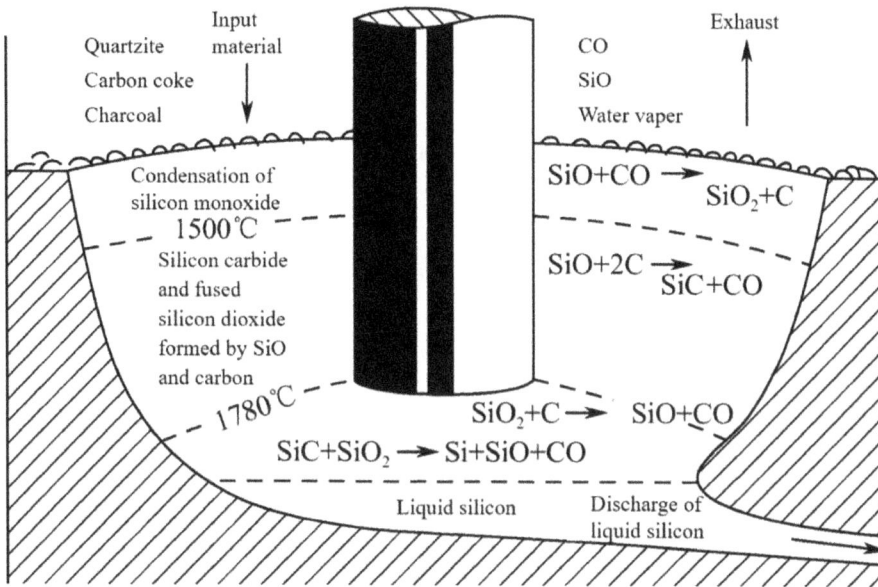

Fig. 6.2: MG-Si production with electric arc furnace.

Although the purity of SG-Si is slightly lower than that of electronic grade poly-silicon, the mainstream method of its production and manufacturing is basically the same as that of electronic grade polysilicon. At present, there are three methods in the market: modified Siemens method, fluidized bed reactor method, and silane (SiH_4) thermal decomposition method.

6.1.2.1 Modified Siemens method

The modified Siemens method is based on the traditional Siemens method. It improves the traditional method and increases the recycle of the reaction gas, thereby increasing the yield of high-purity polysilicon [2–4]. The main recycle and reuse of the reaction gases include H_2, HCl, $SiCl_4$, and $SiHCl_3$. The process of production forms a complete closed loop. Its advantages over the traditional Siemens method are

- Saving energy: As the modified Siemens method uses multiple pairs of rods and large diameter reduction furnaces, it can effectively reduce the power consumption.
- Reduce raw material consumption: The modified Siemens method has effectively recycled all components in the exhaust gas, which can greatly reduce the consumption of raw materials.
- Reduce pollution: Since the modified Siemens process is a closed-loop system, various materials used in the production of polysilicon are fully utilized, and the amount of waste discharged is very limited. Compared with the conventional

Siemens method, pollution is controlled and the environment is protected. At present, most domestic and foreign polysilicon plants use this method to produce electronic grade and solar grade polysilicon.

In addition to trichlorosilane as the main reducing gas, silane, silicon tetrachloride ($SiCl_4$) and silicon dihydrogen dichloride (SiH_2Cl_2) have also been used in the production of modified Siemens process. The flowchart of modified Siemens process is shown in Fig. 6.3.

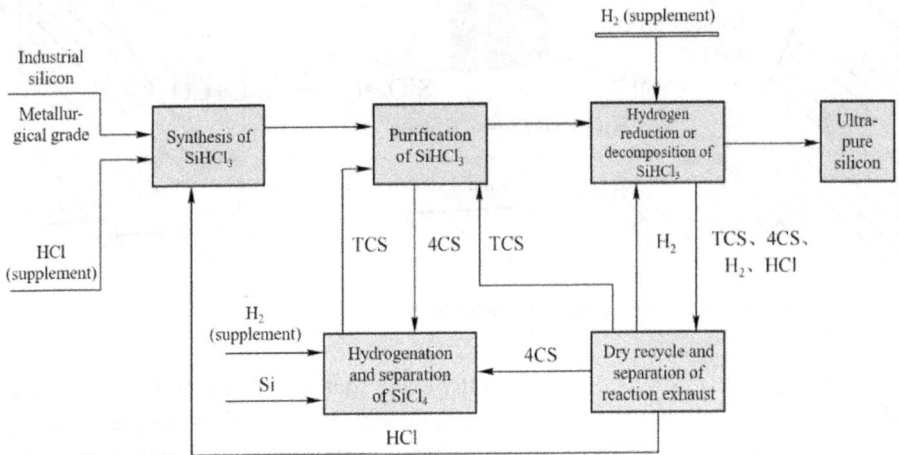

Fig. 6.3: Flowchart of modified Siemens process.

6.1.2.1.1 Synthesis of trichlorosilane
Industrial silicon reacts with hydrogen chloride in a fluidized bed reactor to produce trichlorosilane ($SiHCl_3$). The chemical reaction formula is

$$Si + 3HCl \rightarrow SiHCl_3 + H_2 \uparrow$$

The reaction temperature is 300 °C and the reaction is exothermic. In the process of production, various impurity chlorides are also generated, as shown in Fig. 6.4.

6.1.2.1.2 Purification of trichlorosilane by distillation
Distillation is using difference of boiling points among trichlorosilane, chloride, and hydride impurities to achieve the purpose of purification and excluding impurities. It is achieved in a distillation column. After multistage distillation, the purity of trichlorosilane can reach 9 N level or higher. The vapor pressure of the impurity chlorides contained in the trichlorosilane synthesis product is much lower than the vapor pressure of the trichlorosilane. These metal chlorides belong to the high-boiling

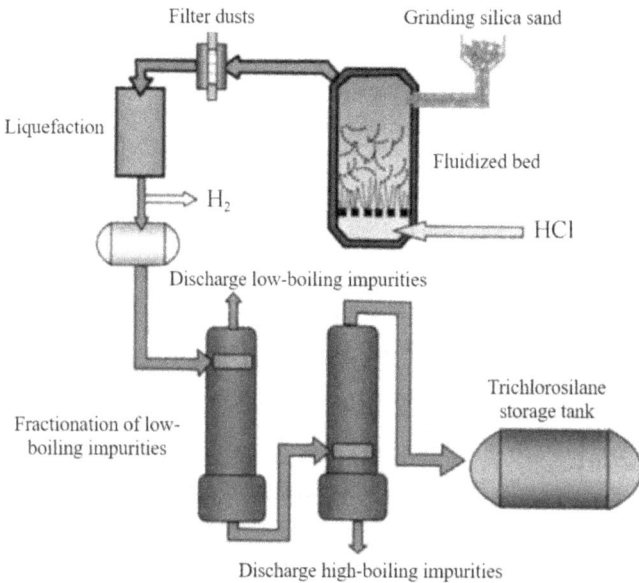

Fig. 6.4: Schematic of the synthesis of trichlorosilane.

components and are easily separated during the distillation. Boron and phosphorus are impurity elements which are difficult to exclude in distillation. Phosphorus mainly exists in high-boiling components and boron is mainly in low-boiling components.

6.1.2.1.3 Reduction of trichlorosilane
High-purity trichlorosilane gas is reacted with hydrogen in a stainless steel bell-type reactor to deposit polysilicon on an inverted U shaped silicon core which is electrically heated to 1,100 °C to form a silicon rod, as shown in Fig. 6.5 [5].
Its chemical reaction equations are

$$2SiHCl_3 \rightarrow SiH_2Cl_2 + SiCl_4$$

$$SiH_2Cl_2 \rightarrow Si + 2HCl$$

$$SiHCl_3 + H_2 \rightarrow Si + 3HCl$$

$$SiHCl_3 + HCl \rightarrow SiCl_4 + H_2$$

Using the silicon core with diameter of 5 to 10 mm and length of 1.5 to 2 m, a silicon ingot produced after the reaction can reach 150 to 200 mm in diameter. The broken silicon ingot is the original polysilicon product. Only 15% of the trichlorosilane is converted to polysilicon. The remaining trichlorosilane companying with H_2, HCl, $SiHCl_3$, $SiCl_4$, and so on are condensed and separated in the reactor to obtain $SiHCl_3$ and $SiCl_4$. $SiHCl_3$ is sent back to the process of reaction. $SiCl_4$ is converted into

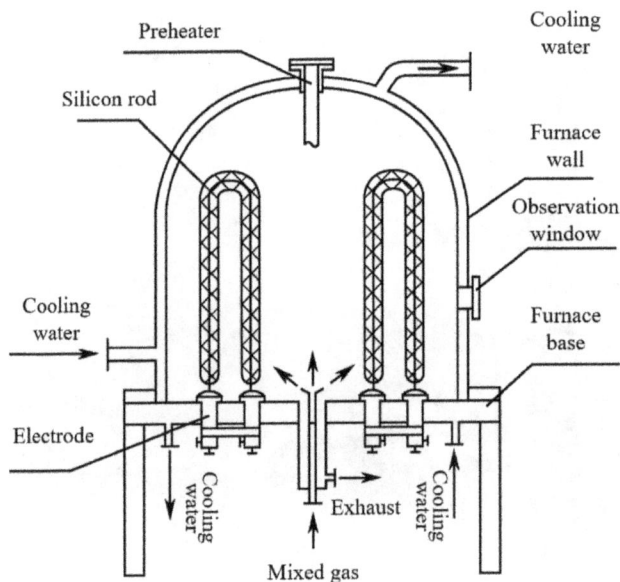

Fig. 6.5: Schematic of Siemens furnace reactor.

SiHCl$_3$ by the subsequent hydrogenation reaction and sent back to the reaction to produce polysilicon. The separation of the gaseous mixture is complex and energy intensive, and to a certain extent, determines the cost of the polysilicon and the competitiveness of the production process.

6.1.2.1.4 Silicon tetrachloride hydrogenation

Silicon tetrachloride is a major by-product of the modified Siemens polysilicon production. 1 kg of polysilicon will produce 10–15 kg of silicon tetrachloride. Silicon tetrachloride can be converted to trichlorosilane by hydrogenation again. Cold hydrogenation is the main method for hydrogenation. The high temperature hydrogenation process widely used before has been phased out due to its high energy consumption and low conversion rate. The reaction equation for cold hydrogenation is

$$3SiCl_4 + Si + 2H_2 \rightarrow 4SiHCl_3$$

The reaction temperature of cold hydrogenation is 500–550 °C, the conversion rate can reach more than 25% with relatively low energy consumption. After the reaction, the gas passes through a dry recovery system to obtain trichlorosilane, which can be used as raw material again after entering the distillation system.

6.1.2.2 Fluidized bed reactor method

The fluidized bed reactor method, as shown in Fig. 6.6, uses $SiCl_4$, H_2, HCl, and industrial silicon as raw materials to generate $SiHCl_3$ in a high-temperature and high-pressure fluidized bed (boiling bed). $SiHCl_3$ is further decomposed and hydrogenated to produce SiH_2Cl_2, which is used to generate silane gas. The produced silane gas is passed into a fluidized bed reactor with small particles of silica fume for continuous thermal decomposition to produce a granular polysilicon product. Polysilicon is deposited on small particles of silica beads floating in a fluidized bed. These silica beads are suspended in the fluidized bed reaction zone in a stream of silane and hydrogen. As the reaction proceeds, the silicon material gradually grows, sinks to the bottom of the fluidized bed, and is collected as granular poly c-Si through the outlet.

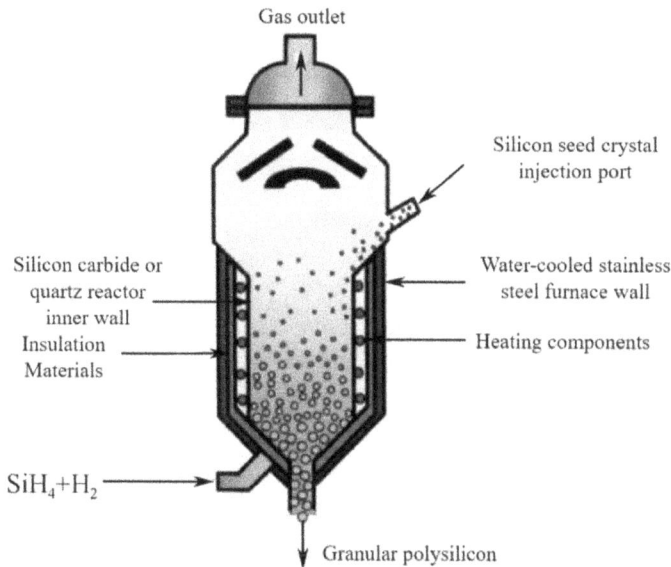

Fig. 6.6: Schematic of fluidized bed reactor.

The surface area of the silicon involved in the reaction in the fluidized bed reactor is large, so the method has high production efficiency, low energy consumption, and low cost. The final product is a millimeter granular poly c-Si that can be used directly for subsequent crystal growth. The disadvantage of this method is its poor safety and danger, and its purity is lower than that of the polysilicon produced by the improved Siemens process.

6.1.2.3 Silane thermal decomposition

In the silane thermal decomposition method, silane (SiH_4) is prepared by methods such as silicon tetrachloride hydrogenation, silicon alloy decomposition, hydride reduction, or direct hydrogenation of silicon [6, 7]. Then, the produced silane gas is purified to produce rod-shaped poly c-Si which has a higher purity in a pyrolysis furnace. Silane is a flammable and explosive gas which requires special measures.

The silane thermal decomposition process is as follows:

(1) Hydrogenation (550 °C, 30 atmospheres)

$$Si + 2H_2 + 3SiCl_4 \rightarrow 4SiHCl_3$$

(2) Disproportionation

$$6SiHCl_3 \rightarrow 3SiH_2Cl_2 + 3SiCl_4$$

$$4SiH_2Cl_2 \rightarrow 2SiH_3Cl + 2SiHCl_3$$

$$3SiH_3Cl \rightarrow SiH_2Cl_2 + SiH_4$$

(3) Thermal decomposition of silane (800 to 1,000 °C)

$$SiH_4 \rightarrow Si + 2H_2$$

The polysilicon production process flowchart of silane thermal decomposition method is shown in Fig. 6.7.

For poly c-Si manufacturing, in addition to abovementioned technologies, such as modified Siemens process, fluidized bed reactor process, and silane thermal decomposition process, there are several other emerging technologies, such as metallurgical processes, gas–liquid deposition processes, heat exchange furnace processes, and chlorine-free technologies.

6.2 c-Si growth technology

c-Si is made of high-purity polysilicon by pulling or directional solidification techniques. Currently used c-Si growth technology includes Czochralski silicon material and cast polysilicon technology. Czochralski silicon has high purity, high energy consumption, high preparation cost, and high PV cell conversion efficiency. Casting polysilicon has low purity, high defect density, low energy consumption and relatively low manufacturing cost, and relatively low conversion efficiency.

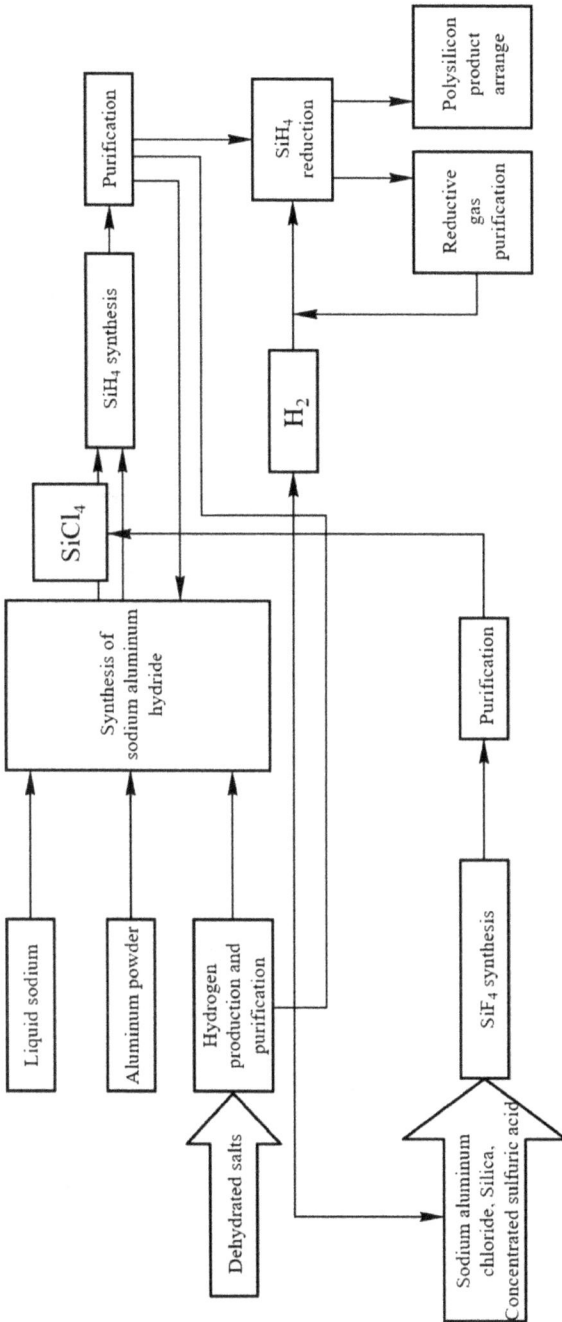

Fig. 6.7: Silane thermal decomposition polysilicon production process.

6.2.1 Mono c-Si growth technology

6.2.1.1 Czochralski method

The method was invented by Polish scientist J. Czochralski in 1918, and was named with his name, also known as CZ method. He used this method to measure the rate of crystallization. In 1950, American scientists G. K. Teal and J. B. Little successfully applied this method to make mono-crystalline germanium, and later, G. K. [8] Teal applied it to make monocrystalline silicon. The equipment used at the beginning was very simple. The loading capacity of a furnace was only a few tens of grams to hundreds of grams. Nowadays, mono c-Si production has a high degree of automation and a high production volume. The amount of material loaded in one furnace can reach hundreds of kilograms. Currently, mono c-Si with a diameter of 16 inches can be produced in a large scale.

In the Czochralski method, first introduce seed crystals as nonuniform crystal nuclei in a crucible filled with molten silicon in a Czochralski crystal furnace, and then when thermal field is controlled, the seed crystal is rotated and pulled up slowly. The mono crystal grows in the same direction as the seed crystal. The schematic diagram of the Czochralski method and the Czochralski furnace are shown in Fig. 6.8.

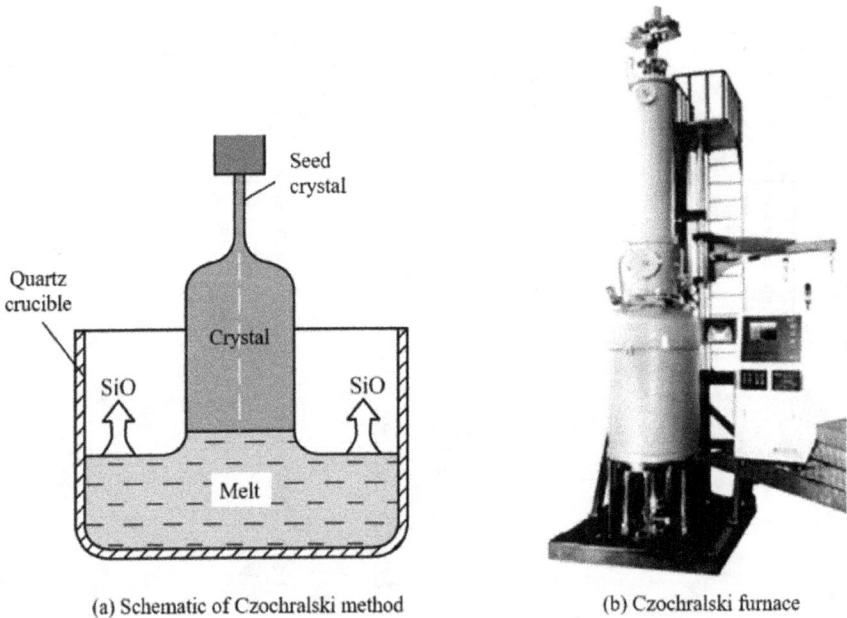

(a) Schematic of Czochralski method   (b) Czochralski furnace

Fig. 6.8: Schematic of Czochralski method and Czochralski furnace.

The steps in Czochralski silicon growth process include melting, seeding, necking, shouldering, equal diameter, ending, and other steps, as shown in Fig. 6.9.

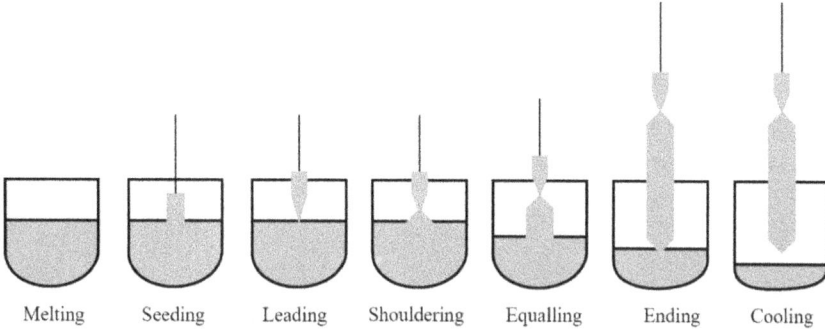

| Melting | Seeding | Leading | Shouldering | Equalling | Ending | Cooling |

Fig. 6.9: Growth process of mono c-Si in Czochralski method.

6.2.1.1.1 Melting

The quartz crucible containing the silicon material and the dopant is placed in a graphite crucible in a Czochralski crystal furnace, and nitrogen gas is introduced as a protective gas under vacuum. By energizing the graphite electrode, the furnace temperature rises. When the temperature in the quartz crucible exceeds the silicon melting point of 1,412 °C, the silicon material starts to melt. After the silicon material is melted, it needs to be kept at this temperature for a period of time so that the temperature and flow of the molten silicon can be stabilized.

6.2.1.1.2 Seeding

The crystal seeds with selected crystal orientation are fixed on the rotating seed crystal axis. The seed crystal slowly descends until the surface is suspended several millimeters above the molten silicon surface to go through "baking crystal." The baking purpose is to make the seed crystal temperature as close as possible to the molten silicon temperature and to reduce thermal shock when seed crystals are contacting the liquid surface and to further avoid the introduction of dislocation defects.

Then the seed crystal is gently immersed in the molten silicon, so that the front part is first dissolved in a small amount. The seed crystal and the molten silicon form a solid–liquid interface, and the seed crystal is gradually raised to leave the solid–liquid interface, and the temperature of the silicon atoms firstly separated from the solid–liquid interface is lowered, thereby mono c-Si is formed.

6.2.1.1.3 Leading

Also called "necking." When the seed crystal just hits the liquid surface, dislocations may be generated in the crystal due to thermal vibration and can extend to the entire crystal. At this point, the crystal diameter is small, so the dislocations are able to slip fast out of the surface of the mono c-Si, ensuring that the mono c-Si grows with no dislocation. After the seeding is complete, the seed crystal will rise rapidly and speed of the crystal growth will increase. The diameter of the newly crystallized mono c-Si is smaller than that of the seed crystal, up to 3 mm, and the length is about 6 to 10 times the diameter, that is the reason for the name "necking."

6.2.1.1.4 Shouldering

After the "necking" is completed, the lifting speed is greatly reduced, so that the silicon crystal is rapidly and directly increased to the required size, and the length of the crystal in the shoulder-releasing stage is generally smaller than the final crystal diameter.

6.2.1.1.5 Equaling diameter

When the shoulder is reached to a predetermined crystal diameter, the lifting speed of the crystal is accelerated, and an almost constant speed is maintained allowing the crystal to maintain a fixed diameter and dislocation-free growth. There are two important factors that may affect the dislocation-free growth of c-Si:
- Thermal stress in the radial direction of c-Si.
- Fine particles in the mono c-Si furnace.

6.2.1.1.6 Ending

At the end of the growth, the growth rate is accelerated again, the molten silicon temperature is increased, the crystal diameter is gradually reduced to form a conical shape, and finally the crystal ingot leaves the liquid surface and the mono c-Si growth is completed.

6.2.1.1.7 Cooling

After the growth of crystal is complete, it is cooled in the furnace until it reaches room temperature. During the cooling process, protective gas must be introduced.

6.2.1.2 Zone melting method

Zone melting method, also known as FZ (Floating Zone) method, as shown in Fig. 6.10. It uses high-purity poly-Si as raw materials which is made in a rod shape and fixed vertically. Put a mono c-Si as a seed on the bottom of the poly c-Si rod. Under the protection of inert gas such as argon or vacuum, the poly-Si rod is heated by a

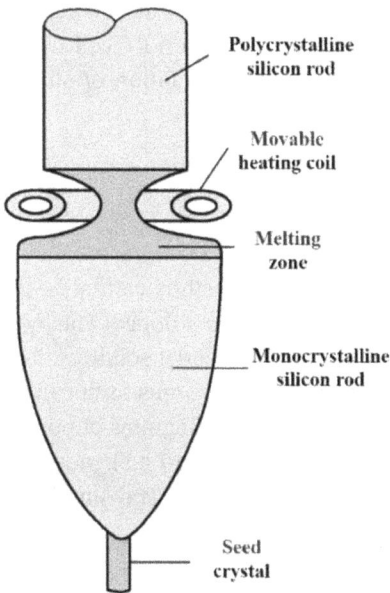

Fig. 6.10: Schematic of zone melting method.

movable high-frequency induction coil to form a melting zone in a middle section. The surface tension of molten silicon and the magnetic buoyancy force of the induction heating coil cause the molten zone to float between the poly c-Si ingot and the mono c-Si formed below, and the poly-Si gradually grows to be mono c-Si through the moving up of the molten zone. Due to the limitation of the induction coil power, the diameter of the mono c-Si ingot made by FZ method cannot be too large.

6.2.1.3 Magnetic field applied Czochralski method

Also known as MCZ method, can be divided into horizontal magnetic field applied Czochralski method and vertical magnetic field applied Czochralski method. In recent years, a combination of a horizontal magnetic field and a vertical magnetic field has been developed to pull the mono c-Si, which is known as cusp (shear transformation) magnetic field method.

Technical principle of MCZ method: the flowing molten silicon is conductive, and the external magnetic field generates Lorentz force on the flow of the conductive molten silicon, thereby suppressing and attenuating the melt flow to achieve the goal of reducing the melt temperature fluctuation and fluctuation of the liquid surface.

The low-oxygen silicon made from MCZ has good performance, which can reduce the degradation of electrical properties of the prepared silicon PV cell under sunlight. However, due to the high production cost and the limitation of silicon wafer area, it is not commonly used.

6.2.2 Casting polycrystalline method

The poly c-Si rods obtained by methods such as the Siemens method cannot be directly applied for the manufacture of PV cells because they are not doped. The molten silicon is injected into the graphite crucible and after directional solidification, a poly c-Si ingot with uniform doping, large crystal grains, and fibrous form can be obtained. Now, with a polysilicon casting furnace, hundreds of kilograms of poly c-Si ingots can be obtained at one time. Compared to pulling mono c-Si rods, the processing cost of ingot polysilicon can be reduced considerably. The simplified production process for poly c-Si ingots is shown in Fig. 6.11.

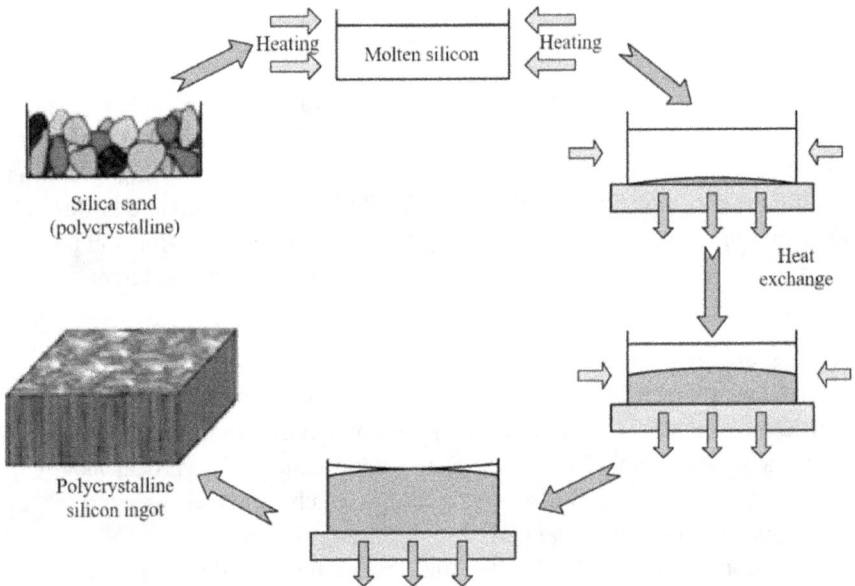

Fig. 6.11: Simplified poly c-Si ingot production process.

According to different means of heating, thermal transfer and crystal surface control, there are four main methods for directional solidification growth of poly c-Si ingots.

6.2.2.1 Bridgeman method

The Bridgeman method is an early directional solidification method used by NEC in Japan, IBM in USA, and so on. The schematic diagram of the Bridgeman crystallization furnace is shown in Fig. 6.12.

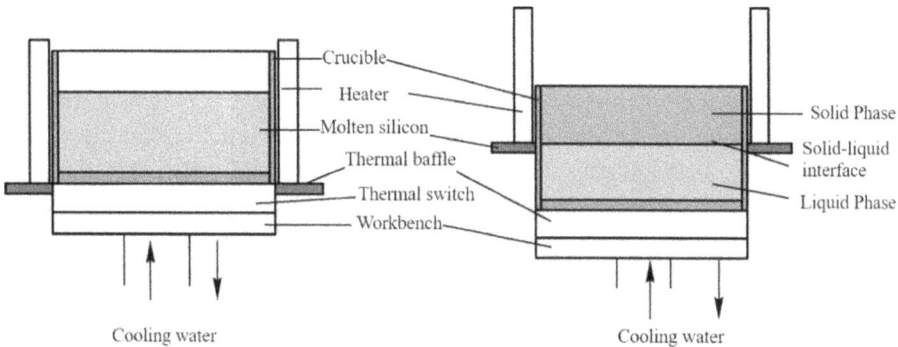

Fig. 6.12: Bridgeman crystallization furnace.

The main features of the process: in order to maintain a relatively fixed solidification and crystallizing plane, the crucible and the heater in the furnace are relatively moved after the starting of solidification, it has two zones which are the liquid phase zone and the crystallized zone, and the heat insulation plate separates the two zones outsides. At the same time, the temperature gradient at the liquid–solid interface must be greater than 0, which means $dT/dx > 0$, and the temperature gradient is close to a constant.

The Bridgeman method's crystal growth speed is controlled by the moving down speed of the crucible table, flowrate, and temperature of the cooling water. The crystal growth speed can be adjusted at any time, and the height of the silicon ingot is mainly limited by the height of the furnace chamber and the crucible. The Bridgeman method generally has a growth rate of 0.8 to 1.0 mm/min.

The main disadvantages of this method are that the structure of the furnace is more complicated, the workbench needs to move up and down, and the down speed must be stable. In addition, the bottom of the workbench needs to be water cooled.

6.2.2.2 Hem-heat exchange method

Hem-Heat exchange method is currently the mainstream method for producing polysilicon ingots in the world, such as GT Solar in the USA, Crystal Systems in the UK, ALD and KR Solar in Germany. The main features of the process are as follows: The crucible and the heater have no relative movement in the whole process of melting and solidification. A thermal switch should be set at the bottom of the workbench. The heat switch is turned off when it is melted to provide insulation. When the solidification starts, the thermal switch is turned on to increase the degree of heat dissipation at the bottom of the crucible and to create a thermal field. Thermal switches have many types, such as flanged, flat, shuttered, and so on. The Hem-Heat exchange method's crystal growth speed is controlled by the heat dissipation intensity at the bottom of the crucible. If it is water-cooled, it is controlled by the flowrate of cooling water (and the temperature difference between the inlet and outlet of water). Since directional solidification can only be a unidirectional heat flow (radiation), radial (crucible side) cannot dissipate heat, that is, the radial temperature gradient tends to be zero, while the crucible and the heater are stationary, so as the solidification proceeds, the isothermal line above the melting point temperature of the thermal field will gradually increase upwards. At the same time, it is necessary to ensure that there is no radial heat flow, so it is difficult to control and adjust the temperature field. From the above analysis, it can be known that the crystal growth speed and temperature gradient of the Hem-Heat exchange method varies, and the height of the silicon ingot is limited. To increase the capacity, the only choice is to increase the cross-section area of the silicon ingot.

An advantage of this method is that the components of the furnace body are relatively simple and not moving, except for the heat switch. The heater in crystallizing furnace of the Hem-Heat exchange method can be vertically and horizontally arranged. The schematic diagrams and an actual photograph are shown in Figs. 6.13 and 6.14 respectively.

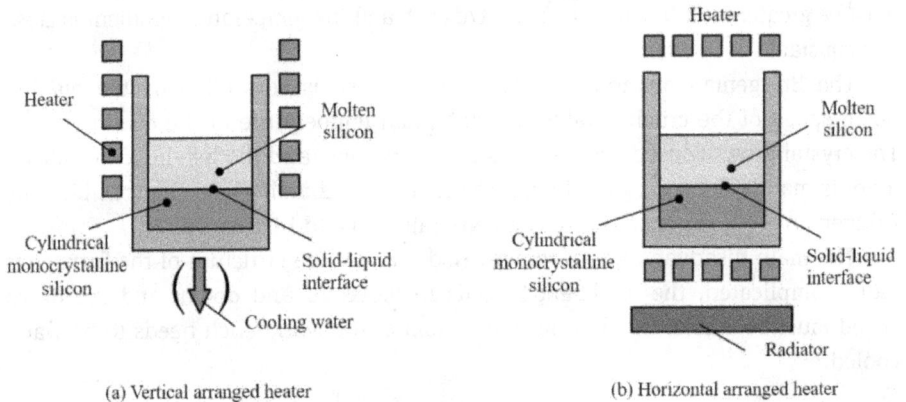

(a) Vertical arranged heater

(b) Horizontal arranged heater

Fig. 6.13: Schematics of two types of heater arrangement.

6.2.2.3 Electromagnetic casting method

Silicon liquid is magnetic in the molten state, and the magnetic field with the opposite polarity generates a strong repulsive force, so that the molten silicon is heated

Fig. 6.14: Hem-heat exchange method polycrystalline furnace.

without contacting the container. In the process of continuous draining, it is cooled by an external water jacket and crystallized [9]. The outside dimensions of the silicon ingot are close to the required dimensions of the silicon wafer, and the general operating cycle is up to 48 h. The features of electromagnetic continuous casting method are there is no need of quartz ceramic crucible, low concentration of oxygen and carbon, relatively small grain size, stable purification effect, and small cross-sectional area of the ingot. The maximum size of the production in Japan was 350 mm × 350 mm, but the height of the ingot can be up to 1 m.

6.2.2.4 Casting technology

The casting method separates melting and crystallization processes. The melting is in an induction furnace lining with quartz and surrounded by an electric resistive heater. Molten silicon is pour-over into a graphite crucible placed on an elevating table, which moving down with a speed of about 1 mm per minute. The crystallization process is based on the Brightman method [10], and started from the bottom of the crucible, as shown in Fig. 6.15. After all silicon in the crucible is crystallized and cold down, the silicon ingot is to be moved out from the crucible.

As shown in the figure, this method can achieve semi-continuous production. The melting, crystallization, and cooling are located in different places, which can effectively increase production efficiency and reduce energy consumption. However, the casting method also has its disadvantage: secondary pollution may be induced by using different crucibles for melting and crystallization. In addition, the equipment is relatively complicated due to the pour-over mechanism and the ingot move-out mechanism.

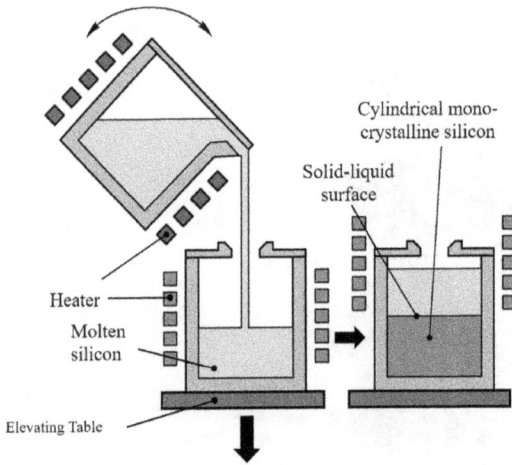

Fig. 6.15: Crystallization furnace of casting method.

6.2.3 Casting mono c-Si method

With rapid development of the PV industry at present, the c-Si used for PV cells is mainly mono c-Si using the Czochralski method and poly c-Si using casting technology. The polysilicon ingot has a large loading amount, simple operation, and low process cost, but the cell conversion efficiency is low and the lifetime is short; the mono c-Si produced by Czochralski method has high conversion efficiency but the output is small with a single loading, and the process is complicate and the cost is high. Therefore, how to combine these two methods and avoid disadvantages and utilizing advantages has become a hot and difficult focus for PV companies all over the world.

The mono c-Si casting method is based on a poly c-Si casting technology in which seed crystals are partially used to obtain poly c-Si wafers whose appearance and electrical properties are similar to mono c-Si wafers at the time of the growth. This technique of forming mono c-Si to an ingot has a slightly higher power consumption than ordinary polysilicon, but the quality of mono c-Si produced is close to that of straight-pulled mono c-Si. This technique is producing mono c-Si with the cost of producing polysilicon. The produced silicon crystal is called mono-like silicon.

The casting mono c-Si method mainly has following two kinds of ingot casting technologies.

6.2.3.1 Seedless crystal ingot

The seedless crystal-guided ingot process has a high requirement for the initial growth control process of nucleation. One method is to use a slotted bottom. The key point of this method is to precisely control the temperature gradient and crystal growth speed during directional solidification to increase the size of the polycrystalline grains. The size of the grooves and the cooling rate determine the size of the grains. The grooves help to increase the grain size. However, because there are too many parameters to control, the seedless crystal ingot process is not easy to apply.

6.2.3.2 Seed crystal ingot

Most of the current quasi-mono crystal technology is seed ingots. This technique first places seed crystals and silicon doping elements in the crucible. The seed crystals are generally located at the bottom of the crucible and then it is heated to melt the silicon raw materials and keep the seed crystals from being completely melted away. Finally, the temperature dropping is controlled and the temperature gradient of solid–liquid phase is adjusted to ensure that the mono c-Si grows from the position of the seed crystals.

Casting mono c-Si products mainly have the following advantages:

- Conversion efficiency of the PV cells produced by casting mono c-Si wafers is higher than that of ordinary polycrystalline wafers and close to Czochralski wafers.
- Compared to ordinary polycrystalline cells, attenuation test is basically at the same level, which means that it has stable performance.
- Compared to ordinary polycrystalline cells, the output power of the cell is significantly increased and the unit cost is reduced.

However, casting mono crystal technology still has many questions, and most companies can only conduct small-scale experiments. Although some enterprises can produce quasi-mono crystals, the control process is complicated and the cost remains high, and industrial production cannot be realized. In short, technology development and mass production of mono c-Si casting problem are the main problems in the field. However, it is certain that the casting mono c-Si technology combines the advantages of mono c-Si and polysilicon, which will help reduce the cost of solar power generation, promote solar power generation, and achieve grid parity grid-parity. Therefore, the casting mono c-Si technology is leading the new trend in the PV industry and the prospect is very broad.

6.3 c-Si processing technology

According to the different growth methods of c-Si, the product generated is a cylindrical silicon ingot or a square silicon ingot. According to different shapes, different processing methods are used to process it into crystal blocks, and the crystal blocks are then cut to get the silicon wafer.

6.3.1 c-Si square cut

The mono c-Si ingot is usually in cylindrical shape, and the silicon ingot is first cut to remove the undesired portion including the head and tail and then cut into a suitable length for processing. In order to increase the effective area of the PV module, the PV cell generally uses a square silicon wafer, and therefore, the silicon ingot needs to be cut into a square like shape. The wire saw is usually used for edge trimming. Along the longitudinal direction of the cylindrical ingot, that is, the crystal growth direction, the silicon ingot is usually cut into round corner quasi-square crystal ingots with a size of 125 mm × 125 mm or 156 mm × 156 mm. Due to the thermal vibration, thermal shock, and uneven crystal growth speed during the growth of the c-Si, the surface of the silicon ingot has many ripple lines. Therefore, the silicon ingot needs to be rounded after the square cutting and the grinding wheel made by diamond grinds the surface of the silicon ingot so that the surface characteristics of the silicon ingot are uniform. In the process of cutting, the surface of silicon wafer will have serious mechanical damage. These damages will cause wafer chipping and microcracking in the subsequent slicing process. Therefore, it is necessary to mechanically grind the surface of the silicon blocks to remove the mechanical damage during the cut.

A poly c-Si ingot is in square shape, and a wire saw is used for cutting it into wafers. The process is shown in Fig. 6.16. First, the surround area of the ingot is cut off in the range of 2 to 3 cm, due to its poor quality. Second, according to the desired wafer size, the silicon ingot is cut into 6 × 6, 7 × 7, or 8 × 8 number of crystal blocks, and the poor quality area on the top and bottom of each blocks are also cut off. Finally, the block is cut into wafers, and the edges of each wafer are then be grinded.

6.3.2 c-Si wafer

The silicon block is usually sliced into wafers by wire-cutting technique. Currently, the thickness of silicon wafers for PV cell is around 180 μm. During wire cutting, a steel wire is wound around two or four guide wheels in series to form a wire mesh. The guide wheel is engraved with a precision wire groove, and the groove pitch determines the thickness of the slice. The nozzles on both sides of the silicon block spray the paste on the wire mesh, and the guide wheel rotates the wire mesh to bring the abrasive slurry into the silicon block. The steel wire presses the grinding sand tightly on the crystal surface and moves forward to perform abrasive cutting, and the silicon block slowly moves down at the same time and is pushed through the wire. The slice process usually takes a few hours. During wire cutting, the speed, pressure of the steel wire, pressure, viscosity, and flow velocity of slurry will affect the cutting quality and speed.

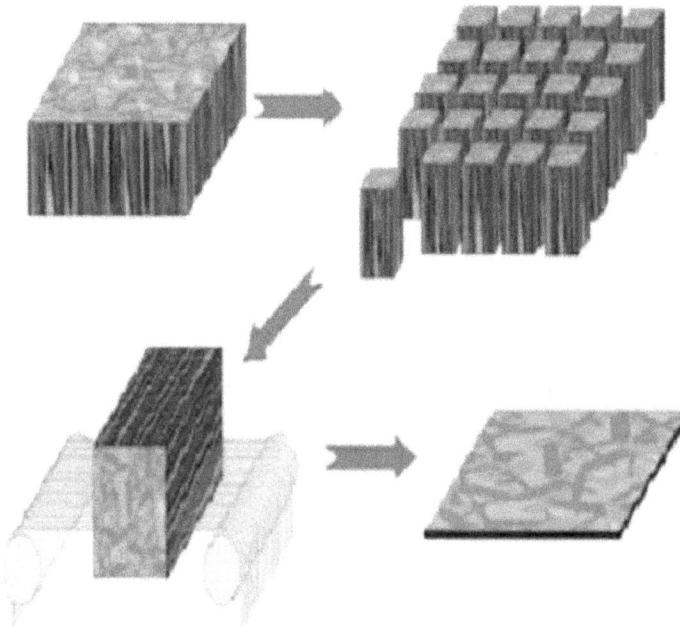

Fig. 6.16: Multicrystalline silicon cutting process.

The diamond wire is a critical change in the wire-cutting process. Diamond wire cutting can further increase the cutting speed while reducing costs by reducing slurry loss. The diamond wire is actually a kind of cutting line with diamond particles embedded on the wire surface. The diamond particles act as an abrasive. This eliminates the need for silicon carbide abrasives, making the entire process cleaner and more environmentally friendly. The size and concentration of abrasive particle vary with the application. At present, it is mainly used for cutting mono c-Si wafers. With the development of poly c-Si surface texturing technology, diamond wire cutting will be gradually applied to the cutting of poly c-Si wafers.

6.4 Crystal silicon PV cell manufacturing technology

The commonly used c-Si PV cells are composed of mono c-Si PV cells (Fig. 6.17) and poly c-Si PV cells (Fig. 6.18). The manufacturing process of these two is basically the same except cleaning and texturing.

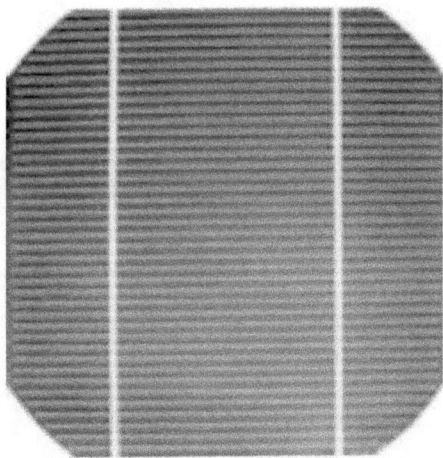

Fig. 6.17: Mono c-Si PV cell.

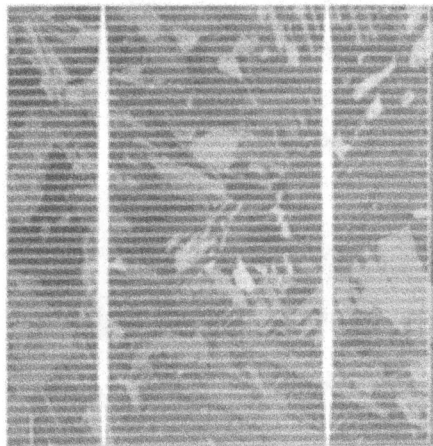

Fig. 6.18: Poly c-Si PV cell.

6.4.1 Surface texturing

In order to increase the sunlight absorption by PV cells, a smooth silicon wafer surface needs to be textured to form a certain geometric shape structure, so that incident light is reflected and refracted on the surface multiple times to increase the light absorption rate [11]. There are several methods for surface texturing, such as mechanical notching, chemical etching, and plasma etching.

Surface texturing can reduce the reflectivity of the PV cell surface to the incident light. The reflectance before and after the wafer texturing is shown in Fig. 6.19. If the antireflection film is added, the reflectivity can be further reduced and even reaches 3% or less. The incident light is repeatedly refracted on the textured surface, which not only changes the incident angle on the silicon but also extend the

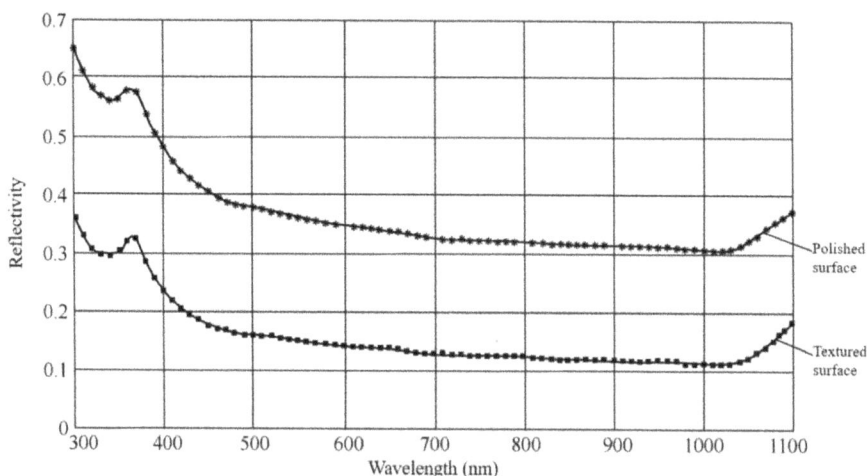

Fig. 6.19: Reflectance before and after texturing.

optical path, increasing the absorption rate of infrared photons and generating more photo carriers by photons near the P–N junction, which increases the collection probability of photo-generated carriers. On silicon wafers with the same size, the area of the P–N junction in the textured PV cell is much larger than that of the smooth one, so that the short-circuit current can be increased, and the conversion efficiency is improved accordingly.

Before making texture, the mechanical damage layer on wafer surface caused by wire cutting has to be removed. There is a high density of cracks in the damage layer, extending from the surface to the wafer body. The defect at the crack damage is a strong recombination center of the electron hole pair, which has a great influence on the cell efficiency, so the damage layer must be removed. For mono c-Si wafers, the damage layer is generally removed by a 10% to 20% concentration of NaOH solution and the surface of the silicon wafer is uniformly etched a certain thickness at a temperature of 75 to 100 °C. At present, many companies are omitting this process in order to increase production and complete this part of work at the same time as the texturing process. For polysilicon wafers, HF and HNO_3 mix solution is used to remove the damage layer while texturing.

The etching principle and etching solution used in mono c-Si and poly c-Si wafers are quite different. Due to the different atomic densities of different crystal planes, the alkaline solution has different etching rates for different crystal planes. The (100) crystal plane has the fastest etching rate, the (110) crystal plane is the second, and the (111) crystal plane has the slowest etching rate. The mono c-Si wafers used for PV cells are mostly < 100 > crystallographic. Using anisotropic alkaline etching characteristics, a (111) pyramidal structure are formed over the surface. This structure is densely covered on the surface of the cell and makes the cell appear as a layer of velvet to naked eye. Therefore, it is called "textured surface" and the pyramidal morphology of the mono c-Si surface observed by scanning electron microscopy is shown in Fig. 6.20. In the industry production process, most of them use a mixed solution of 1% to 2% NaOH and isopropyl alcohol (IPA) at an etching temperature of 70 to 85 °C. Because of the randomness of the etching process, the sizes of the square cones are not exactly the same but are usually 1 to 4 μm.

Fig. 6.20: Morphology of the texturing surface on mono c-Si.

The effect of adding IPA to the solution is to form a top of a certain number of pyramids and protect the tops of these pyramids during erosion. In recent years, the universal application of texturing additives has completely replaced IPA, which not only reduces the cost of chemicals and waste in processing but also makes the size of the pyramids smaller and more perfect.

Mono c-Si texturing is usually made with a trough type texturing and cleaning machine, as shown in Fig. 6.21. The process is as follows: removing damage layer – alkali texturing – water washing – HCl soaking cleaning – water washing – HF soaking cleaning – water washing – drying. Soaking with HCl solution (10–20%) is to neutralize the residual alkali liquor on the wafer surface. Simultaneously, Cl$^-$ in HCl reacts with the metal ions in the silicon wafer and further removes metal ion on the surface of the silicon wafer. Soaking with HF solution (5–10%) is to remove the oxide layer on the surface of the silicon wafer and form a hydrophobic surface.

The crystal orientation on the surface of poly c-Si is randomly distributed, so the anisotropic etching of alkaline solution is not ideal for poly c-Si wafers, and because the reaction speed between different grains on the polysilicon surface is not the same, the etching will produce bumps and cracks instead of forming a uniform textured surface. Therefore, polysilicon texturing is usually acid-etched by using a mixed solution of HF and HNO$_3$ and isotropically etched on the surface at a low temperature of 5 to 10 °C. It is based on the principle that the surface damage caused by slicing (micro-cracks) is etched faster, forming a groove-like textured

Fig. 6.21: Trough type texturing and cleaning machine.

structure on the surface. The poly c-Si crystal orientation is randomly distributed. After etching, irregular pits appear on the surface. These pits are like "worms" on the surface of the wafer. When viewed under a microscope, they look like oval spheres, which can also be called "textured surface." The morphology of the textured poly c-Si surface is shown on Fig. 6.22.

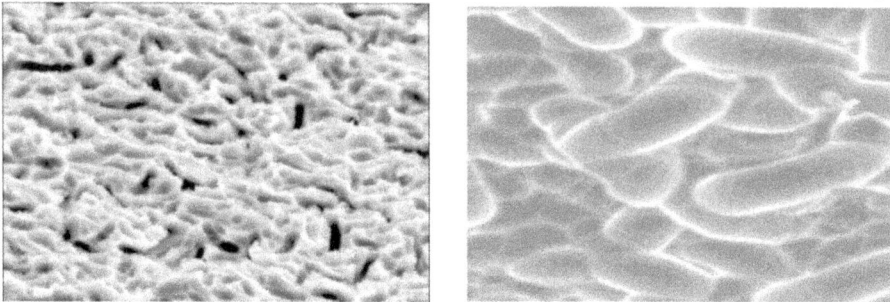

Fig. 6.22: Morphology of polysilicon texturing surface.

Poly c-Si texturing is usually produced with a chain-type texturing and cleaning machine, as shown in the left figure of Fig. 6.23. The figure on the right shows the internal details of the machine. The process is as follows: acid texturing – water washing – alkaline washing – water washing – HCl + HF soak cleaning – water washing – air knife drying. The ratio of $HNO_3:HF:H_2O$ is usually 8:1:11 and the reaction process is

Fig. 6.23: Acid texturing cleaning machine (left) and internal details (right).

- oxidation of silicon: oxidation of silicon to silicon dioxide by nitric acid.
- dissolution of silica: reaction of silica with HF to produce soluble hexafluorosilicic acid.

The reaction is repeated and the silicon wafer is constantly etched to form textured structure. Reaction equations are as follows:

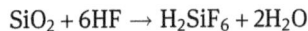

$$3Si + 4HNO_3 \rightarrow 3SiO_2 + 2H_2O + 4NO \uparrow$$

$$SiO_2 + 6HF \rightarrow H_2SiF_6 + 2H_2O$$

After acid texturing and water washing, an alkaline washing is performed. It mainly uses NaOH or KOH (5%) to remove metastable porous silicon formed on the wafer surface during texturing. Porous silicon can result in higher recombination rates, though it is beneficial for reducing surface reflectivity. The soak cleaning with a mixed solution of HCl (10%) and HF (8%) is to remove the residual alkaline solution and metal impurities on the surface of the silicon wafer, and also to remove the oxide layer on the surface.

6.4.2 Diffusion junction process

The P–N junction is the core of the PV cell and is also the key process in the PV cell manufacturing process. Thermal diffusion is the most common method for preparing P–N junction. By heating, pentavalent impurities are incorporated into P-type silicon or trivalent impurities are incorporated into N-type silicon. The commonly used pentavalent impurity element in c-Si PV cells is phosphorus, and the trivalent impurity element is boron. In addition, the method of preparing a N^+-N back field by using phosphorus ion implantation for an N-type silicon substrate has gradually been applied to the preparation process of an N-type bifacial PV cells in recent years.

The requirement for diffusion is to obtain appropriate junction depth and sheet resistance of diffusion layer for PV cell P–N junctions. With shallow junctions and low surface concentration, the short-wavelength response of the cell will be good but will increase the series resistance. The fill factor of the cell can only be increased by adding density of the electrode lines in this case but that will make the process more difficult. If the surface concentration is too high, a heavily doping effect will be induced, causing the decrease of both open-circuit voltage and short-circuit current of the PV cells. In the actual PV cell production, various factors should be considered comprehensively so the junction depth of PV cells is generally controlled at 0.2–0.5 μm.

Whether the diffusion quality meets the requirements of manufacturing can be obtained by measuring the sheet resistance. The sheet resistance also called volume resistance, is the resistance exhibited by per unit area of a thin semiconductor layer, with the common unit of "ohms per square", denoted as "Ω/sq". It can reflect the total amount of impurities doped during the diffusion process. The sheet resistance is closely related to the diffusion time, diffusion temperature, and gas flow rate. In general, the higher the concentration of the diffusion impurity, the stronger the conductivity and the smaller the sheet resistance. In recent years, in order to improve the conversion efficiency of PV cells, the diffused sheet resistance has been continuously improved. At present, the sheet resistance of P-type silicon cells is about 105 Ω/sq. With the development of the silver paste technology, the volume resistance will further increase. The high volume resistance can increase the purity of the junction region and reduce the recombination rate of the junction region in the cell, thereby increasing the short circuit current of the PV cell.

PV cells usually use $POCl_3$ (oxychloride phosphorus) liquid source diffusion method to prepare the P–N junction, and the liquid $POCl_3$ is brought into the diffusion furnace by N_2 gas to achieve diffusion. $POCl_3$ decomposes at high temperature (> 600 °C) to produce phosphorus pentachloride (PCl_5) and phosphorus pentoxide (P_2O_5). The reaction equation is as follows:

$$5POCl_3 \xrightarrow{>600\,°C} 3PCl_5 + P_2O_5$$

With oxygen supplement, PCl_5 reacts with oxygen and further decomposes into P_2O_5. The chemical reaction equation is

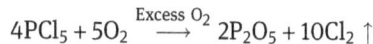

$$4PCl_5 + 5O_2 \xrightarrow{Excess\ O_2} 2P_2O_5 + 10Cl_2 \uparrow$$

The generated P_2O_5 reacts with silicon at the diffusion temperature to form SiO_2 and P atoms and forms a phosphosilicate glass (PSG) on the surface of the silicon wafer. P atoms then diffuse from the PSG into the silicon:

$$2P_2O_5 + 5Si \rightarrow 5SiO_2 + 4P \downarrow$$

At present, in the production of PV cells, the thermal diffusion method is mainly soft landing tube-type diffusion, as shown in Fig. 6.24.

(a) Boat enter state: The paddle is in high position and send the quartz boat with wafers into the furnace tube

(b) Paddle transportation state: The paddle descends to a low position to place the quartz boat in the furnace tube

(c) Process state: The paddle exits furnace tube and gate seals

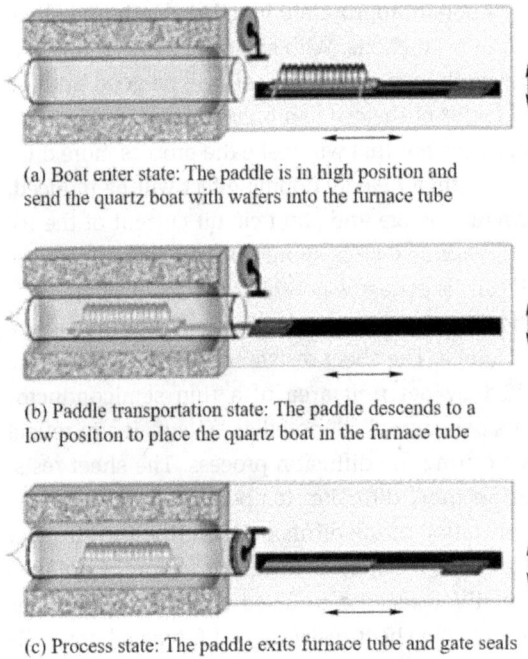

Fig. 6.24: Soft landing diffusion schematic.

A silicon carbide paddle feeds a quartz boat into the furnace and then the paddle exits. After the furnace gate closed, the diffusing process starts. The diffusion source enters and exits from the end of furnace, which ensures the sealing and temperature uniformity of the furnace gate and improves the uniformity of diffusion. The low-pressure diffusion method has been gradually introduced to the market in recent years. This method can greatly improve the uniformity of the sheet resistance and greatly increase the production rate.

6.4.3 Edge isolation

During the diffusion process, diffusion layers are also formed on the edges and backside of the silicon wafer. This diffusion layer causes a partial short circuit between the front and back electrodes of the PV cell, reducing the shunt resistance of the PV cell, and reducing the conversion efficiency of the PV cell. Therefore, the diffusion layer at the edge must be removed.

At present, a chemical etching method is mainly used. This method utilizes a mixed solution of HNO_3, HF, and H_2SO_4 to remove about 1 μm of the back and edges of the silicon wafer at the same time. With a chain cleaner, the wafer moves on the rollers and only the back and edges contact the etching solution and the front remains intact. This method is widely used for its high degree of automation and high shunt resistance for PV cells.

The older method is plasma etching, which alternately acts on the silicon wafer with fluorine and oxygen under glow discharge to remove the edge containing the diffusion layer. The reaction gas is CF_4 and O_2. The main working principle is the use of high-frequency glow discharge reaction to activate the reactive gas into active particles, such as atoms or radicals. These active particles diffuse to the locations which needed etching, where they react with the etched material and is removed by forming volatile products. Hundreds of silicon wafers are stacked together during the etching process and only the edge diffusion layer is removed. A small amount of diffusion layer is still left on the back surface, which is fused through the subsequent sintering process by aluminum–silicon eutectic. Therefore, the shunt resistance of the PV cells using this method is not as high as the chemical etching method, and it has been used less frequently in industrial production now.

6.4.4 Remove phosphosilicate glass

During the diffusion process, a layer of phosphorus-containing SiO_2 is formed on the surface of the silicon wafer, which is called PSG. This PSG is relatively loose and insulative, and will have an adverse effect on the subsequent coating process and the electrical properties of the cell, so it should be removed before coating.

This PSG on the surface of the silicon wafer is usually removed by the reaction of HF and SiO_2. The reaction equation is

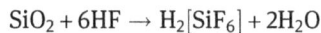

$$SiO_2 + 6HF \rightarrow H_2[SiF_6] + 2H_2O$$

At present, trough washing machine is mainly used to remove PSG on a production line of mono c-Si PV cells, and on the poly c-Si PV cell production line, the PSG has been removed before or after the edge isolation.

6.4.5 Deposition antireflection film

If light incidents on the surface of a smooth silicon wafer, about one-third of the light is reflected. After the surface of the silicon wafer is textured, the reflection loss is reduced to less than 10%. In order to further increase the utilization of light, an antireflection film can be deposited on the surface of the silicon texturing surface, and the reflectivity can be reduced to less than 5%, as shown in Fig. 6.25.

The commonly used antireflection film materials and their refractive indices for PV cells are shown in Tab. 6.1. Among them, the most commonly used antireflection films for c-Si PV cells are Si_3N_4, SiO_2, and TiO_2 [12].

Before coating   After coating

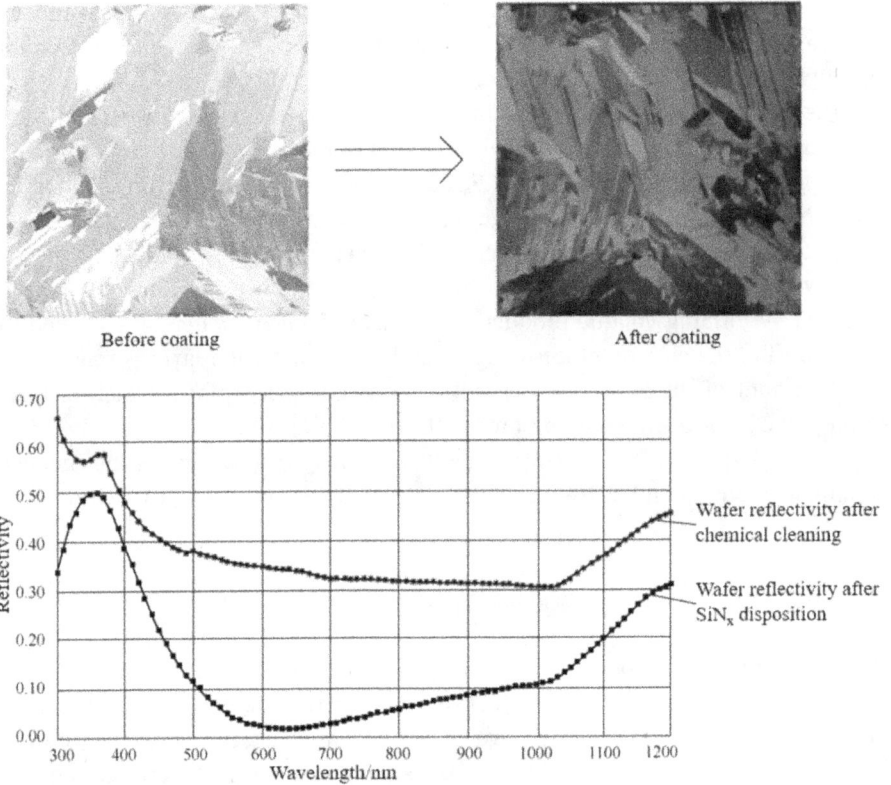

Fig. 6.25: Surface characteristics and reflectivity of poly c-Si before and after coating.

Tab. 6.1: Refractive index of materials used for antireflection film.

Material	MgF_2	SiO_2	Al_2O_3	SiO_2	Si_3N_4	TiO_2	Ta_2O_5	ZnS
Refractive index	1.3–1.4	1.4–1.5	1.8–1.9	1.8–1.9	About 1.9	About 2.3	2.1–2.3	2.3–2.4

The commonly used antireflection film preparation method for c-Si PV cells is to use plasma enhanced chemical vapor deposition (PECVD) to prepare SiN_x thin films. SiN_x thin film has good optical properties, which can reduce the reflection of light and increase the light absorption rate. The active hydrogen atoms produced during deposition process can passivate the dangling bonds and grain boundary defects on the wafer surface, reducing the surface and interface recombination rates, increasing the minority carrier lifetime, and thus increasing the open circuit voltage and short circuit current. At the same time, it also has excellent antioxidation and insulation properties and good ability to block sodium ions, masking metal and water vapor diffusion, and its chemical stability is also very good. Most acids can hardly etch the SiN$_x$ film, except hydrofluoric acid and hot phosphoric acid. The PECVD equipment currently used for depositing SiN$_x$ thin films is mainly divided into flat type and tube type, as shown in Fig. 6.26. The gases used are NH$_3$ and SiH$_4$. The reaction equation is as follows:

$$SiH_4 + NH_3 \rightarrow SiN_X : H + H_2 \uparrow$$

(a) Flat type   (b) Tube type

Fig. 6.26: Flat type and tube type PECVD equipment.

6.4.6 Screen printing electrode

When PV cells are placed under the sunlight, positive and negative charges are accumulated on both sides of the P–N junction, thus generating a photovoltaic electromotive force. In practical applications, it is necessary to collect the generated current through the front and back electrodes. Usually, the electrodes on the light-receiving surface of PV cell are called as front electrodes, and the ones on the back of cells are called as back electrodes. The electrode material should generally meet the following requirements: it can form a strong ohmic contact with silicon, should have excellent conductivity, high collection efficiency, strong solderability, low cost, small bulk resistance, and should be suitable for processing. At present, electrodes are generally prepared by screen printing and then sintered in a high-temperature atmosphere to form ohmic contacts.

The front electrode is usually made of metallic silver paste. The main reason is that silver has good conductivity, solderability, and low diffusion in silicon. The conductivity of the metal layer formed by screen printing and sintering depends on the chemical compound of the paste, the content of the glass frit, the roughness of the screen, the sintering conditions, the thickness of the screen plate, and so on. The front electrode consists of main grid (called cell busbar), and fine grid (called cell finger). In order to reduce the light-shielding area, the finger's width should be as small as possible, and it has reached 60 μm or less at present. The main function of fingers is to collect photo-generated current. The function of busbar is to transmit the current collected by fingers to the external circuit. The width is generally 1.5 to 2 mm. Most cells have three or four busbars. With the maturity of technology, five or six busbars and no busbars technology are emerging in the market.

The back electrode usually uses a low solid content silver paste as the backside soldering area and collect current. Other areas use aluminum paste to fill the back of the PV cell as much as possible to reduce the series resistance of the PV cell, and it can also form an aluminum back field to reduce the back recombination [13].

The preparation process of the electrodes includes three printings: print back silver electrodes, aluminum back field, and front silver electrodes, and the conventional process is as follow:

print back silver electrodes – drying – print aluminum back field – drying – print front silver electrodes – drying – sintering

Common industrial screen printing equipment and rapid sintering furnaces are shown in Figs. 6.27 and 6.28.

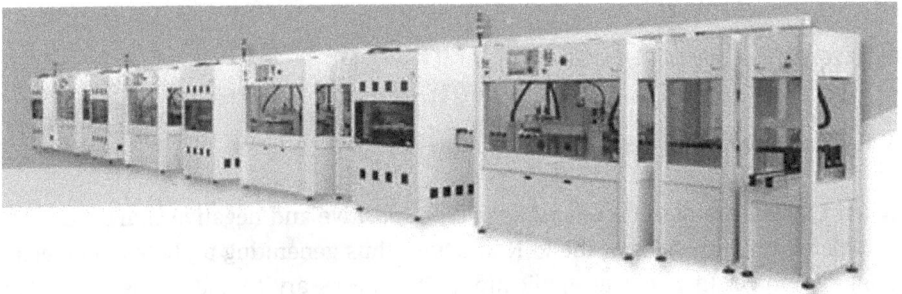

Fig. 6.27: Screen printing equipment.

6.4.7 High temperature sintering

The printed PV cell is rapidly sintered at a high temperature (usually at a heating rate of 150–200 °C/s) so that the front silver paste acts with the SiN_x film and forms a good ohmic contact with the emitter region. The eutectic reaction of the aluminum paste on the back surface with silicon forms a P^+ layer, thus forming a P/P^+ back surface field and ohmic contacts. The back field can prevent the minority carrier (electrons) from recombining by diffusing into the back surface, thereby reducing the back surface recombination loss and increasing the current density of cells.

Fig. 6.28: Rapid sintering furnace.

The sintering furnace usually adopts a chain structure (Fig. 6.28) with infrared lamp heating and short sintering time which is usually only ten seconds passing through the high temperature area. The typical sintering process is roughly divided into four stages.

- The first stage is the drying zone. When drying at 150 °C, all the solvents in the paste are volatilized first. Otherwise, the bubbles generated from the solvent will cause cracks during sintering at high temperature.
- The second stage is the combustion zone, which is performed at 300–400 °C. Under the action of oxygen, the organic binder in the paste is burned off.
- The third stage is the sintering zone, which is an important step for electrode formation. At 700–800 °C, the glass frit in the paste burns through the SiN_x film to fuse the silver electrode with silicon and forms a good ohmic contact.
- The fourth stage is the cooling stage, forming positive and negative electrode contact and aluminum back field.

6.4.8 PV cell I–V characteristics test

After the PV cell is manufactured, its performance parameters must be measured by a test instrument in which the volt-ampere (I–V) curve is the most important parameter. Through the I–V curve test of the PV cell, the following parameters can be obtained: optimum operating voltage, optimum operating current, maximum power (also called peak power), conversion efficiency, open circuit voltage, short circuit current, fill factor, and so on.

The I–V tester of a PV cell is roughly composed of a light source, enclosure, cell holding equipment, measuring instrument, and display part. The requirement of the light source requires that the spectrum of the emitted light beam is as close as possible to the ground solar spectrum AM1.5 and the light intensity is uniform and stable in working area, and the intensity can be adjusted within a certain range. Because of the large power of the light source, most testers use pulsed flash to save energy and avoid temperature rise in the test area. The cell holding equipment need to be firm and reliable with convenient operation. The ohmic contact between the probe and the PV cell and the table should be as good as possible because the size of the PV cell develops in the direction of larger area and larger current, which requires testing must be in good contact. For example, the current of a 156 mm × 156 mm silicon PV cell is about 8 A, and the voltage of a single PV cell is only about 0.6 V. If the contact is not good during the test, even if a 0.01 Ω series resistance is generated, $0.01 \times 8 = 0.08$ V voltage drop will be caused. This is absolutely unacceptable for PV cell testing. Therefore, measuring a large area PV cell must use a Kelvin electrode, which is commonly referred as a four-wire system, to ensure the accuracy of the measurement. A single PV cell test system is shown in Fig. 6.29.

Fig. 6.29: Single PV cell tester.

Modern PV cell test equipment system mainly includes three parts, which contains solar simulator, test circuit, and a computer controlling and processing system. The solar simulator also includes three major parts: an electric light source circuit, an optical path mechanical device, and a filter device. The test circuit uses a clamped voltage electronic load connected to the computer. The computer test controller mainly completes the control of the flash pulse of the electric light source circuit, data collection, automatic processing, and display of the I–V data. The block diagram of the PV cell test equipment system is shown in Fig. 6.30.

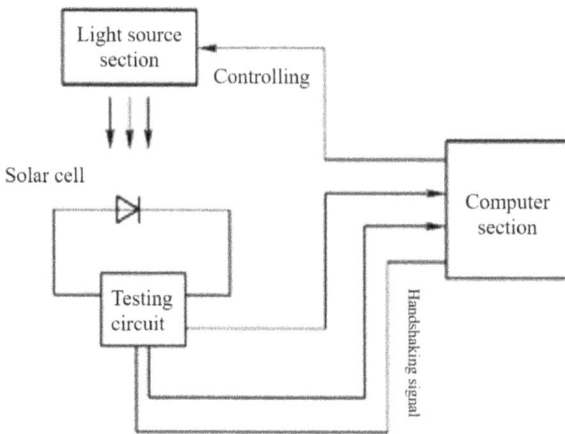

Fig. 6.30: Block diagram of PV cell test equipment.

6.5 PV module packaging

Bare PV cell usually cannot be used for power generation directly, because PV cells are thin and brittle, have poor mechanical strength and are prone to cracking. The moisture and corrosive gases in the atmosphere will gradually oxidize and rust the electrodes, so the bare PV cell cannot withstand the harsh conditions of open air. At the same time, the working voltage of a bare PV cell is usually only about 0.6 V, and the power is too small to meet the actual needs of general electrical equipment. Therefore, PV cells must be provided with mechanical, electrical, and chemical protection, packaged into PV modules, to generate electricity to the load.

The PV module is defined as a complete and environmentally protected assembly of interconnected PV cells.

Before PV module packaging, according to the power and voltage requirements, calculations are needed for the size, number, layout, connection and junction box position. The general requirements for terrestrial c-Si PV modules are working life of 25 years or more, good insulation, good sealing, sufficient mechanical strength, impact resistance, good UV radiation stability, and low cost.

The structure and the packaging method of the mono c-Si PV module (Fig. 6.31) and the poly c-Si PV module (Fig. 6.32) are basically the same. The sealing material ethylene vinyl acetate (EVA) is usually used to bond the upper cover plate (PV glass), the PV cell, and the lower cover (PV back sheet) together, and the periphery is reinforced with an aluminum frame, and the junction box is mounted on the back. The structure of the module is shown in Fig. 6.33.

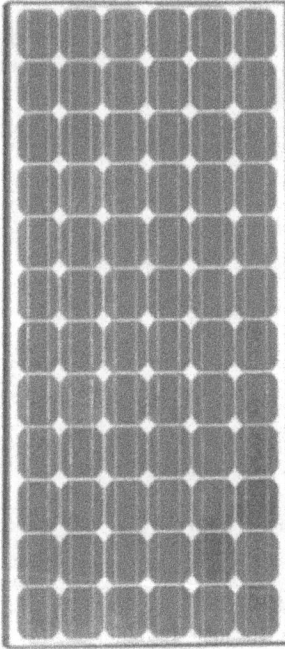

Fig. 6.31: Mono c-Si PV module.

Since off-grid PV systems mostly use lead–acid batteries as energy storage devices, the most commonly used battery voltage is 12 V. For the convenience of use, early c-Si PV modules were typically made of 36 PV cells in series. The optimum operating voltage was about 17.5 V, which took into account the general blocking diode and line losses. In the case if the operating temperature was not too high, the battery can be charged normally. However, special attention should be paid: in the grid-connected PV system, the assembly of 36 PV cells in series does not have a working voltage of 12 V but about 17.5 V.

At present, a module with 60 or 72 pieces of 156 mm × 156 mm PV cells in series is commonly produced. However, due to the demand for large PV power plant and lower system cost, there is a tendency to make the single module power rating higher and higher. As of the first half of 2020, single module power rating reached 580W.

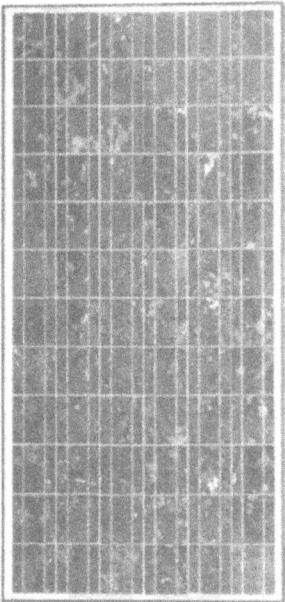

Fig. 6.32: Poly c-Si PV module.

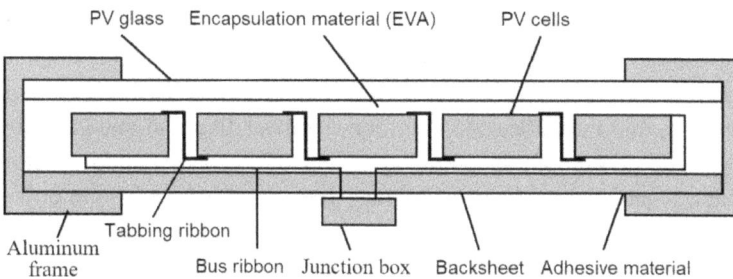

Fig. 6.33: PV module structure diagram.

6.5.1 PV module packaging materials

6.5.1.1 PV glass

PV glass covers the front of the PV module and constitutes the outermost layer of the module. It is the main part of PV cell protection. Therefore, solid and firm, strong impact resistance and long service life are required. At the same time, the light transmittance should be high and the loss of incident light is minimized. At present, commonly used PV glass for module encapsulation is mainly low-iron and ultra-white tempered glass. The common thickness is 3.2 mm, and the light transmittance is required to be over 90%. Since it is subjected to impacts such as hail in practical applications, it must have a considerable strength, and it usually applies chemical or physical toughening on the glass. In order to further reduce the loss in reflection, some also perform antireflection process on the glass surface, such as embossed glass or coated glass, to reduce the reflectivity of the glass and increase the light transmittance.

In addition to glass, some use polycarbonate, polyacrylic glass, and others. As the packaging material, these materials have good light transmission, light weight, can be applied to any irregular shape, and easy to process. However, some materials are not resistant to aging, appear yellow for a short time and the light transmittance is severely degraded, the temperature tolerance is also poor, and the surface is easily scratched, so the application is quite limited. Now it is mainly used for some mini modules.

6.5.1.2 Encapsulation material

Between the glass and the PV cells, and between the PV cells and the back plate, it is necessary to use an encapsulation material for adhesion. At present, in a most common PV module, an EVA film is generally used, which is heated and melted in a vacuum lamination process and solidified after cooling thereby bonding the glass, the PV cells, and the backplane material together. EVA film is a copolymer of ethylene and vinyl acetate. It has properties such as transparent, soft, low melting temperature, good fluidity, and hot-melt adhesive properties, which are in line with the requirements of PV module packaging, but its poor heat resistance, low cohesive strength, and easy thermal shrinkage cause the PV cell to break or delamination. In addition, there are disadvantages such as aging, cracking, and discoloration when it is used outdoors for a long time. Therefore, it is necessary to modify the EVA film before actual use. The main modification is to add additives that can stabilize the polymer during the preparation of the EVA film, such as UV absorbers and heat stabilizers. In addition, organic peroxide cross-linking agents need to be added in order to increase the degree of cross-linking and avoid excessive heat shrinkage. Finally, an EVA film suitable for PV module encapsulation applications is obtained through extrusion molding. Its main performance indicators are

– Light transmittance should be higher than 90%.
– The degree of cross-linking is (70 ± 10)%.
– Peel strength: glass/EVA film is larger than 30 N/cm, TPT/ EVA film is larger than 15 N/cm.
– The properties is stable under the operating temperature range from −40 to 90 °C.
– Antiaging with high UV resistance and thermal stability.

EVA film used for PV module packaging should also have good electrical insulation properties. Usually, EVA thickness should be in the range from 0.3 to 0.8 mm with the width of 1,100, 800, 600 mm, and meet other specifications.

6.5.1.3 Backsheet material

PV module backsheet material can have a variety of options, depending on application condition. A small number of mini modules used in solar lawn lamps and many others often use temperature-resistant plastics or glass-reinforced plastics as the backsheet, according to the requirements of the module structure. For general PV modules, a special Tedlar Polyester Tedlar (TPT) backsheet is commonly used. The sandwich structure is composed of Tedlar and polyester aluminum film or iron film. It is used as a protective layer on the back side of the module and has the properties of moisture resistance, sealing, flame retardancy, weather resistance, and so on. TPT composite material is white, can reflect sunlight, and increase the efficiency of the module. It has a high infrared reflectivity and can reduce the operating temperature of the module.

Bifacial modules and BIPV modules that make the modules into PV curtain walls or PV roofs still use glass as the backsheet material and become double-sided glass modules, which can increase the strength, but the weight has to be increased a lot.

6.5.1.4 Frame

In order to increase the mechanical strength of the modules, it is usually necessary to add a frame around the module laminator, and installation holes on appropriate locations of the frame are required for fixing bolts and nuts. The frame material is mainly made of aluminum alloy, stainless steel, and reinforced plastic. Aluminum alloy frames are usually surface-oxidized to give them good corrosion resistance. Silicone is a commonly used adhesive between the frame and the laminator.

Depending on the requirements of the place of use, frameless module may also be used.

6.5.1.5 Junction Box

The junction box is used to connect external circuit and transmit current. It is usually bonded to the backsheet with silicone. The junction box is equipped with a bypass diode and the diode has a unidirectional conduction characteristic and forms a parallel connection with module circuit. When the module is shaded or damaged, the bypass diode plays an important role in the PV junction box in bypassing current generated by normal modules in series with the affected ones. Without bypass diode, the affected module will have harmful hot spots.

6.5.1.6 Solder ribbon

Solder ribbon has the following requirements: conductive, otherwise it will reduce the current collection probability which will reduce module output power; good solderability, to make a solid soldering to avoid pretend and false soldering phenomenon; moderate hardness, considering the cell thickness and soldering methods to avoid cracks. Solder ribbons are usually tin-coated copper strips, divided into two groups based on their size: tabbing ribbon (solder to cell busbar) and bus ribbon (solder to multiple tabbing ribbons at the end of the module). According to the tin coating layer, there are two types, as lead and lead-free, and the solder temperature of these two types of tapes are different.

6.5.1.7 Other materials
PV module packaging, in addition to the above materials, also needs soldering flux, silicone adhesives, connecting cables and so on.

6.5.2 PV module package process

Figure 6.34 illustrates the PV module packaging process.

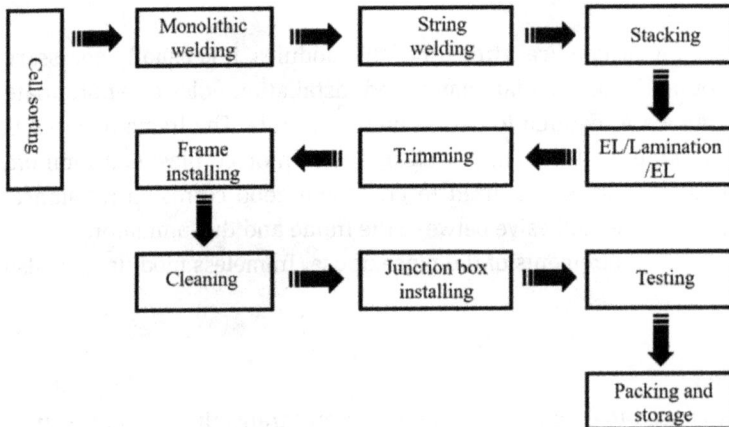

Fig. 6.34: PV module packaging process.

6.5.2.1 Cell sorting
If the PV cells with different operating currents are connected in series, the total current of the cell string is the same as the minimum operating current of all cells, which obviously will cause great waste. Therefore, before PV modules are packaged, PV cells with different performance parameters need to be sorted, and then PV cells with similar parameters are grouped for use.

In industrialized production lines, automatic sorting equipment is usually used to sort PV cells with different performance parameters into several groups and put them into respective bins to prepare PV modules with different powers.

6.5.2.2 String soldering

The front and back electrodes of the PV cell are sequentially soldered in series according to design requirements by using several tabbing ribbons (depends on how many cell busbars) to form a cell string. These tabbing ribbons are soldered with a bus ribbon to form positive and negative electrodes. The soldering process has a great influence on the performance of the module. Soldering requires a firm connection, good contact, consistent spacing, uniform solder joints, and a smooth surface and bad soldering, such as a large difference in height, can affect the quality of lamination and increases cell crack rate. At present, automatic soldering machine (Fig. 6.35) has been widely used, which helps to reduce the cell crack rate and improve the reliability and consistency of soldering.

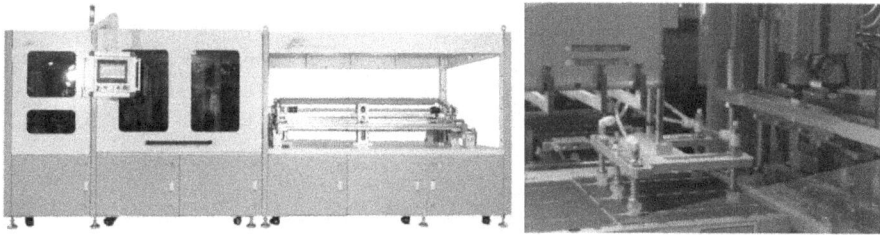

Fig. 6.35: Automatic soldering machine (left) and details (right).

6.5.2.3 Stacks

Place the arranged cell strings in the order shown in Fig. 6.36, laying PV glass, EVA film, cell strings, EVA film, and backsheet (or another glass) in order from bottom to top to form a PV module. Once laid, an electroluminescence (EL) test is used to check for problems such as missing solder, virtual soldering, cracks, and black spots.

6.5.2.4 Lamination

Module lamination is a key process for module production. Lamination has a direct impact on the useful life of the module. The stacking module is placed in a laminator machine, the air in the module is extracted by vacuum pump, and then the EVA is heated to melt. The molten EVA flows under pressure and fills the space among the glass, PV cells, and the backsheet. At the same time, the air bubbles are discharged in the middle, and cells, glass, and backsheet are bonded together.

The lamination time, heating temperature, and vacuum degree are main parameters of the lamination process. If the heating temperature is too high or the vacuum time is too short, bubbles may appear in the laminated module and affect the module quality. After the lamination, the module is also subjected to an EL test. A laminator machine is shown in Fig. 6.37.

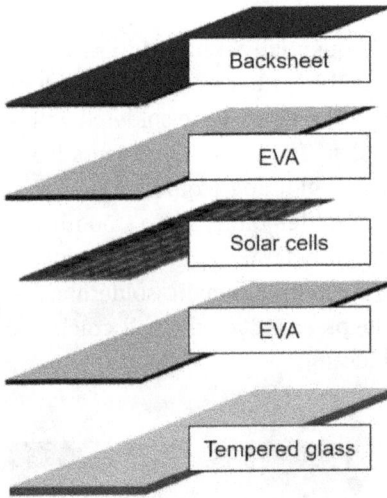

Fig. 6.36: Stacking sequence diagram.

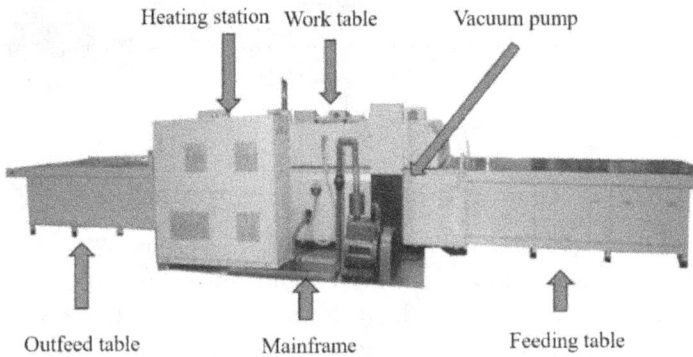

Fig. 6.37: Laminator machine.

6.5.2.5 Trimming

After the EVA is laminated, the EVA melts and flow outwards due to the pressure and solidifies to form burrs around the edge of the module. This will reduce the quality of frame installation, so it should be removed after lamination.

6.5.2.6 Install frame

The gap between the frame and the module is filled with a rubber sealing band or silicone, which bond the frame and the module together. The installation of an aluminum frame to the module can increase the overall strength of the module, further seal the module, extend the working life of the PV cells, and easy for installation.

6.5.2.7 Install junction box

The positive and negative bus ribbons are connected to the terminals inside the junction box. The junction box is fixed to the back side of the module with adhesive. For some BIPV modules, junction box may be placed on the edge of the module for the convenience of installation.

6.5.2.8 Testing

The purpose of the test is to measure the appearance, insulation properties, and electrical performance of the module.

The high voltage test is to apply a certain voltage between the module frame and the terminal leads to test the insulation and dielectric stress of the module and to ensure that it is not damaged under harsh natural conditions (lightning strike, etc.).

A solar simulator is used to measure the electrical properties of the PV modules under the standard conditions, namely AM 1.5, 25 °C, and 1,000 W/m^2.

After testing, a label is put on the module, which has product model and major parameters. The major parameters include the maximum power, open-circuit voltage, short-circuit current, maximum power voltage and current, fill factor, and efficiency, as well as the manufacturer's name and the date of manufacture. Then it can be packed and enter the warehouse.

References

[1] The corresponding writing group, practical industrial silicon technology[M]. Beijing: Beijing Industrial Press, 2005.

[2] Method for producing highest-purity silicon for electric semiconductor devices: USA, US 3042494[P]. 1962.

[3] Method for producing pure silicon: USA, US 3146123[P]. 1964.

[4] Method of producing hyperpure silicon: USA, US 3200009[P]. 1965.

[5] Method and apparatus for manufacturing high-purity silicon rods: USA, US 4150168[P]. 1979.

[6] Kai He. Preparation of crystalline silicon from silicone. [J] China Non-Metallic Mining Industry Guidebook, 2005, 51, 48–50.

[7] Chuanbin Tang. Overview of granular polysilicon production, Rare Rare-earth Metals, 2001, (3): 29–31.

[8] E. Teal and G. K. Buehler. Growth of silicon single crystals and of single crystal silicon P-N junction. [J] Physical Review, 1952, 87, 190.

[9] K. Kaneko, R. Kawamura, H. Mizumoto. et al., Manufacturing of polycrystalline silicon ingot by electromagnetic casting. In 11th European PVSEC, 1997: 1070–1074.

[10] J. M. Kim and Y. K. Kim. Growth and characterization of 240 kg multicrystalline silicon ingot grown by directional solidification. [J] Solar Energy Materials & Solar Cells, 2004, 81, 217.

[11] S. Zhong, W. Wang and M. Tan. et al., Realization of quasi-omnidirectional solar cells with superior electrical performance by all-solution-processed Si nanopyramids. [J] Advanced Science, 2017, 4(11), 1700200.

[12]   Y. F. Zhuang, S. H. Zhong and X. J. Liang. et al., Application of SiO2 passivation technique in mass production of silicon solar cells. [J] Solar Energy Materials and Solar Cells, 2019, 193, 379–386.
[13]   X. Gu, X. Yu and J. Xu. et al., Towards thinner and low bowing silicon solar cells: form the boron and aluminum co-doped back surface field with thinner metallization film. [J] Progress in Photovoltaics: Research and Applications, 2013, 21(4), 456–461.

Exercises

6.1   What are the methods for producing solar grade polysilicon materials?

6.2   Briefly describe the main steps of the Czochralski process.

6.3   What are the methods of directional solidification growth of polysilicon?

6.4   Summarize the production process steps of c-Si PV cells.

6.5   What is the purpose and method of texturing in the production of mono c-Si and poly c-Si PV cells?

6.6   Give brief description of the method and principle of PV cell diffusion bonding.

6.7   What is the purpose and method of edge etching in the preparation of c-Si PV cells?

6.8   Why should we make antireflection film on the surface of PV cells? What are the common materials for antireflection coatings?

6.9   What are the materials used for the front grid and back field of c-Si PV cells? And what are their roles?

6.10 What performance parameters do you need to measure for a typical PV cell performance test?

6.11 Why PV cells need to be encapsulated into module for practical application?

6.12 Give brief description of the process for the preparation of c-Si PV modules.

6.13 What are the parameters for the performance test of PV modules?

Chapter 7
Components of a PV system

A PV system requires a variety of components to cooperate in order to work properly [1]. Among them, PV modules are the most important components. In addition, all the equipment and devices and other components in the system are often collectively referred to as balance of system (BOS). Common BOSs mainly include controller, diode, inverter, energy storage, circuit breakers, transformers, protection switches, electricity meters, combiner box; connecting cable and sleeve; mounting rack, support structure, tracking device, fastener for component installation, system AC and DC grounding, lightning protection device, solar, wind, and other environmental monitoring and data acquisition system.

Under different application conditions, the specific content of the BOS will be different. This chapter only discusses some major system components.

7.1 PV array

Normally, a single PV module cannot meet the voltage or power requirements of the load. It is necessary to combine several modules to form a PV array through series and parallel connections before they can work properly.

PV arrays are DC power generation units that consist of mechanical and electrical assembly of PV module, PV panel, or PV subarray and their support structure.

If there are different modules connected differently in an array, the parts with the same structure and connection are called subarray (portion of a PV array that can be considered as a unit).

The specific connection method of PV modules needs to choose series or parallel method according to the requirements of system voltage and current. The modules with the similar optimum operating currents should be connected in series. For example, when connecting the water pipes, generally, the water pipes of different lengths can be connected together, but the inner diameters should be approximately the same; otherwise, the flow will be limited by the minimum inner diameter. Similarly, if PV modules with different operating currents are connected in series, the total current will be limited by the minimum output current of the module. This must be taken into account.

At present, the common method used in PV arrays is in series connection first and then in parallel.

When the power of a single string of PV modules cannot meet the demand, multiple strings of PV modules need to be connected in parallel to supply power to the loads. In this case, it is necessary to ensure that the number of series elements of each string of PV modules is consistent as much as possible, that is, the working voltage of each string of PV modules has to be the same, to minimize the loss due to the parallel mismatch.

https://doi.org/10.1515/9783110524833-007

7.2 Diode

In a PV array, the diode is a very important component. There are two types of commonly used diodes.

7.2.1 Blocking (anti-backflow) diodes

A blocking diode is often required in series between the PV array and the storage battery or the inverter. A solar cell is equivalent to a diode with a P–N junction. When at night or rainy days, the working voltage of the PV array may be lower than the DC bus voltage of its power supply, thus battery or inverter will reversely send power to the PV array, which will consume energy and cause heating of the array, and even affect the life of the module. The blocking diodes are connected in series in the circuit of the PV array and limiting the current to only one direction.

Since the blocking diode would cause voltage drop in the circuit, a certain amount of power is consumed by the blocking diode. The silicon rectifier diodes generally have a voltage drop of 0.6–0.8 V, and a voltage drop of a large capacity silicon rectifier diode could reach 1–2 V. When the voltage of the system is low (such as below 100 V DC), Schottky diodes may also be used, whose voltage drop is 0.2–0.3 V. However, it should be noticed that the reverse voltage withstands and current capacity of Schottky diodes are relatively small.

In the actual system design, if it can be ensured that the DC bus will not send electricity back to the PV array, a blocking diode may not be required, therefore to reduce system cost and energy loss.

7.2.2 Bypass diode

When PV cells are connected in series to form a PV module, it is necessary to connect one or more diodes in parallel on both sides of each cell string. In this way, when one of the cells is shaded or malfunctioned to stop generating power, a forward bias may be formed across the diode to bypass the current so as not to affect the power generation of the other normal cells in that module and also protect the shaded cell against high forward bias or damage due to "hot spot effect." This type of diode connected in parallel with a section of cell string is called a bypass diode. At present, the bypass diodes are generally encapsulated in the junction box of the PV module and become part of the PV module, as shown in Fig. 7.1.

Schottky diode is one type of diode widely used as bypass diode now. However, when selecting a model of Schottky diode, one should be extremely careful to ensure that there is adequate margin in the capacity to prevent breakdown damage. Its reverse voltage withstand capacity should normally be twice the maximum open-circuit voltage

Fig. 7.1: PV module junction box with two bypass diodes.

of the parallel-connected cell string, and the current capacity should also be twice the expected maximum operating current.

7.3 Energy storage equipment

A PV system is affected by climatic conditions. It can only generate electricity when there is sunlight during the day, and it can be changed at any time due to the influence of solar irradiance. Usually, energy storage equipment needs to be equipped for off-grid PV systems to store excess electricity generated in the presence of sunlight for use in the evening or on rainy days [2]. Even if it is a grid-connect PV power generation system, if the grid power supply in the region is unstable and the load is very important where power supply cannot be interrupted (e.g., military, communications, and hospitals), energy storage equipment should also be installed.

In the long run, the share of solar power in electrical grid will be gradually increased, and it will occupy an important position in the energy supply by the end of this century. To overcome the limited operating characteristics of solar power generation and make it a reliable and stable alternative energy source, large-scale application of energy storage should be installed in the system.

7.3.1 Mainstream energy storage technology

Large-scale energy storage technologies are classified according to the medium they are stored in. Commonly, they are divided into three categories: mechanical energy storage, electrochemical energy storage, and electromagnetic energy storage. The energy storage characteristics of various energy storage methods are all different. In the following sections, we will briefly introduce the basic principles and current status of various energy storage technologies.

7.3.1.1 Mechanical energy storage

Mechanical energy storage converts electrical energy into mechanical energy and converts mechanical energy into electrical energy when needed. At present, there are several practical applications.

7.3.1.1.1 Pumped energy storage

Pumped energy storage is currently one of the most widely used energy storage technologies in power systems. Its main application areas include energy management, frequency control, and provision of system backup capacity. By the end of 2016, the total installed capacity in the world reached 168 GW, accounting for 96% of the total energy storage capacity. The largest one is the Bath County Pumped Storage Station, USA, with a capacity of 3,003 MW.

Pumped storage power station needs to be equipped with two reservoirs: high reservoir and low reservoir. During the low load period of the power grid, the system works in charging state, that is to pump water from the low reservoir to high reservoir, which converts electrical energy to gravitational potential energy of water. When the grid load is high, the system works in generating state, releasing water from high reservoir to run the water turbine and generate electricity.

The storage time of pumped station can range from a few hours to several days, or even months. Some high dam hydropower station with large water storage capacity can be used as pumped storage power station for power management. The use of mines or other caves to achieve underground pumped water storage is technically feasible. The ocean can sometimes be used as a low reservoir. In 1999, Japan built the first pumped storage power station that uses seawater.

The advantages of a pumped storage are: such technology is well developed and reliable, and the capacity can be very large which is only limited by the reservoir capacity. The disadvantages are as follows: the construction of high and low reservoirs is subject to geographical conditions, and there are certain requirements for the fall distance between the two reservoirs; there is a considerable amount of energy loss in pumping and generating cycles, with the overall efficiency from 70% to 80%, the energy density is low, and the construction period is long with huge initial investment.

7.3.1.1.2 Flywheel energy storage

Flywheel energy storage is the storage of energy in the form of kinetic energy in a high-speed rotating flywheel. The entire system consists of high-strength alloy and composite rotors, high-speed bearings, double-fed motors, power converters, and vacuum container. The electric energy drives the flywheel to rotate at a high speed, and the electric energy is converted into the kinetic energy of the flywheel. When required, the flywheel decelerates and the electric motor operates as a generator. The acceleration and deceleration of the flywheel enable charging and discharging. The flywheel energy storage principle is shown in Fig. 7.2.

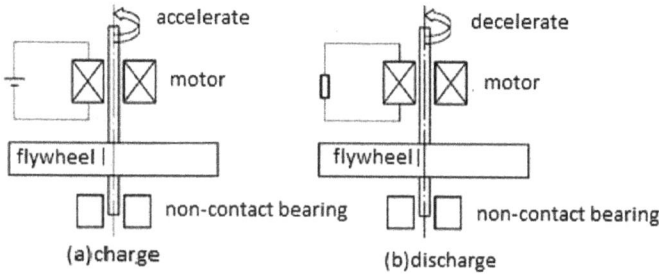

Fig. 7.2: Illustration of flywheel energy storage.

The flywheel system operates in a high-vacuum environment. Its characteristics are almost no friction loss, small air resistance, long lifetime, and low impact on the environment. It is suitable for power grid frequency regulation and power quality assurance. Outstanding advantages of flywheel energy storage are said to be free of maintenance, long equipment life (cycles 10^5–10^7), high specific energy (100–130 Wh/kg), high efficiency (up to 90%), and no impact to environment. The flywheel has excellent cycle usage and load tracking performance and can be used between short-term energy storage applications and long-term energy storage applications. The disadvantage of flywheel energy storage is that the energy density is relatively low, with relatively high cost for ensuring the safety of the system, and its advantages cannot be demonstrated in small-scale applications.

The USA, Germany, Japan, and other developed countries have conducted a lot of R&D on flywheel energy storage technologies. Japan has produced a variable frequency speed control flywheel energy storage system with the world's largest capacity (26.5 MVA, 1,100 V, speed 510,690 rpm, moment of inertia of 710 tm^2).

The University of Maryland in the USA has also developed a 24 kWh electromagnetic levitation flywheel system for electric power peak shaving. The flywheel weighs 172.8 kg, and the working speed ranges from 11 610 to 46 345 rpm. The system output has a constant voltage from 110 to 240 V, and the overall efficiency is 81%. Economic analysis shows that all costs can be recovered after running for several years.

Flywheels with large capacity generally use superconducting magnetic bearing technology. Currently, there are many research institutes working on it, including the French National Center for Scientific Research, SISE in Italy, Mitsubishi Heavy

Industries in Japan, and the Argonne National Laboratory in the USA. The largest one is located at the Fraunhofer ISE, Germany. They are developing 5 MWh/100 MW superconducting flywheel energy storage power stations. Each flywheel weighs 12 t and the efficiency of the system is as high as 96%.

7.3.1.1.3 Compressed air energy storage

Compressed air energy storage is a method of energy storage proposed in the 1950s. It consists of two cycles: one is an air compression cycle and the other is an expansion cycle. During the compression, the doubly fed electric motor acts as an electric motor and uses the excess electric power from grid to drive the compressor and push the high-pressure air into an air storage tank. When the grid demand is high, the doubly fed electric motor acts as a generator. Stored compressed air is first reheated by regenerator, and then fuel is combusted in the combustion chamber, driving power into the expansion system to perform work such as driving a combustion turbine to generate electricity. The construction of the compressed air energy storage power station is restricted by the topography and has special requirements for the geological structure. Demonstration plants have been built in Germany, the USA, Japan, and Israel.

At present, the forms of the compressed air energy storage system also vary a lot. According to the working medium, storage medium, and heat source, it can be divided into a traditional compressed air energy storage system which needs to burn fossil fuels, compressed air energy storage system with a heat storage device, and liquid gas compression energy storage system.

7.3.1.2 Electrochemical energy storage

Electrochemical energy storage, or battery, realizes the interconversion between electrical energy and chemical energy through electrochemical reactions, thereby realizing the storage and release of electrical energy. Battery technology has developed rapidly since the introduction of Daniel's battery in 1836. Room temperature batteries include lead-acid battery, nickel-cadmium battery, nickel-hydrogen battery, lithium-ion battery, and flow battery. Sodium-sulfur battery is high-temperature battery.

7.3.1.2.1 Lead-acid battery

The lead-acid battery is a rechargeable battery with lead and its oxide as the electrode and sulfuric acid solution as the electrolyte. It has been developed before 150 years and is the earliest commercially used secondary battery. Lead-acid batteries have low energy storage costs, high reliability, and high efficiency (70–90%). They are currently one of the most mature and widely used power technologies. However, lead-acid batteries have a short cycle life (500–1,000 cycles), low energy density (30 to 50 Wh/kg), a narrow operating temperature range, slow charging speed, and gas evolution when overcharged, and lead is a heavy metal which has a large impact on the environment. Therefore, its application and development have been greatly limited. At present, many energy storage systems based on lead-acid batteries have been established around the world. For example, the 8.8 MW/8.5 MWh battery energy storage system of BEWAG in Berlin, Germany, is used for peak shaving and frequency regulation. The battery energy storage system in Puerto Rico can stabilize island power of 20 MW for 15 minutes (5 MWh).

In recent years, many companies around the world have been devoting themselves to the development of improved lead-acid batteries that have better performance and can meet various application requirements. An outstanding example is a lead-carbon super battery. The lead-carbon super battery was invented by the Commonwealth Scientific and Industrial Research Organization (CSIRO) in Australia. It replaces part or all of the lead anode with commonly used ultracapacitor carbon electrode materials. It is a combination of lead-acid battery and supercapacitor (SC), and has a fast charge and discharge speed, high energy density, and long service life. It can be used for hybrid electric vehicles, uninterruptible power supply (UPS), and so on.

7.3.1.2.2 Sodium-sulfur battery

The sodium-sulfur battery was first invented by Ford, USA, in 1967, and it has been only 40 years old. The sodium-sulfur battery is a secondary battery that uses metal sodium as a negative electrode, sulfur as a positive electrode, and a ceramic tube as an electrolyte membrane. At certain operating temperature, sodium ions undergo a reversible reaction between the electrolyte membrane and sulfur, resulting in the release and storage of electrical energy. Regular lead-acid batteries, nickel-cadmium batteries, and so on are composed of solid electrodes and liquid electrolytes, while sodium-sulfur batteries are different. It is composed of molten liquid electrode and solid electrolyte. The negative electrode active material is molten metal sodium. The active materials for the positive electrode are sulfur and sodium polysulfide molten salt. Since sulfur is an insulator, sulfur is generally filled in a conductive porous carbon or graphite material. The solid electrolyte membrane is a ceramic material called Al_2O_3, and the cell container is generally made of a metal such as stainless steel. It has high specific energy and can discharge at high current and high power.

Tokyo Electric Power Company Holdings, Inc. and NGK Insulators, Ltd. jointly developed sodium-sulfur batteries for energy storage. Its application targets at power plant load frequency regulation, UPS emergency power supply, and grid electric quality regulator, and it began to enter the practical stage of commercialization in 2002. According to statistics of 2007, Japan's annual production of sodium-sulfur battery capacity has exceeded 100 MW and began to export overseas. The disadvantage of this type of battery is the high cost of the materials compared with lead-acid battery. In addition, the operating temperature of the sodium-sulfur battery is 300–350 °C. Therefore, when the battery is working, high-performance vacuum insulation and heat insulation technology is necessary, which increases the operating cost and reduces the system reliability.

Sodium-sulfur batteries have been successfully used to cut peaks and fill valleys in power grids, emergency power supplies, PV wind power, and other renewable energy sources, and to improve power quality. There are already hundreds of sodium-sulfur battery energy storage power stations in operation, which has broad application prospects.

7.3.1.2.3 Flow battery

The flow battery is generally called a redox flow battery and is a high-performance battery in which the positive and negative electrolytes are circulated separately. It was first funded by NASA and applied for a patent in 1974. At present, most of the applications are all vanadium flow batteries, and its working mechanism is to use vanadium ion solutions with different valences as active materials for positive and negative electrodes, and store them in their respective electrolyte storage tanks. When the battery is charged and discharged, the electrolyte pumped between positive and negative electrode chambers, and oxidation and reduction reactions take place at the surface of the electrode to achieve charging and discharging of the battery.

7.3.1.2.4 Sodium/nickel chloride battery

The sodium/nickel chloride battery is a new type of energy storage battery developed on the basis of sodium-sulfur battery. It is very similar with sodium-sulfur battery: both of them use metal sodium as the negative electrode, and β-Al_2O_3 is the solid electrolyte. What is different is that its positive electrode is a molten transition metal chloride ($NiCl_2$, $FeCl_2$) and sodium chloroaluminate instead of sulfur. The battery operates at a slightly lower temperature (250–350 °C) and also has a high specific energy and long life, no self-discharge, simple operation and maintenance, and so on. Due to the ability to assemble under discharge and withstand overcharge and overdischarge, sodium/nickel chloride battery is safer than sodium-sulfur battery, while the only disadvantage is lower specific power.

7.3.1.2.5 Lithium-ion battery

Lithium-ion batteries are improved products for lithium batteries. Lithium batteries exist long ago, but lithium is a highly active metal. It often burns and bursts when it is charged. Later on, components that can inhibit the activity of lithium (such as cobalt and manganese) are added into lithium batteries, so that the lithium battery is more safe, efficient, and convenient, and the old lithium battery is basically phased out.

When the battery is being charged, lithium ions are generated on the positive electrode of the battery, and the generated lithium ions move to the negative electrode through the electrolyte. The carbon as a negative electrode has a layered structure, and it has many micropores, and lithium ions reaching the negative electrode are embedded in the micropores of the carbon layer. The more lithium ions are embedded, the higher the charging capacity is. Similarly, when the battery is discharged, lithium ions embedded in the negative carbon layer escape and move back to the positive electrode. The more lithium ions return to the positive electrode, the higher the discharge capacity is. Charge and discharge process of lithium-ion battery is the intercalation and deintercalation of lithium ions. During the charging and discharging process, lithium ions are moving back and forth between positive and negative electrodes, and are vividly referred to as "rocking chair batteries."

Lithium-ion batteries have the advantages of high energy density, high energy storage efficiency, and long cycle life. A noticeable drawback is that there is a recession phenomenon. Unlike other rechargeable batteries, the capacity of a lithium-ion battery will slowly decline, regardless of the number of uses, and it is related to the temperature.

At present, the standard cycle life of single-cell lithium-ion battery has exceeded 6,000 times, and the specific energy and cycle life of lithium-ion battery have basically met the requirements of energy storage applications.

Regarding lithium-ion battery's application in power systems, BYD and Lishen of China, Samsung and LG of Korea are in the leading position.

7.3.1.3 Electromagnetic energy storage

7.3.1.3.1 Superconducting magnetic energy storage

The superconducting magnetic energy storage (SMES) uses a coil made of a superconductor placed in a low-temperature environment to store magnetic energy. The low temperature is provided by a cryogenic apparatus that includes a liquid nitrogen or liquid helium vessel. The power conversion/adjustment system connects the superconducting coil with the AC power system and can charge and discharge the energy storage coil according to the needs of the power system. Two power conversion systems are commonly used to connect the energy storage coils to the AC power system: one is a current source converter; the other is a voltage source converter. Power transfer does not require the conversion of energy forms, since only electricity is involved. It has the advantages of fast response (ms level), high conversion efficiency (≥96%), high specific capacity (1–10 Wh/kg), and specific power (104–105 kW/kg). Real-time large-capacity energy exchange and power compensation with the power system can be realized. However, SMES technology is still very expensive and has not yet been commercialized.

7.3.1.3.2 Supercapacitor energy storage

SC is a new type of energy storage unit that has been developed in recent decades which is similar to a conventional capacitor and an electrochemical battery. It has the discharge power of traditional capacitors and also has the ability to store energy-like electrochemical batteries. In the power system, it is mostly used for short-time, high-power load smoothing and meeting the requirements of peak power, such as high-power DC motor start-up support and dynamic voltage restorer, to increase the power supply level during voltage drop and transient interference.

7.3.1.3.3 High-energy capacitor energy storage

As we all know, a capacitor is also an energy storage unit. The stored energy is proportional to the square of its own capacitance and terminal voltage [3]. Capacitor energy storage is easy to maintain without superconductors. The important feature of capacitive energy storage is to provide instantaneous high power, which is very suitable for lasers, flashlights, and other applications.

The various energy storage technologies mentioned above have different performances in terms of energy density and power density. The power system also puts forward different technical requirements for different applications. Not a few energy storage technologies can fully satisfy the power system requirements. For various applications, it is necessary to take into consideration the needs of all factors and choose a matching energy storage method [4].

In 2011, the USA installed a variety of energy storage technologies with different capacities: pumped energy storage 22,000 MW; compressed air energy storage 115 MW; lithium-ion battery 54 MW; flywheel energy storage 28 MW; nickel-cadmium battery 25 MW; sodium-sulfur battery 18 MW; and others.

According to the characteristics of various energy storage technologies, pumped energy storage, compressed air energy storage, and electrochemical battery energy storage are suitable for large-scale and large-capacity applications such as system peak shaving, large-scale emergency power supply, and renewable energy access. The energy storage such as superconductors, flywheels, and SCs is suitable for applications where short time high-quality power is required, such as dealing with sudden voltage drop and momentary interruption, improving the quality of users' power, suppressing low-frequency oscillations in power systems, and improving system stability.

Although lead-acid battery is still the most used battery in the world, it cannot meet the requirements for large-scale and high energy storage of power systems in the future. The characteristics of sodium-sulfur battery make it possible for future large-scale electrochemical energy storage. In particular, flow battery is expected to gradually replace lead-acid battery in the next 10 to 20 years. The lithium-ion battery is also expected to be a rising star with emerging electric vehicles. The comparison of these energy storage technologies is shown in Tab. 7.1.

Tab. 7.1: Comparison of different energy storage technologies.

	Type	Typical nominal power	Rated capacity	Characteristics	Application
Mechanical energy	Pumped storage	100–2,000 MW	4–10 h	Applicable to large scale, mature technology, slow response, need geographic resources	Daily load regulation, frequency control, and system standby
	Compressed air	10300 MW	1–20 h	Applicable to large-scale, slow response, need geographic resources	Peak shaving, frequency control, and system standby
	Flywheel	5 kW to 10 MW	1 s to 30 min	High specific power, high cost, high noise	Peak shaving, frequency control, UPS, and power quality
Magnetic energy	Superconducting magnetic	10 kW to 50 MW	2 s to 5 min	Fast response, high specific power, high cost, difficult maintenance	Transmission and distribution stability, suppress oscillation
	High-energy capacitor	1–10 MW	1–10 s	Fast response, high specific power, low specific energy	Transmission system stability, power quality control
	Supercapacitor	10 kW to 1 MW	1–30 s	Fast response, high specific power, high cost, low specific energy	Can be applied to custom power and FACTS
Electrochemical energy	Lead-acid battery	1 kW to 50 MW	Several min to several h	The technology is mature and the cost is low. Short life, environmental issues	Power quality, power station standby, black start
	Flow battery	5 kW to 100 MW	1–20 h	Long life, deep discharge, high efficiency, environmental-friendly, low energy density	Power quality, backup power, peak load adjustment, energy management, renewable energy storage, EPS
	Sodium-sulfur battery	100 kW to 100 MW	Several h	Specific energy and specific power are higher. High temperature conditions, transportation safety issues need to be improved	Power quality, backup power, peak load adjustment, energy management, renewable energy storage, EPS
	Lithium-ion battery	kW to MW	Several min to several h	High specific energy and safety issues need to be improved	Power quality, backup power, UPS

Source: www.bjx.com.cn

At present, all of these energy storage technologies are still unable to meet the large-scale application needs of solar energy and wind power. The development of electric energy storage technology still has a long way to go.

7.3.2 Batteries for PV systems

Currently, the most commonly used energy storage device in off-grid PV system is battery, in which the batteries perform alternately floating charge and discharge. Generally, it can be considered this way: in summer when solar irradiation is high, in addition to supplying electricity to the load, there is excess electric energy that can charge the battery. In winter when the amount of solar irradiation is low, the amount of power generated by PV array is insufficient and electric energy stored in the battery is discharged. On the basis of this seasonal cycle, a much smaller daily cycle will be added: the battery will be charged when there is excess electricity during the day and the load will be supplied by the battery at night. Therefore, the battery is required to have long cycle life, small self-discharge, resistance to overcharge and discharge, low temperature influence, and high charge and discharge efficiency. Of course, factors such as low cost, convenience, and easy maintenance are also considered. Currently used batteries are lead-acid batteries and lithium batteries.

In a selection and design of a PV battery system, the factors considered include following aspects.

- Electrical characteristics: voltage, capacity, charge rate, and discharge rate;
- Performance: relationship between cycle life and depth of discharge (DOD), system self-discharge, and so on;
- Physical properties: size and weight;
- Maintenance needs: immersion or VR seals;
- Installation: location, structural requirements, environmental conditions;
- Safety and auxiliary systems: racks, trays, fire protection;
- Cost: warranty terms and availability.

7.3.2.1 Lead-acid battery

The lead-acid battery was invented by G. Plante in 1859. Since its invention, it has become the world's largest and most widely used battery due to its low price, easy availability of raw materials, reliable performance, easy recycle, and high current discharge. After more than 100 years of development, lead-acid battery has been continuously improving and used in various fields, such as automobiles, communications, electric power, railways, electric vehicles, and UPS. In recent years, applications in the field of solar power generation are also increasing [5]. The appearance of the lead-acid battery is shown in Fig. 7.3.

Fig. 7.3: Appearance of lead-acid battery.

7.3.2.1.1 Basic structure

The lead-acid battery consists of positive and negative plates, separators, electrolyte, battery case, and terminal. The active material of the positive plate is lead dioxide (PbO_2), and the active material of the negative plate is gray spongy lead (Pb). The electrolyte is dilute sulfuric acid (H_2SO_4). The basic structure of a lead-acid battery is shown in Fig. 7.4.

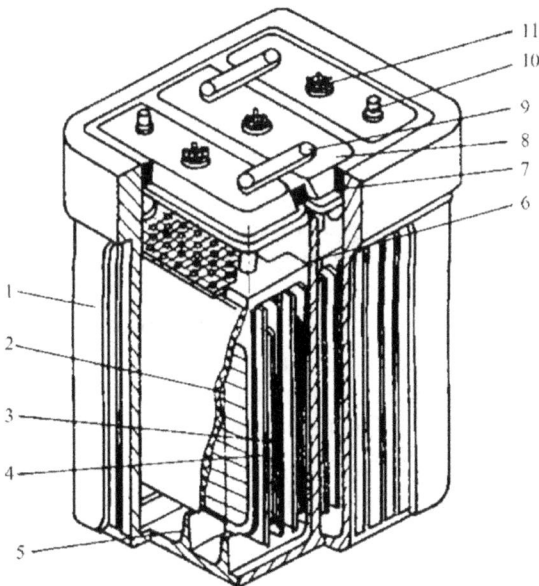

Fig. 7.4: Basic structure of lead-acid battery.
1, Hard rubber case; 2, negative plate; 3, positive plate; 4, separator; 5, strap; 6, bus bar;
7, sealing glue; 8, case cover; 9, connection bar; 10, terminal; 11, vent plug.

7.3.2.1.2 Model code

According to the China Mechanical Industry Standard JB/T 2599, Name, Type Organization and Naming Convention of Lead-Acid Battery, the model code is divided into three segments:

The first segment is a number, indicating the number of single cells in a battery, and the nominal voltage of each single cell is 2 V. When there is only one cell, the first segment can be omitted. The 12 V battery normally used is made up of six cells in series.

The second segment contains two to four letters, indicating the type and characteristics of the battery. The type of the battery is mainly divided according to its use. The characteristic of the battery is an additional part. Type and characteristic codes are shown in Tab. 7.2.

Tab. 7.2: Battery types and feature codes.

Battery type	Code	Battery characteristics	Code
Start-up	Q	Sealed	M
Stationary	G	Maintenance-free	W
(Electric) driving	D	Dry	A
Diesel locomotive	N	Wet	H
Train passenger car	T	Antiacid	F
Motorcycle	M	With liquid	Y
Beacon	B		
Boats and ships	C		
Valve controlled	F		
Energy storage	U		

The third segment shows the rated capacity of the battery. When needed, other codes can be added after the rated capacity. For example:

- GFM-600 is represented as a single cell with a nominal voltage of 2 V, fixed valve regulated sealed type battery, rated capacity at 20 h is 600 Ah;
- 6-QA-120 indicates that there are six cells in series, nominal voltage is 12 V, start-up battery, equipped with dry type electrode plates, with a rated capacity of 120 Ah at 20 h.

Although there are different explanations for the product models of each battery manufacturer, the basic meanings in the product models do not change, and they are usually represented by the above methods.

7.3.2.2 Maintenance-free lead-acid battery

The electrolyte in traditional lead-acid battery will reduce with use. This is because the Sb inside will contaminate the sponge-like pure lead on the negative plate, weakening the back electromotive force (back EMF) in the battery after fully charging and causing excessive water decomposition. A large amount of oxygen and hydrogen escape from the positive and negative plates, respectively, to reduce the electrolyte. Therefore, it is necessary to regularly replenish the electrolyte; this will increase the maintenance workload. The maintenance-free lead-acid battery adopts a lead-calcium alloy grid, and the amount of water decomposed during charging is small, which reduced vaporization velocity of the liquid and lower water evaporation. The oxygen produced on the positive plate during charging is passed through a recombination reaction on the negative plate. It forms back to water; therefore, it is not necessary to add water for maintenance within the specified lifetime.

Other measures are also taken, such as the sealed case, a labyrinth-structured gas chamber in the case cover, and a rich fluid design adopted by the special design of fluorine plastic rubber porous vent plug, with 20% more electrolyte than the average lead-acid battery. Using porous low-resistance polyethylene (PE) separators, fills acid solution between plate pack and the tank to increase heat capacity and heat dissipation, does not accumulate heat and prevents thermal runaway, makes this new type of battery less affected by temperature than conventional batteries. As a result, the lead-acid battery dry-up failure has been avoided. Since the maintenance-free sealed lead-acid battery has very little sulfuric acid gas released, compared with conventional batteries, it does not need to add any liquid, has less corrosion to wiring terminals and wires, has small internal resistance, good low-temperature start-up performance, resistance to overcharging, large starting current, long battery storage time, no leakage of acid and convenient for transportation and maintenance, and it has been widely used in small- and medium-sized PV systems. The appearance of the sealed lead-acid battery is shown in Fig. 7.5.

7.3.2.3 Valve-regulated sealed lead-acid battery

Sealed valve-regulated lead-acid batteries (VRLA) can be divided into absorbed glass mat (AGM) and gel-sealed (GEL) types according to different electrolytes and separators. AGM battery mainly uses a glass fiber separator, and the electrolyte is adsorbed in the voids of the separator. GEL battery mainly uses a PVC-SiO$_2$ separator. The electrolytes are gelified electrolytes. They are SiO$_2$ gels and certain concentrations of sulfuric acid. They are mixed together in an appropriate ratio to form a porous, multichannel polymer. Both H$_2$O and H$_2$SO$_4$ are adsorbed therein to form a solid electrolyte.

The main features of the VRLA battery are: safe and sealed, during normal operation, the electrolyte will not leak from the battery terminal or case, special liquid-absorbing separator keeps the acid inside, and there is no free acid inside the

Fig. 7.5: Appearance of sealed lead-acid battery.

battery, so the battery can be placed anywhere. With a gas release system, when the internal pressure of the battery exceeds normal level, the VRLA battery will emit excess gas and automatically reseal, the maintenance is simple and the gas generated by the unique gas composite system is converted into liquid. There is no need to add water during VRLA battery operation and long service life.

VRLA battery is widely used in electric tools, emergency lights, UPS, electric wheelchairs, computers, communications equipment, and renewable energy.

7.3.2.4 Lead-acid battery performance parameters

The following discussion is mainly focused on the most used lead-acid battery.

7.3.2.4.1 Battery voltage

The nominal voltage of a single lead-acid cell is 2 V. Lead-acid batteries with capacity less than 200 Ah are usually consisting only one cell and have a rated voltage of 2 V, and with capacity more than 200 Ah are generally containing six cells in series and have a rated voltage of 12 V. The actual battery voltage varies with the charge and discharge conditions. Taking a 2 V cell as an example, the float voltage at the end of normal charging is set at 2.6–2.7 V. During discharge, the voltage slowly drops and when it reaches 1.75 V, it cannot continue to discharge. Overcharging and overdischarging can damage the battery's plate, affecting the battery life.

7.3.2.4.2 Battery capacity

The capacity of the battery is the battery's ability to store electrical energy. A fully charged lead-acid battery is discharged under a certain discharge condition (communication battery is generally specified to be discharged at a rate of 10 h at 25 °C), and the amount of electricity that can be given when discharging to the specified termination voltage is called battery capacity, represented by the symbol C, with the unit of Amp-hour, or Ah. When the battery is discharged with a constant current, its capacity is equal to the product of the discharge current value and the discharge time. The discharge rate is usually indicated at the lower corner of C.

The capacity of the battery is not a constant, it is related to the degree of charging, discharge current, discharge time, ambient temperature, age of the battery, and so on. Usually during an application, the battery discharge rate and electrolyte temperature are the most important factors affecting capacity.

7.3.2.4.2.1 Relationship between battery capacity and discharge rate

The same battery has different capacity at different discharge rate, as shown in Tab. 7.3. Discharge rate can be expressed in two different ways: hourly rate and current rate.

- Hour rate: A discharge rate, in amperes, will deliver the specified hours of service until drop to a given cutoff voltage.

Tab. 7.3: Effect of discharge rates on lead-acid battery capacity (Ah).

Battery specification	Capacity of hourly rate (cutoff voltage)				
	20 h (10.8 V)	10 h (10.8 V)	5 h (10.5 V)	3 h (10.5 V)	1 h (10.02 V)
12 V/40A h	43.4	40	36	32.7	25.6
12 V/50 Ah	54	50	45	41.1	32
12 V/65 Ah	70.5	65	58.5	53.3	41.6
12 V/75 Ah	82	75	67.5	61.5	48.5
12 V/90 Ah	98	90	80	73.8	57.6
12 V/100 Ah	108	100	90	83.1	65
12 V/150 Ah	162	150	135	123	97.5
12 V/200 Ah	216	200	180	165	130

Source: Zhongshi Li, Design, Construction and Application of Solar Photovoltaic Power System.

A 12 V battery, if discharged at 2 A, and dropped to 10.5 V in 5 h, then the capacity is:

$$C_5 = 2\,A \times 5\,h = 10\,Ah$$

With the same battery, if discharged at 1.5 A, and dropped to 10.5 V in 10 h, the capacity is

$$C_1 0 = 1.2\,A \times 10\,h = 12\,Ah$$

The former is called 5 h discharge rate, noted as C_5; the latter is called 10 h discharge rate, noted as C_{10}, C's subscript is the hour rate.

- Current rate: expressed as a multiple times of the rated capacity of battery in watt-hours. For example, if a battery with a capacity of 100 Ah is discharged at 100 A, then all the power will be discharged in 1 h, and the discharge rate will be $1C_1$ and if a battery is discharged at 10 A, and all the power discharged after 10 h, the rate is $0.1C_{10}$. Other discharge rates can be obtained similarly.

For a given battery, the larger the discharge current, the smaller the battery capacity. According to the different conditions of application, the rated capacity of a car battery is usually using 20 h rate capacity C_{20}, stationary or motorcycle batteries using 10 h rate capacity C_{10}, and traction and electric vehicle batteries using 5 h rate capacity C_5. Lead-acid batteries for general PV applications use 10 h capacity C_{10}.

7.3.2.4.2.2 The relationship between battery capacity and temperature
The temperature of a lead-acid battery electrolyte has a significant effect on the battery capacity. When the temperature is high, the viscosity of the electrolyte decreases, resistance decreases, diffusion rate increases, and reaction rate increases. These all increase the capacity. However, when the temperature rises, the battery's self-discharge will increase and the electrolyte consumption will increase.

The battery capacity drops rapidly at low temperatures. When the temperature of a universal battery drops to 5 °C, the capacity drops to about 70%. Below –15 °C, capacity will drop to less than 60%. The charging reaction is very slow below –10 °C, which may make it difficult to recover after discharge. If it cannot be charged in time after discharging, there is a danger of freezing when the temperature is lower than –30 °C. Therefore, in the practical application of lead-acid battery, ambient temperature correction should be added to more accurately determine the battery overcharge and overdischarge voltage values.

7.3.2.4.3 Battery efficiency
In off-grid PV systems, batteries are used as common energy storage devices. When charging, the electrical energy generated by PV array is converted into chemical energy. When it is discharged, chemical energy is converted back to electrical energy to supply the load.

In reality, battery is not a perfect energy storage device. There must be a certain amount of energy loss in the working process, and it is usually represented in terms of energy efficiency and Ah efficiency.

(1) Energy efficiency: The ratio of the energy output during discharging to the energy input during charging. The main factor affecting energy efficiency is the internal resistance of the battery.

(2) Charging efficiency (also called Coulomb efficiency): The ratio of the quantity of electricity charges of output when discharging to the quantity of electricity charges of input when charging. The main factors affecting the charging efficiency are various internal negative reactions in the battery, such as self-discharge.

For an off-grid PV system, the average charging efficiency is about 80% to 85%, and it can increase from 90% to 95% in winter. This is because:

- The battery has a higher charging efficiency when it is in a relatively low state of charge (SOC, 85–90%).
- If most of the electricity generated by PV array is directly used by the load, which is more efficient than charging the battery (the experimentally measured charging efficiency is 95%).

7.3.2.4.4 Battery self-discharge

When the battery is not in use, the storage capacity of the battery will automatically decrease with the extension of the time. This phenomenon is called self-discharge. The relationship between self-discharge and storage time is shown in Fig. 7.6.

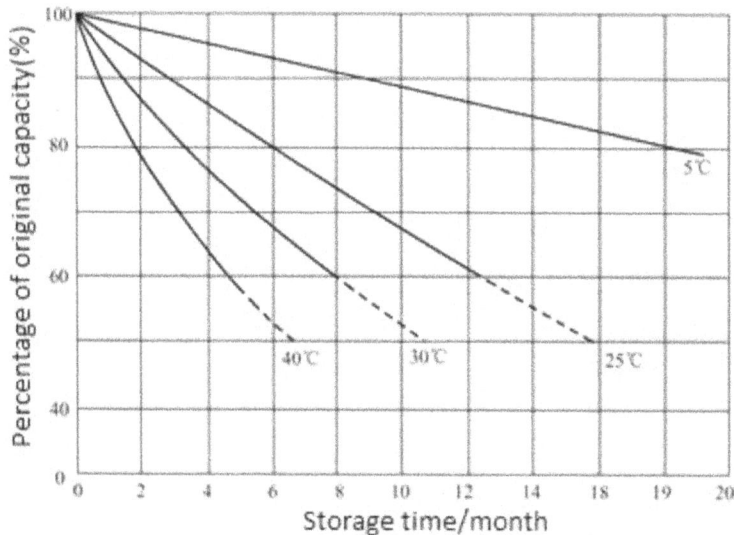

Fig. 7.6: Relationship between self-discharge and storage time.

7.3.2.4.5 Battery discharge depth and charge state

DOD refers to the percentage of ampere-hours discharged to its nominal capacity. DOD will cause sulfation on internal plate surface of the battery and lead to an increase of battery internal resistance. In severe cases, individual battery will experience "reverse polarity" and permanent damage. Therefore, excessive DOD will affect battery service life seriously. If not necessary, do not leave the battery in deep discharge. Under normal circumstances, in the PV system, the DOD of the battery is 30% to 80%.

Another important parameter to measure the degree of battery charge is the SOC. Usually, the state where the battery is charged until it can no longer absorb energy at a certain temperature is defined as SOC = 100%, or fully charged, and the state where the battery can no longer discharge energy is called SOC = 0%.

Therefore, the general lead-acid battery SOC is defined as

$$SOC = \frac{C_r}{C_t} \times 100\% \qquad (7.1)$$

where C_r and C_t represent the remaining energy and total energy of the battery at a certain moment.

The relationship between the SOC and DOD is

$$SOC = 1 - DOD \qquad (7.2)$$

As the battery discharges with its SOC gradually decreasing, the corresponding electrolyte's specific gravity and open-circuit voltage will also become smaller, and the freezing point of the electrolyte will increase, as shown in Tab. 7.4. The SOC of the battery and the specific gravity of the electrolyte can be generally determined by measuring the open-circuit voltage of the battery. If it is found that the battery has been overdischarged, the electrolyte may be added if necessary, and charging under small current for an extended time may be performed. So it is possible to recover SOC partially in this way.

Tab. 7.4: Characteristics of typical deep-cycle lead-acid batteries.

SOC	Specific gravity	Voltage of single cell (V)	Voltage of 12 V battery (V)	Freeze point (°C)
Full	1.265	2.12	12.70	−57.2
75%	1.225	2.10	12.60	−37.2
50%	1.190	2.08	12.45	−23.3
25%	1.155	2.03	12.20	−16.1
0%	1.120	1.95	11.70	−8.3

Source: Sandia PV Architectural Energy Corporation. December 1991

7.3.2.4.6 Battery life

In off-grid PV systems, lead-acid batteries are usually the component with shortest lifetime.

Depending on the battery usage and method of application, the method of assessment for lifetime is not the same. For lead-acid battery, it can be divided into three evaluation methods: charge and discharge cycle life, float charge life, and constant current charge life. The batteries used in the renewable energy field are mainly concerned with the former two.

The charge and discharge cycle life of a battery is measured by the number of charge and discharge cycles, while the float charge service life is measured by the battery's working life. According to regulations, the charge/discharge cycle life of fixed (open type) lead-acid battery should not be less than 1,000, and the service life (float charge) should not be less than 10 years.

In fact, the battery life has a lot to do with the battery quality and working conditions, DOD, and maintenance. Figure 7.7 shows battery capacity versus DOD and number of cycles.

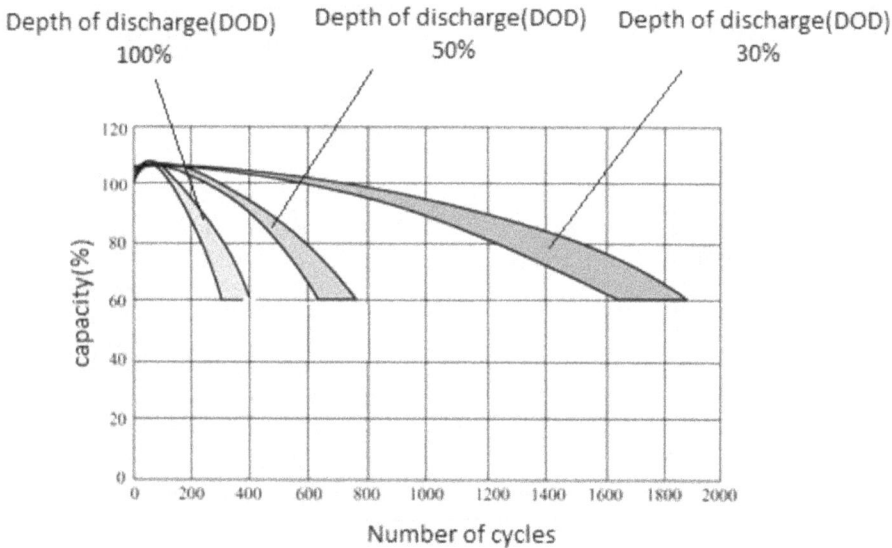

Fig. 7.7: Battery capacity versus DOD and number of cycles.

In summary, lead-acid battery is one of the most critical components in off-grid PV systems and often has the shortest lifetime in the entire system. It is necessary to design battery in a PV system with proper type and specifications, to select a right model with sufficient capacity. Careful installation and maintenance can ensure long-term stable operation of off-grid PV systems.

7.4 Controller

The controller in a PV system is an important device for managing and controlling system [6]. In different types of PV systems, the controllers are different, with various functions and complexity. A controller consists of electronic component, meter, relay, switch, and others. In a grid-connect PV system, the controller is often integrated with an inverter. Naturally, its function also becomes a part of the inverter function. Therefore, this section mainly discusses the case of the off-grid PV system [7].

In an off-grid PV system, the core function of the controller is to manage and automatically control the entire PV system to ensure the safe and reliable operation of components, batteries, and loads. The controller shall provide the battery with the best charging current and voltage, charge the battery efficiently, quickly and stably, to reduce the losses during the charging process, and to extend the battery service life. At the same time, it has battery protection function of disconnecting and reconnecting to avoid overcharging and overdischarging. If the user uses a DC load, it can also have a voltage regulation function when needed to provide a stable DC power supply for the load. The shape of different PV controllers is shown in Fig. 7.8.

Fig. 7.8: Different PV controllers.

7.4.1 Types of controllers

7.4.1.1 Controller classification
Controllers can be roughly divided into three types.

7.4.1.1.1 Pulse-width modulation (PWM) controller in parallel connection
This controller features an electronic switch in parallel with the battery. When the battery is fully charged, the electronic switch in the controller diverts the current

from PV array to its internal resistors or power modules, and then consumes them in the form of heat. Because this method consumes thermal energy, it is generally used only in small, low-power systems. This type of controller does not have mechanical parts such as relays, thus could work reliably.

Figure 7.9 shows the circuit diagram of a common PWM controller. D is a PV blocking diode, and Q is a PWM-controlled switch. As can be seen from the figure, when the battery is full, switch Q is completely turned on and all power generated from PV consumed by the controller, not to charge the battery.

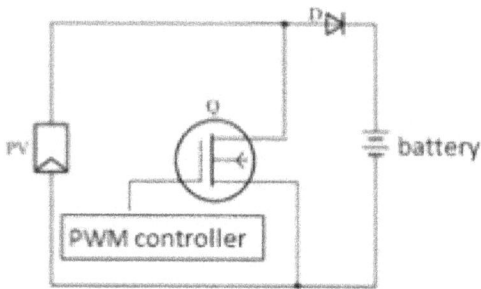

Fig. 7.9: Circuit diagram of a PWM controller in parallel.

7.4.1.1.2 PWM controller in series connection

As shown in Fig. 7.10, for a PWM controller in series connection, Q_2 is a PWM control switch. When charging the battery, Q_2 is turned on. When the battery is full, the switch Q_2 turned off, and there is no power to the battery. And also, the PV array is under open circuit, solved hot spot problem with short circuit on parallel controllers.

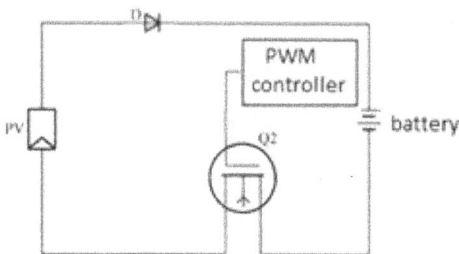

Fig. 7.10: Circuit diagram of a PWM controller in series.

7.4.1.1.3 Maximum power point tracking controller

When the PV array is at the maximum power point, the maximum power can be output. This type of controller automatically tracks the maximum power point of the array. It multiplies the voltage and current of the PV array to obtain power, and then determine whether this power reaches the maximum at a given time. If it is not at the maximum power point, it adjusts the pulse width to change the output duty cycle, therefore change the PV output current, until the maximum power reaches. The current is sampled again in real time and a duty cycle adjustment is made to keep the PV output on maximum power. Through this optimization process, the PV array can always be operated at the maximum power point. At the same time, the PWM method is used to make the battery charging current as a pulse current to reduce the polarization of the battery and improve the charging efficiency.

7.4.1.2 Comparison of three controllers
The advantages and disadvantages of PWM controllers in parallel or in series, and maximum power point tracking (MPPT) controllers are compared as follows.

7.4.1.2.1 PWM controller
Advantages: Simple structure, low cost; controller can be made very small, and is easy to install and use.

Disadvantages: The system is inefficient. The output voltage of the PV array is controlled by the battery voltage and cannot track the maximum power point of; due to the influence of the battery pack, accurate balanced floating charge management cannot be implemented.

7.4.1.2.2 Maximum power tracking controller
Advantages: High system efficiency; regulated modular design, high system reliability.

Disadvantages: The circuit structure is complex, and the cost is relatively high; the volume and weight are relatively large.

7.4.2 The main functions of the controller

7.4.2.1 Battery charge and discharge management
The controller should have an ability to disconnect from power source when the battery is fully charged and re-connect to power source as required [8]. For a battery with rated voltage of 12 V, the reference values of disconnection and re-connection are:
- Start-up lead-acid battery: disconnect at fully charged at 15.0–15.2 V; re-connect for charging at 13.7 V;
- Fixed lead-acid battery: disconnect at fully charged at 14.8–15.0 V; re-connect for charging at 13.7 V;
- The sealed lead-acid battery: disconnect at fully charged at 14.1–14.5 V; re-connect for charging at 13.2 V.

The controller should also have low voltage disconnection to load and restoring supply when sufficient recharging obtained.

When the voltage of the 12 V battery drops to the overdischarge point of 10.8 V, the controller should be able to automatically cut off the load; when the voltage of the 12 V battery recharged to the recovery point of 13.2 V, the controller should be able to automatically or manually restore power supply to the load.

Considering the environment and operating temperature dependence of the battery, the controller should be equipped with a temperature compensation function. The temperature compensation function is mainly to set a more reasonable charging voltage for the battery under different working environment temperatures to prevent overcharge or undercharge which may cause charge and discharge capacity of the battery to drop prematurely or even damaged. The general specification for the temperature coefficient is $-(3-7)$ mV/°C.

7.4.2.2 Equipment protection functions

7.4.2.2.1 Load short-circuit/overload protection
To protect the charge controller and battery under load short circuit and overload.

7.4.2.2.2 Internal short-circuit protection
Can withstand the internal short circuits of charge controllers, inverters, and other devices.

7.4.2.2.3 Reverse discharge protection
Circuit protection that prevents the battery reversely discharging to PV module.

7.4.2.2.4 Reverse polarity protection
Can withstand the reverse polarity of load, PV module, or battery.

7.4.2.2.5 Lightning protection
Can withstand breakdown caused by lightning strikes, especially in the area with frequent lightning activities.

7.4.2.3 PV system status display
The controller should be able to display the working status of the system. For the controller of a small PV system, the SOC of the battery can be indicated by the color of LEDs as follows:
  - Green indicates that the battery has sufficient power and can work normally.
  - Yellow indicates that the battery has insufficient power.

– Red indicates that the battery has very low SOC and must be charged immediately; otherwise, it will damage the battery. At this time, of course, the output of the controller to the load has also been automatically disconnected.

For large- and medium-sized PV systems, the basic technical parameters such as voltage, current, power, and ampere-hour should be displayed by meters or digital display system.

7.4.2.4 PV system data and information storage
Especially for large-scale PV systems, data and information storage devices should be provided, and if necessary, analysis should be performed for system improvement.

7.4.2.5 PV system fault handling and alarm
When the system fails, it can automatically take protective means, or use sound, light and other alarms so that operators can deal with it in time to avoid system damage.

7.4.2.6 PV system remote monitoring and control
For large-scale PV systems, remote monitoring and control and other devices may be required by electric grid operators or local utility authorities [9].

Of course, the functions of a controller are not the more the better. It needs to be balanced by regulator's requirement, system operation, safety, reliability, and cost.

7.4.3 Main technical specifications of controllers

In order to make a controller working more effectively, there are also certain requirements for its own performance.

7.4.3.1 Quiescent current
The no-load loss of the controller should be reduced as much as possible to improve the efficiency of the PV system. The quiescent current of the controller should be as low as possible, and it should not exceed 1% of its rated charging current.

7.4.3.2 Loop voltage drop
The voltage drops of the controller during charging or discharging should not exceed 5% of the rated voltage of the system.

7.4.3.3 Vibration resistance

After a vibration test at 10–55 Hz, amplitude of 0.35 mm, and 30 min in each of the three axial directions, the controller should still work normally.

7.4.3.4 High voltage resistance

When the battery is removed from the circuit, the controller must be able to withstand 1.25 times of the nominal open-circuit voltage of the solar module for 7 h.

7.4.3.5 High current resistance

The controller must be able to withstand 1.25 times of the nominal short circuit current of the solar module for 1 h. Switching components in the controller must be able to switch this current without damaging itself.

The controller generally has three pairs of terminals, which are connected to PV array, battery, and DC load or inverter, respectively. When connecting note that the three positive poles of the controller are connected to the positive poles of the PV array, the battery, and the DC load, and three negative poles are connected to the negative poles of the corresponding component. The polarity cannot be reversed.

7.5 Grid-connect PV inverters

7.5.1 Type of grid-connect PV inverters

In recent years, with the rapid decline in the cost of all PV system components, grid-connect PV systems have become the mainstream of PV applications. Grid-connect PV inverters convert DC power output from PV array into AC power that feed into the electric grid [10].

As a bridge between the PV array and the power grid, the grid-connect PV inverter needs to have three basic functions [11]:

- The first is to efficiently convert power from DC to AC, including MPPT control and inverter functions.
- The second is that the energy output from the PV system is properly fed into the grid, and the harmonic current is required to be low, with high power quality, and to follow grid voltage and frequency within a certain range; In addition, the inverter also needs to have the ability to support the stability of the power grid, such as meeting the fault ride-through capability of the power grid, delivery of the active and reactive power to the grids in real-time, and implement virtual synchronous generator (VSG) in combination with energy storage system.
- The third is to have various protection functions for PV systems, such as islanding protection, insulation monitoring, and PV module potential-induced degradation (PID) protection/recovery.

Grid-connect PV inverters involve many disciplines such as power electronics, control theory, PV power generation systems, and electric power systems [11]. The control system usually uses high-performance control chips such as DSP/ARM/CPLD as the core, and the controller accepts power from the DC side as well as the current and voltage sampling signals of the grid side through analysis and processing, maximum power optimization, voltage and current adjustment, and PWM waveform generation, a control command is sent to the power conversion main circuit to realize the grid-connect power generation.

In recent years, various new devices, new topologies, new control chips, and new control algorithms have emerged. The technical indicators of the inverter have been continuously improved. The maximum conversion efficiency of the inverter has reached 99%, and it will reach 99.5% or more in the future. MPPT efficiency exceeds 99%, and it will reach 99.9% or more in the future. The power density of the inverter is continuously increasing and the power of the single unit is also increasing. The history of its development is shown in Tab. 7.5.

Tab. 7.5: Summery of inverter development.

	First generation	**Second generation**	**Modern**	**Next generation**
Time	2002–2007	2008–2011	2012–2016	2017–2020
Power devices	Low-power IGBTs, MOSFETs	Third-generation IGBT	Fourth-generation and fifth-generation IGBTs	SIC, GAN device
Topology	Two-level	Two-level	Two-level, three-level	Triple level, multilevel
Capacitor	Electrolytic capacitor	Electrolytic capacitor	Electrolytic capacitor, film capacitor	Thin-film capacitors, no electrolytic capacitors, low capacitance design
Processor	MCU (96 series)	DSP24 series	DSP28 series	Double DSP, 200 M frequency
Maximum conversion efficiency	95%	97%	98.70%	99.50%
European efficiency	94%	96%	98.50%	99%
MPPT efficiency	97%	98%	99%	99.90%
Lifetime	5–10 years	10–15 years	15–25 years	25–30 years

According to the power of the grid-connect PV inverters, it can be divided into three main types: centralized inverter, string inverter, and microinverter.

The power rating of a single unit in a centralized inverter varies from several hundred kilowatts to several megawatts. The system topology generally adopts DC-AC single-stage conversion, and power devices generally use high-current IGBTs. The main features of a centralized inverter are high rated power per inverter, low number of MPPT, and low cost per watt. It is commonly used in large ground, floating, and roof-top PV systems.

The power rating of a single string inverter is 2.5–100 kW, and the system topology uses DC-DC-AC two-level conversion. The main features of string inverters are small rated power per inverter, flexible application, and a large number of MPPTs. Usually, one or several strings per MPPT can partially resolve mismatch issues due to inconsistent mounting orientation and various shadings. String inverters are mainly used for complex hill mounted, small- and medium-sized distributed PV systems.

The power of the microinverter is generally a few hundred watts. Usually, a single PV module can be equipped with a microinverter. Its characteristic is that it can perform MPPT on a single PV module, which is small size and weight, and easy to install. Because it does not require components to be connected in series, the system has low DC voltage and high safety. The disadvantages are that the system cost is high, and low-rated power per inverter leads to a large number of inverters in system configurations. The main application is in the residential system. Various inverters are shown in Fig. 7.11.

| Centralized inverter | String inverter | Micro-inverter |

Fig. 7.11: Various inverters (not in the same scale).

7.5.2 Main requirements for grid-connect PV inverters

The operation of a grid-connect PV system has higher technical requirements for inverters, which mainly include [12]:

(1) The inverter output is required to meet the power quality requirements. The feed-in electricity from PV systems to utility grid, and harmonics and DC components of the output current must meet the specifications of the grid.

(2) The inverter should operate efficiently under a large range of irradiance and temperature variations. The power of PV generation system comes from solar energy, and the irradiance and temperature change continuously, so the DC power input during operation changes a lot, which requires the inverter can operate efficiently under different irradiance and temperature conditions.

(3) The inverter is required to track PV array at the maximum power point. The output power of PV modules has nonlinear relationships. The inverter is required to have the maximum power tracking function, that is, regardless of changes in irradiance, temperature, or other factors, the maximum power output of the PV array can be achieved through the automatic adjustment of I–V point by the inverter.

(4) The inverter is required to have good grid connection performance, adapt to a variety of complex grid environments, such as voltage amplitude and frequency anomalies, and have the capability of grid fault (including low voltage, zero voltage, high voltage) ride-through capability, not only ensuring that the network is not disconnected during the power grid failure, but also providing certain active and reactive power during the fault to help the local power grid to recover.

(5) The inverter is required to have a certain reactive power output capability to support the stable operation of the power grid and can quickly respond to active and reactive power dispatches of the power grid in real time.

(6) It is required that the inverter can constitute a multienergy complementary system with other energy sources such as energy storage. For example, in some cases without power grids, PV systems constitute microgrid with matched energy stored to supply power to selected load. The PV power generation system smooths the output power of the PV system to increase the grid-friendliness and combines the energy storage to achieve VSG functions.

(7) The inverter is required to have features such as high power density, strong environmental adaptability, and high reliability. It can operate stably and reliably under various harsh environments such as high temperature, low temperature, damp heat, and windy and dusty conditions.

7.5.3 Working principle of grid-connect PV inverters

The DC output from the PV array must be converted to AC before it can be connected to the grid. Figure 7.12 shows the schematic diagram of a typical single-phase grid-connect inverter. The system consists of a grid-connect AC inductor L, power insulated-gate bipolar transistors (IGBTs) (T_1–T_4), DC storage capacitor C, DSP controller, and

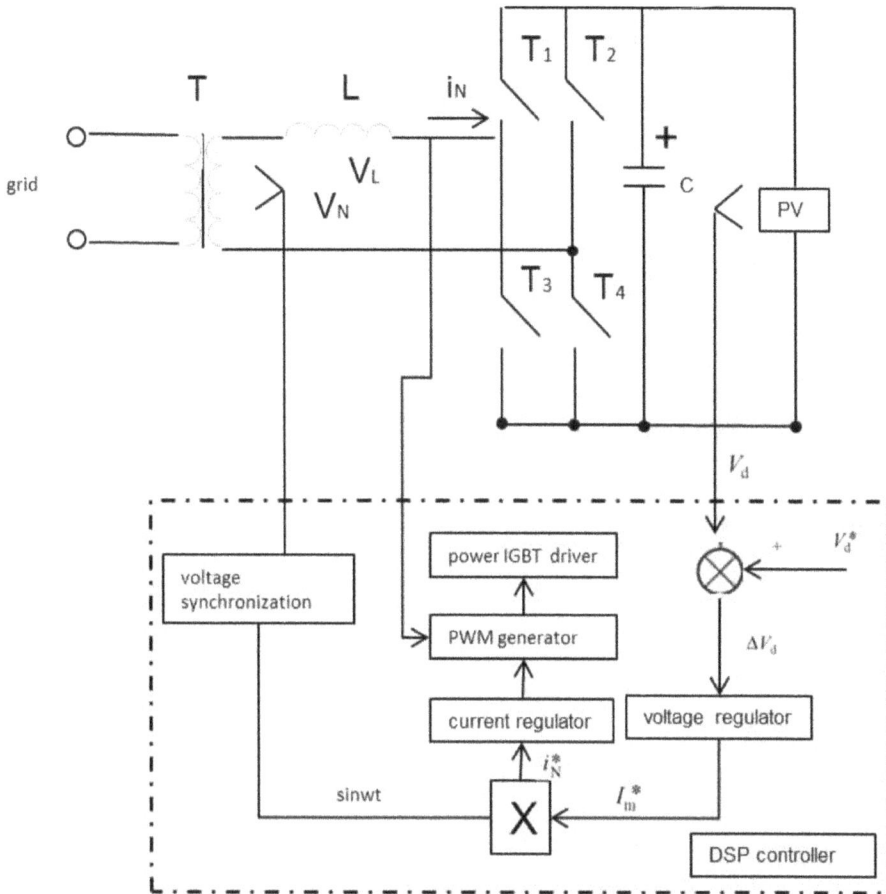

Fig. 7.12: Schematic of a single-phase grid-connect inverter.

other components. In grid-connect operation, the grid-side current sinusoidal control process is as follows: at first, the DC set voltage V^*_d is compared with the feedback voltage V_d to obtain the voltage difference signal ΔV_d, after adjusted by ΔV_d, an output current I^*_m command is generated, and its phase is obtained from the unit sine wave signal $\sin\omega t$ synchronized with the grid voltage. The two are multiplied and the sinusoidal current command signal i^*_N can be obtained. After being controlled by the current regulator, the PWM mode generator outputs a control signal to force the output current to track the input current. When i_N and V_N are inverted, electrical energy will be fed from the PV array to the grid.

Generally, for large- and mid-scale grid-connect PV systems of 100 kW or more, three-phase inverters are usually connected to the grid. The system principle is shown in Fig. 7.13.

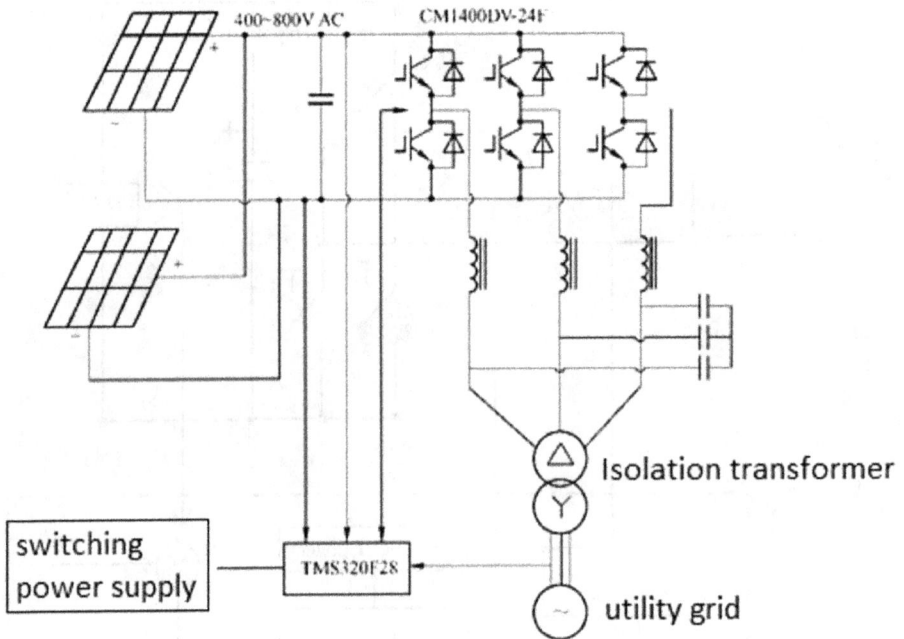

400–800V AC CM1400DV-24F

switching
power supply

TMS320F28

Isolation transformer

utility grid

Fig. 7.13: Schematic of a three-phase grid-connect inverter.

7.5.4 Main functions of grid-connect PV inverters

7.5.4.1 MPPT control

The output of PV modules has nonlinear characteristics. In addition to the internal characteristics of PV cells, the output power is related to changes in irradiance, temperature, and load. Under different external conditions, PV modules can operate at different and unique maximum power point values, as shown in Fig. 7.14. Therefore, the inverter needs to find the maximum power point of the PV module in order to maximize the conversion of solar energy into electrical energy.

The PV inverter detects the output power of the PV array in real time and predicts the possible maximum power output of the PV array through a certain control algorithm to change the current impedance value so that the PV array outputs the maximum power.

The MPPT algorithm is shown in Fig. 7.15. Assume that curve 1 and curve 2 in the figure are two output characteristic curves of the same PV arrays under different irradiance, and points A and B are corresponding to their maximum power points. Assume that the system operates at point A at a certain moment, when the irradiance changes, the output characteristic of the PV array rises from curve 1 to curve 2. At this point, if the load 1 remains unchanged, the system will operate at point A', which deviates from the maximum power point at curve. In order to continue to

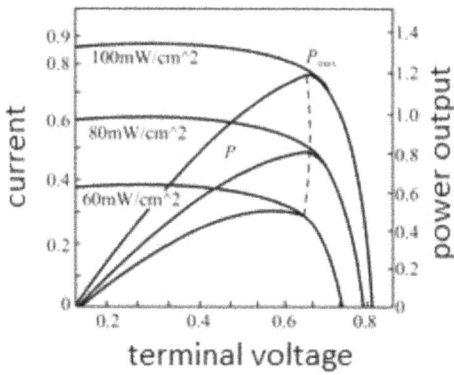

Fig. 7.14: *I–V* and *P–V* characteristics of a PV module.

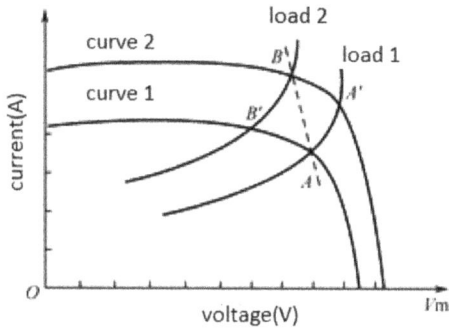

Fig. 7.15: MPPT algorithm analysis.

track the maximum power point, the load characteristics of the system should be changed from load 1 to load 2 to ensure that the system operates at the new maximum power point B.

There have been many studies on the MPPT control technology for PV system, and various control algorithms have also been developed. Commonly used MPPT algorithms include constant voltage tracking, disturbance observation, conductance incremental methods, and fuzzy logic control algorithms. These methods have their own advantages and disadvantages. In practical applications, suitable control algorithm can be selected according to specific conditions and requirements.

7.5.4.2 Grid connection

In addition to transform the DC output from the PV array into a high-quality AC power with the same amplitude, frequency, and phase angle to the grid, grid-connect inverter also has to adapt to changes in a certain range of the grid voltage and frequency variations, including power grid fault ride-through capability, reactive power support capability, real-time dynamic response to the grid's active and reactive power dispatching of PV systems, and VSG functions combined with energy storage [13].

Grid low-voltage ride-through (LVRT) means that when grid voltage drops a lot to abnormal status, the PV system will continue to maintain operation without disconnecting from the grid until reached a preset time period, or the fault been recovered and the system returns to normal and stable operation. During grid faults, inverters are also required to be able to output certain active and reactive power to help restore local grid faults and improve grid stability. Except LVRT, the inverter may also have zero-voltage ride-through and high-voltage ride-through functions. The specific requirements are shown in Figs. 7.16 and 7.17.

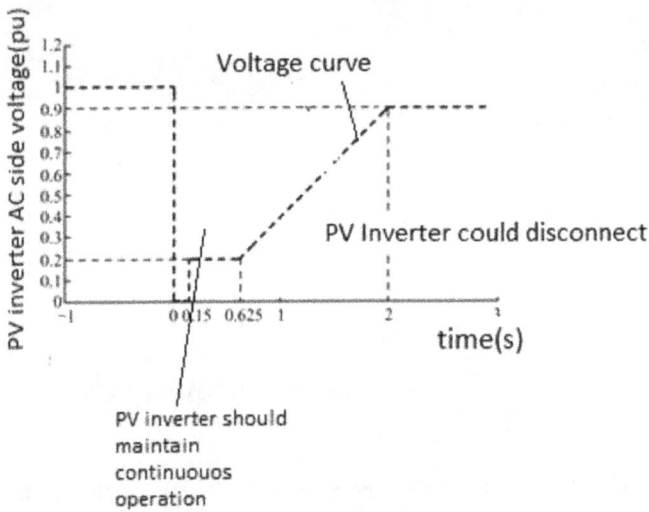

Fig. 7.16: Low-voltage ride-through requirement for PV inverters.

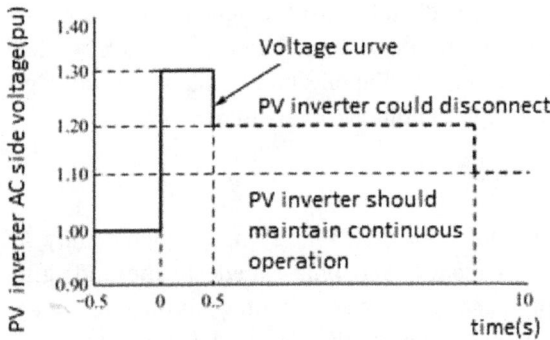

Fig. 7.17: High-voltage ride-through requirement for PV inverters.

With the increase of the installed capacity of PV power and its percentage in the power system, the power system not only requires the PV power plant to have active power output capability, but also needs to have a certain reactive power output capability and can quickly respond to the active and reactive power dispatching of the power grid. It plays a certain supporting role on the power grid and improves the stability of the power grid operation. The inverter reactive power capacity and dynamic response time requirements are shown in Fig. 7.18. When the inverter outputs the rated active power P_n, it also needs to output 0.48 P_n of reactive power, and the dynamic response time should be less than 30 ms.

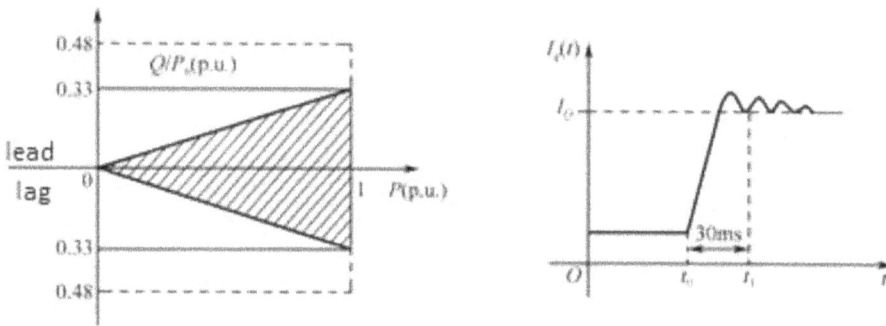

Fig. 7.18: Inverter reactive power capacity and dynamic response time.

PV system can also realize the function of VSG by combining energy storage, that is, by simulating the characteristics of the synchronous generator model, active power frequency regulation, and reactive power regulation, the grid-connect inverter can be comparable to the traditional synchronous generator in terms of operating mechanism and external characteristics, to further enhance the grid-friendliness of the PV system. The system structure is shown in Fig. 7.19.

7.5.5 System protection function

In addition to its own protection functions such as lightning protection, overcurrent, overheating, short circuit, reverse connection, abnormal DC voltage, and abnormal grid voltage, the grid-connect inverter needs to have the PID protection function and anti-islanding function.

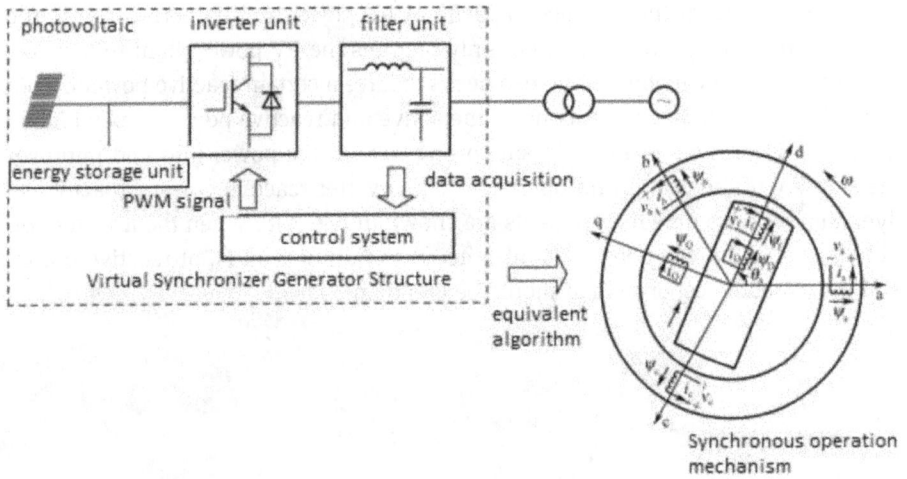

Fig. 7.19: Virtual synchronization control (VSG) system.

7.5.5.1 PID protection

The high voltage between the solar cell and its metal frame in the crystalline silicon PV module may cause the continuous degradation of the performance of the module. This phenomenon is called the PID effect. The PID problem has become one of the most important factors affecting the power generation of PV power plants. Especially in high-temperature, high-humidity floating PV power systems, and rooftop PV systems, the probability of PID greatly increases. Therefore, in addition to the continuous improvement of the protection of the components themselves, it is also required that the inverter has a PID protection function.

The common PID protection methods mainly include the negative pole grounding and the negative pole virtual grounding of PV systems. The negative grounding method means that the negative pole of the PV array or the inverter is directly grounded through a resistor or fuse, so that the voltage between the negative electrode of the solar panel and ground is kept at the same potential as the grounded metal frame to eliminate the influence of the negative bias voltage. The negative virtual grounding is to adjust the virtual neutral point potential of the alternating current to the ground by detecting the negative potential of the PV system to the ground, so as to increase the negative pole potential to the ground and ensure that the negative pole potential to the ground is greater than or equal to the ground potential.

In addition, it is possible to apply a reverse voltage to the PV modules during PV system shutdown at night to recover components PID potential that has occurred during the daytime.

7.5.5.2 Anti-islanding function

The islanding effect is a phenomenon that when a PV system is connected to the power grid in parallel and supplies power to the load, in the condition that the power grid fails or is interrupted, the PV system continues to independently supply power to the load. When the power output from the PV system is balanced with the load, the load current will be completely provided by the PV system. At this time, even if the power grid is disconnected, the voltage and frequency at the output end of the PV system will not change rapidly, so that the system cannot correctly determine whether the power grid has a fault or is interrupted, thus resulting in an islanding operation condition.

The islanding effect will have certain adverse effects on the system, threaten the personal safety of the maintenance personnel, damage some of the loads that are sensitive to frequency changes, and damage the equipment due to the large transient current caused by the power supply phase recovery. Therefore, at this condition, the inverter should also stop its operation. This called anti-islanding function, and it is an important function for a grid-connect inverter.

Common anti-islanding function methods include passive detection method and active detection methods:

- Passive detection method is generally used to detect the voltage and frequency of the public power grid, as a basis for judging whether the public power grid is faulty or interrupted. Commonly used methods include voltage and frequency protection relay detection method, phase jump detection method, voltage harmonic detection method, frequency change rate detection method, and output power change rate detection method.
- Active detection method means that the voltage or frequency of the system is actively disturbed periodically at the output of the inverter, and observe whether or not the grid is affected, as a basis for judging whether there is a failure or interruption in the public power grid. Methods include output power fluctuations, addition of inductors and capacitors, and frequency offset method.

7.5.6 Other aspects

Grid-connect PV inverters generally have the functions of data acquisition, recording, and display, and can display the working status and power generation status of the PV system at any time. They usually have RS-485, Ethernet, and other communication interfaces, can perform data transmission through the communication line, and are capable of convenient remote communication and monitoring.

In grid-connect PV power stations, electrical equipment such as disconnectors, leakage protectors, and surge protectors are generally required to be installed, depending on the specific conditions. Some large-scale PV grid-connect PV stations also have equipment such as step-up transformers.

In an off-grid PV system, if there is an AC load, it must be equipped with an off-grid inverter. This kind of inverter does not require grid connection and anti-islanding functions, but it still needs functions such as inverter and MPPT. Under normal circumstances, the off-grid inverter also has the function of the controller, so the performance also meets the main technical requirements of controllers in Section 7.4.3.

Overall, modern grid-connect PV inverters are a combination of microelectronics, power electronics and PV, and have all necessary functions for grid connection. With the advancement of electronic technology and the continuous improvement of components, its efficiency, capacity, power density, and reliability will continue to be improved, creating better conditions for large-scale PV grid connection applications.

References

[1] V Vernon Risser, et al. Stand-alone photovoltaic systems a handbook of recommended design practices. SAND, 1995, 87–7023.

[2] Bella Espinar and Didier Mayer. The role of energy storage for mini-grid stabilization. Report IEA-PVPS T11-02: 2011, July 2011.

[3] R Kötz and M. Carlen. Principles and applications of electrochemical capacitors. Electrochimica Acta, 2000, 45(15–16), 2483–2498.

[4] Rachel Carnegie, et al. Utility scale energy storage systems. State Utility Forecasting Group, June 2013.

[5] B Andersson, et al. Lead-acid battery guide for stand-alone photovoltaic systems. IEA Task III, Report IEA-PVPS 3-06: 1999.

[6] J. Dunlop, W Bower and S Harrington. Performance of battery charge controllers. 22nd IEEE Photovoltaic Specialist Conference, Las Vegas, 640–645, 1991.

[7] W. J. Kaszeta, et al. Handbook of secondary storage batteries and charge regulators in photovoltaic systems; final report. SAND81-7135.2002.

[8] J. P. Dunlop. Batteries and charge control in stand-alone photovoltaic systems fundamentals and application. Florida Solar Energy Center, prepared for Sandia National Laboratories, Photovoltaic Systems Applications Dept, January 1997.

[9] E Koutroulis and K. Kalaitzakis. Novel battery charging regulation system for photovoltaic applications. IEE Proceedings – Electric Power Applications, 2004, 151(2).

[10] Fengjun Liu. Sine wave inverter. Beijing : Science Press, 2002.

[11] Xing Zhang and Renxian Cao. Solar photovoltaic grid-connected power generation and its inverter control. Beijing : China Machine Press, 2012.

[12] J. Rodriguez, Jih-Sheng Lai and Fang zheng Peng. Multilevel inverters: A survey of topologies, controls and applications. IEEE Transactions on industrial Electronics, 2002, 49(4), 724–738.

[13] Johan H.R. Enslin, Interconnection of distributed power inverters with the distribution network. IEEE Power Electronics Society Newsletter, FOURTH Quarter 2003.

Exercises

7.1   What are the main components of the PV power generation system? What are their roles?

7.2   What are the diodes used in the PV system? What are their roles? How to connect them?

7.3   Which PV systems need energy storage devices?

7.4   Briefly describe the types of major energy storage technologies currently in use.

7.5   What does the battery capacity mean? Battery capacity is determined by what factors?

7.6   What is the specific meaning of the battery's energy efficiency, charging efficiency, and DOD?

7.7   Please explain the major functions of a controller in an off-grid PV system and how many types of controllers.

7.8   What is the major function of grid-connect inverter?

7.9   What is PID effect and how to prevent it?

7.10 What is islanding effect and how to deal with it?

Chapter 8
Photovoltaic power generation system design

In recent years, the PV power generation system (also simply called PV system) installation has developed rapidly. The emergence of various types of PV system marks that the PV power system has entered a brand-new stage of large-scale application [1]. However, there are still a large amount of PV products and projects, even the large- and medium-sized PV power stations, have not undergone the design optimization yet. Those power stations work but with unsatisfied systematic performance and efficiency. Actually, while people spent lots of efforts to increase module efficiency by 1%, it is so easy and often to decrease system efficiency by 10% due to a poor system design, or even causes whole system breakdown. Therefore, more attention should be paid to optimal design of PV systems.

8.1 Main target of PV system design

8.1.1 Definition and classification of PV systems

The PV system and PV power station have distinct meanings. The PV system converts solar radiation energy directly to electricity according to photovoltaic effect, while the PV power station consists not only the PV system, but also other auxiliary facilities and buildings for operation and maintenance people.

Based on whether the PV system connects to a public electrical grid, the PV system can be divided into grid-connected and off-grid PV systems. The grid-connected PV system can be further classified into user-side and grid-side systems.

There is no absolute criterion for PV classification. The current method used by International Energy Agency is: (1) small-scale: power < 100 kW; (2) medium-scale: 100 kW to 1 MW; (3) large-scale: 1 MW to 10 MW; and (4) super-large scale: >100 MW.

According to a Chinese National Standard, GB 50797 *Code for Design of Photovoltaic Power Station*, the PV power station is classified into three types based on capacity of power transmission and distribution, quality of electricity and other technological requirements: (1) small-scale power station: installation power ≤ 1 MW, commonly used voltage 0.4 to 10 kV; (2) medium-scale power station: 1 MW < installation power ≤ 30 MW, commonly used voltage 10 to 35 kV; and (3) large-scale power station: installation power > 30 MW, commonly used voltage 35 kV or higher.

Small-scale power stations can also be categorized into feed-in or non-feed-in, based on if power is deliver into grid at point of common coupling (PCC).

https://doi.org/10.1515/9783110524833-008

8.1.2 Target and requirement of PV system design

The main target and requirement are different for different types of PV systems [2]. The grid-connected PV systems are required to generate electrical power as much as possible. Therefore, the target of system design is to lower the energy loss and increase the annual generated power.

However, in terms of off-grid PV system, the application relates to local climate and the requirements are distinct in different areas even for the same load. The maximum power generation is not always the optimal choice. In addition, the cost of PV power generation is relatively high. Therefore, in order to fully meet the needs of electricity and achieve the most rational distribution of PV modules and batteries, a scientific optimal design for off-grid PV system is needed [3].

PV and wind hybrid systems are mainly used as off-grid power source. Besides the PV, it is also equipped with a wind power system due to higher cost of PV system. The hybrid system with optimal design, that rationally allocates suitable capacity of PV and wind power generation, is able to lower the system cost, guarantee the power supply, and produce the best possible complementary of solar and wind power.

As for PV and diesel engine hybrid system with diesel engine as backup power during long period of rainy days or lower irradiation in winters, an enhanced optimal design is also necessary to fully make use of advantages of PV system and decrease the cost of power generation. The PV system is also a significant component of microgrid, hybrid system, and electricity purchase projects.

8.2 Design of grid-connected PV system

A series of design steps are necessary for the construction of large- and medium-scale PV systems, and the design of capacity is the most important factor [4].

8.2.1 The capacity design of grid-connected PV system

8.2.1.1 Factors that determine the electric energy production
The factors that affect the generation capacity of grid-connected PV systems can be classified into installed capacity, system efficiency, and solar irradiance.

8.2.1.1.1 Installed capacity
The installed capacity of PV system is the sum of rated power for all PV modules in the system.

It is obvious that the generated power output depends on the installed capacity of PV system. Under the circumstance of the same other conditions, the higher the installed capacity, the greater the power can be generated.

8.2.1.1.2 Performance Ratio (PR)

The system efficiency of PV system is decided by the transfer efficiency of PV array itself and the PR. PR is an important criterion for evaluating PV system and is defined as the ratio of generated AC electrical energy to the solar energy received by the PV array. It has no relationship with capacity of system, solar radiation, tilt angle, or direction of module array.

$$PR = Y_f/Y_r \tag{8.1}$$

where Y_f is the electricity generated by every unit power of PV power generation, and

$$Y_f = E_{pv}/P_0 \ (kWh/kW) \tag{8.2}$$

where E_{pv} is annually (or monthly) generated electricity (kWh) by PV; P_0 is the installed capacity of PV system (kW).

And Y_r is a full sun exposure time of the module in hours, that is, the equivalent time with irradiance of 1 kW/m^2 to total irradiation dose received by the PV module array:

$$Y_r = H/G \ (h) \tag{8.3}$$

where H presents annual (or monthly) solar irradiation and G is irradiance under standard test condition, $G = 1,000$ W/m^2.

The PR of a grid-connected PV system relates to the quality of system design, construction, installation, equipment, accessories, and efficiency of BOS (balance of system) which includes inverter, control instruments and so on, and the loss due to interconnections, as well as other factors, such as maintenance. Those factors can be divided into several aspects shown below:

(1) Loss due to mismatched modules

The PV array of grid-connected PV system consists of a large number of PV modules. The optimal working voltage and current is different from each other even though the rated power of all modules is the same. In theory, modules should be sorted before connecting into strings, and those with similar working currents should be connected in series. Then, the strings with similar working voltage will be parallelly connected. However, there is limited time for module selection due to the large number, and modules are connected randomly, which consequently causes the mismatch of modules and lower total power than the sum of individual module power.

(2) Loss due to electrical cable

The electrical cable loss includes DC current loss and AC current loss. A module is connected to other modules, combination boxes or inverters, with electrical cables. The cable lose is mainly from the cable resistance. In the condition of small diameter of electrical cable and numerous connecting points, the poor connection due to careless installation also causes extra lose.

(3) Loss due to shading

The module will cumulate dust on its surface, and it usually cannot be cleaned by rain due to small tilt angle. The power generation can be compromised by soiling or shading of trees or buildings nearby the module. Besides, the short distance between two arrays also causes the shading loss.

At present, the size of PV modules increases, and there are more cells in series on a module. In this way, the output power is deeply reduced even only a small area is shaded. As shown in Fig. 8.1, the output power of two modules on the left decreases half of their initial value, while the output power of module on the right will decrease about 100%. Therefore, it is necessary to reduce shading of PV modules.

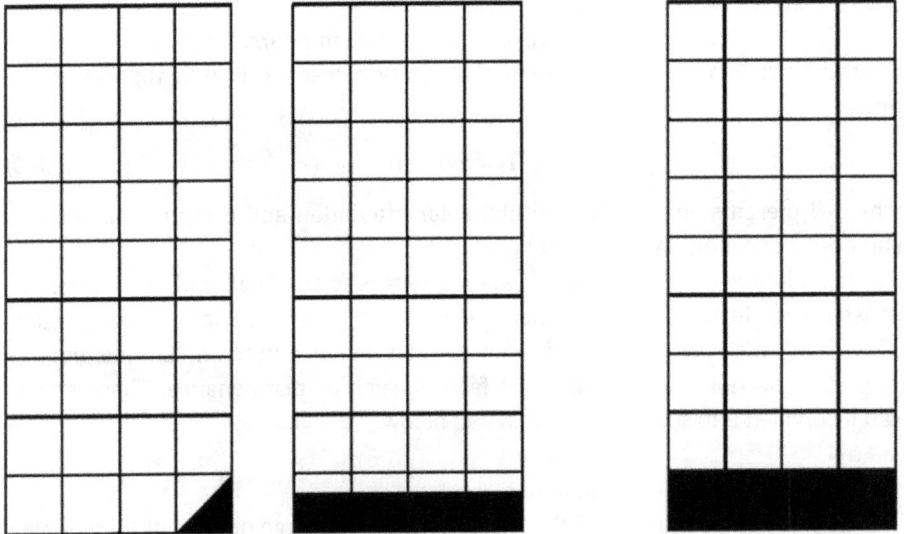

Fig. 8.1: Effect of shading on module output power.

(4) Temperature effect

The rated power of PV module is measured under the standard test condition (STC). The real output power is lower than rated power when module temperature is higher than 25 °C.

(5) Efficiency of balance of system (BOS)

A PV system includes not only the PV arrays, but also controller, inverter, combining box, transformer, and other components, which also consume energy. The lower the efficiency of those components, the greater the energy loss.

M,.Halting

Due to system maintenance, instrument breakdown or misoperation, the PR of PV system also decreases.

8.2.1.1.3 Solar irradiation

Another determining factor of generated electricity is solar irradiation. The PV module array with same rated power generates different amount of electricity in different locations.

The generated electricity depends on the solar irradiation received by the PV module array during working period. The generating capacity can be decided by solar irradiation. However, it is hard to accurately predict the daily solar irradiation due to the randomness. Therefore, the solar irradiation should be evaluated with local average solar irradiation over years (at least 10 years). China started to record the solar irradiation since 1953, and there were 66 meteorological observation stations that had measured total and diffuse solar irradiations before 1993. However, after 1993, there were only 17 meteorological observation stations that continued measuring data. The unit of solar irradiation used before 1985 was cal/m^2 and was changed to MJ/m^2 after 1985. There are no existing long-time solar irradiation data for most of areas which can be used directly. It is the biggest problem for PV and thermal application design. The commonly used data for forecasting is from meteorological database (such as Meteonorm, NASA) built-in the PV power generation calculation software.

What needs to be pointed out is: there is difference between solar irradiation values measured by our meteorological observation stations and that provided by NASA. Those data from NASA is relatively higher for most areas except for Qinghai-Tibet Plateau.

Due to the fact that the PV modules are installed with tilt angle, the correct solar irradiation value should be modified to the number of tilted surface followed by the procedures in the Chapter 2.

For the percentage of solar irradiation on PV array with different orientation and tilt angle, one method was introduced in literature as a polyhedron shown in Fig. 8.2, which actually is not correct. The percentage of each face on this polyhedron is only a schematic diagram for specific location, while due to the different latitude and ratio of direct light to diffuse light in different locations, the percentage of solar irradiation is different. The calculation needs to be adjusted based on different circumstances.

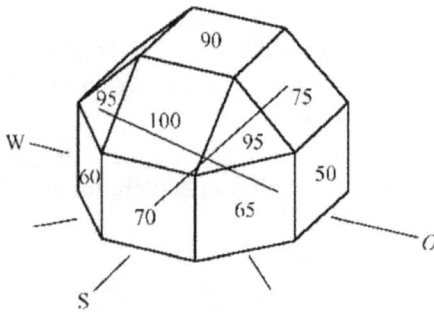

Fig. 8.2: Illustration of solar irradiation on different orientation and tilt angle (%).

Some other design even uses the horizontal radiation data from meteorological observation instead of irradiation on the incline surface or calculates the irradiation by multiplying the randomly chosen "tilt angle," which are both not accurate.

Based on different tilt angle, the monthly solar irradiation on PV array is different. Therefore, under the same local solar irradiation, the electric energy production can be improved by adjusting the tilt angle of PV array.

With regard to above negative factors, most of them can be eliminated through selecting high-quality components, optimal design, careful installation, appropriate maintenance, or other methods. Even the solar irradiation cannot be controlled, it is still possible to find the optimal tilt angle by calculation or experiments to maximize the irradiation on the PV arrays. For a PV system with MW capacity, a small mismatch of tilt angle could cause thousands kWh loss every year.

8.2.1.2 Estimate energy production for grid-connected PV system

8.2.1.2.1 Confirm the site parameters
The site parameters include rated power of PV power station, local climatic and geographical conditions, and multi-year monthly average of horizontal solar radiation information.

8.2.1.2.2 Obtained the optimal tilt angel
The design of grid-connected PV system must consider that how to increase the received solar irradiation on the module array as much as possible. The optimal tilt angle is the one that enables module array to receive the most solar irradiation during the whole year.

There are several methods that can be applied to calculate the optimal tilt angle. For example, Klein and Theilacker mentioned in 1981 that to calculate and compare the solar radiation on different tilted planes and obtain the maximum solar irradiation H_t and corresponding optimal tilt angle, the monthly average solar irradiation could also be calculated at the same time [5]. With the equation of monthly average solar irradiation on tilted plane, the optimal tilt angle for specific area can be decided by local latitude, total solar irradiation, and direct solar radiation (or diffuse solar radiation). The optimal tilt angle for grid-connected PV system facing equator in different regions in China is listed in Tab. 8.1. The local monthly

Tab. 8.1: Optimal tilt angle of grid-connected PV system in China.

Region	φ (°)	β_{opt} (°)	$\bar{H}_T$(kWh/m²/d)	Region	φ (°)	β_{opt} (°)	$\bar{H}_T$(kWh/m²/d)
Haikou	20.02	10	3.892	Xi`an	34.18	21	3.318
Zhongshan	22.32	15	3.065	Zhengzhou	34.43	25	3.881
Nanning	22.38	13	3.453	Houma	35.39	26	3.949
Guangzhou	23.10	18	3.106	Lanzhou	36.03	25	4.077
Mengzi	23.23	21	4.362	Ge`ermu	36.25	33	5.997
Shantou	23.24	18	3.847	Jinan	36.36	28	3.824
Shaoguan	24.48	17	2.993	Xining	36.43	31	4.558
Kunming	25.01	25	4.424	Yushu	33.01	31	4.937
Tengchong	25.01	28	4.436	Hetian	37.08	31	4.867
Guilin	25.19	16	2.983	Yantai	37.30	30	4.225
Ganzhou	25.51	15	3.421	Taiyuan	37.47	30	4.196
Fuzhou	26.05	16	3.377	Yinchuan	38.29	33	5.098
Guiyang	26.35	12	2.653	Minqin	38.38	35	5.353
Lijiang	26.52	28	5.020	Dalian	38.54	31	4.311
Zunyi	27.42	10	2.325	Ruoqiang	39.02	33	5.222
Changshu	28.13	15	3.068	Tianjin	39.06	31	4.074
Nanchang	28.36	18	3.276	Kashi	39.28	29	4.630
Luzhou	28.53	9	2.528	Beijing	39.56	33	4.228
Emei	29.31	28	3.711	Datong	40.06	34	4.633
Chongqing	29.35	10	2.452	Dunhuang	40.09	35	5.566
Lasa	29.40	30	5.863	Shenyang	41.44	35	4.083
Hangzhou	30.14	20	3.183	Hami	42.49	37	5.522
Wuhan	30.37	19	3.145	Yanji	42.53	37	4.054

Tab. 8.1 (continued)

Region	φ (°)	β_{opt} (°)	$\overline{H}_T$(kWh/m²/d)	Region	φ (°)	β_{opt} (°)	$\overline{H}_T$(kWh/m²/d)
Chengdu	30.40	11	2.454	Tongliao	43.26	39	4.456
Yichang	30.42	17	2.906	Erlianhaote	43.39	40	5.762
Changdu	31.09	30	4.830	Wulumuqi	43.47	31	4.208
Shanghai	31.17	22	3.600	Changchun	43.54	38	4.470
Mianyang	31.27	13	2.739	Yining	43.57	36	4.740
Hefei	31.52	22	3.344	Ha`erbin	45.45	38	4.231
Nanjing	32.00	23	3.377	Jiamusi	46.49	40	4.047
Gushi	32.10	22	3.504	A`letai	47.44	39	4.938
Ge`er	32.30	33	6.348	Hailaer	49.13	44	4.769
Nanyang	33.02	23	3.587	Heihe	50.15	45	4.276

average solar irradiation is summarized from the 1981–2000 "Annual register of meteorological radiation in China" published by National Meteorological Center. The solar irradiation on tilted plane is calculated based on equation that proposed by Klein and Theilacker. The φ in the table is local latitude; β_{opt} is the optimal tilt angle for PV array; $\overline{H}_T$ is the annual average solar irradiation on PV array.

For some PV systems, especially the building-integrated PV systems, the PV arrays cannot face the equator all the time due to constrain of buildings. Instead, the sub-arrays have different orientations. The optimal tilt angle for different azimuth angle is also different. The optimal tilt angle should be calculated based on azimuth angle of each sub-arrays in order to receive the largest solar irradiation during the whole year.

The annual peak sun hours can be calculated by converting the unit of solar irradiation to kWh/m²/y, which is not always the sunlight time provided by local meteorological stations.

8.2.1.2.3 Set the PR

The PR is different for each individual PV power station due to various real operation conditions. IEA-PVPS has conducted plenty of investigations and analysis and pointed out that PR for early stage grid-connected PV systems were around 60% – 80%. At present, the PR values for high-quality PV power stations are greater than 80%, some stations even have PRs larger than 85%. The recommended PR is 80% for grid-connected PV systems with optimal tilt angle.

8.2.1.2.4 Calculation equation

The annual energy generation for grid-connected PV system can be calculated with equation below:

$$E_{out} = H_t \cdot P_0 \cdot PR \tag{8.4}$$

where E_{out} is the exported electric energy of grid-connected PV system (kWh/y); H_t is sunlight hours, which is calculated as total solar irradiation received by PV arrays (kWh/m²/y) divided by full solar irradiance ($G = 1,000$ W/m²); P_0 is actual power of PV system (kW); And PR is performance ratio.

8.2.1.2.5 Obtain the energy production of grid-connected PV system

The annual energy production can be obtained by substituting numbers into eq. (8.4). Similarly, the monthly energy production can be also calculated with monthly average solar irradiation data.

If the PV system contains several sub-arrays with different orientation, the optimal tilt angle and solar irradiation should be confirmed for each sub-array, respectively. Then the energy generation of every sub-array can be calculated with equation above, and the sum of energy for all sub-arrays is the annual energy generation for this system.

Example 8.1 A PV power station with capacity of 1 MW is planned to be built on a factory rooftop in Shanghai, calculate the maximum energy production in each month.

Solution According to the meteorological data of Shanghai, the solar irradiation on different tilted planes can be calculated and compared. Then the tilt angle of annual maximum solar irradiation is confirmed as 22° with orientation of south, which is the optimal tilt angle. Therefore, the monthly average solar irradiation on tilted plane can be calculated. Meanwhile, the monthly energy production is calculated with equation above and assume PR of 80%, the result is shown in Tab. 8.2.

Tab. 8.2: Energy production in Shanghai with a 1 MW PV power station.

Month	Horizontal solar irradiation H_0 (kWh/m²/d)	Solar radiation with tilt angle of 22° H_t (kWh/m²/d)	Daily average production E_d (kWh/d)	Monthly average production E_m (kWh/m)
January	2.079	2.481	1,985	61,529
February	2.598	2.926	2,341	65,542
March	2.974	3.138	2,510	77,822
April	4.036	4.050	3,240	97,200
May	4.652	4.485	3,588	111,228

Tab. 8.2 (continued)

Month	Horizontal solar irradiation H_0 (kWh/m^2/d)	Solar radiation with tilt angle of 22° H_t (kWh/m^2/d)	Daily average production E_d (kWh/d)	Monthly average production E_m (kWh/m)
June	4.164	3.963	3,170	95,112
July	4.864	4.635	3,708	114,948
August	4.611	4.550	3,640	112,840
September	3.816	3.962	3,170	95,088
October	3.144	3.483	2,786	86,378
November	2.442	2.907	2,326	69,768
December	2.114	2.620	2,096	64,976

The sum of monthly energy production gives total produced energy of 1,052,431 kWh/y. It indicates that the PV power station with average rated power of 1 W is able to generate 1 kWh electric energy every year with optimal tilt angle in Shanghai.

However, it needs to point out that the discussion above does not mention the degradation of PV modules. The output power of PV modules decreases after a certain period due to degradation of cell properties. Besides, the degradation of encapsulant materials also decreases the output power of PV modules. It is generally assumed that the yearly degradation rate "r" for crystal silicon PV module is around 0.8%, and the value for amorphous silicon PV module is higher. Therefore, to calculate the energy generation of PV system during n years in its life cycle, if the build-up time is regarded as the 0th year, then the equation of total energy production is:

$$E_{out,\,n} = \sum_{0}^{n-1} H_t \cdot P_0 \cdot PR(1-r)^n \qquad (8.5)$$

Example 8.2 A PV power station with crystal silicon cell was built in Yinchuan. If the performance degradation rate "r" = 1%, PR = 80%, life cycle = 20 year, and the module arrays were installed with the optimal tilt angle. Calculate the yearly and total energy production.

Solution The optimal tilt angle for Yinchuan is 33° (Tab. 8.1). The daily irradiation on tilted plane of each month is 5,098 Wh/m^2, that is, the daily full sun time is 5.098 h. Therefore, the total full sun time in a year is 1860.77 h. Substitute the PR and n into eq. (8.5):

$$E_{out,n} = \sum_{0}^{n-1} H_t \cdot P_0 \cdot PR(1-r)^n = \sum_{0}^{19} 1860.77 \times 10 \times 0.8(1-0.01)^n$$

The yearly energy generation is different n from 1 to 20, as shown on Tab. 8.3.

Tab. 8.3: Yearly energy generation of a 10 MW PV station in Yinchuan (MWh).

n	$E_{out,n}$	n	$E_{out,n}$	n	$E_{out,n}$	n	$E_{out,n}$
1	14,886.2	6	14,156.6	11	13,462.8	16	12,803
2	14,737.3	7	14,015	12	13,328.2	17	12,674.9
3	14,589.9	8	13,874.9	13	13,194.9	18	12,548.2
4	14,444	9	13,736.1	14	13,062.9	19	12,422.7
5	14,299.6	10	13,598.8	15	12,932.3	20	12,298.5

And the sum of energy production during 20 years gives total energy generation as 271,066 MWh.

8.2.1.3 Introduction of common design software

Various elements are involved in design of PV power stations and have complex relationship between each other. Several design software has been developed to solve this problem. At present, the commonly used software for PV power station design are PVSystem, RETScreen, and so on. The comparison and characteristic analysis is listed in Tab. 8.4. The corresponding meteorological database includes Meteonorm, NASA, and others. The comparison of each meteorological database is summarized in Tab. 8.5.

Two types of common PV design software, PVSystem and RETScreen have similar calculation principles. Given the same solar energy data, two software get almost same results. However, due to the fact that the built-in database for these two software are different, the results also have large discrepancy with the built-in database.

There are three solar irradiance databases in China: (1) CMA solar irradiance database (actual measured data); (2) CWERA solar energy resource evaluation database (calculated data with meteorological method); and (3) CWERA solar energy resource evaluation database (satellite data).

By comparing NASA data with data from China's meteorological stations, it is found that there are different deviations in different regions. For the central and eastern regions, the cloud volume is relatively large, some districts are also affected by water body, snow, and mountains. Therefore, the real ground irradiance has relatively large discrepancy with NASA database, especially for the areas with more rainy days, the NASA data is more than 10%, which is greater than actual measured data.

Tab. 8.4: Comparison of common PV design software.

Design Software	Software Features	Major Function	Meteorological Database
PVSystem	PVSystem is a professional and comprehensive software for PV system design with expandable model database. It contains four main parts of preliminary design, detailed design, database, and tools. This software is also equipped with extensive meteorological database, PV system module database, and common solar tools. PVsyst6.1 integrated MN6.1 data，while the latest PVsyst version 6.3 integrated MN7.1 data.	(1) Select the type of PV system: grid-connected, independent, PV pump, DC power grid, and so on. (2) Select the configuration parameters of PV modules: fixed mode, tilt angle, array spacing, orientation angle, and so on. (3) Evaluate shading of architecture building on PV systems; calculate shading time and shading ratios. (4) Simulate energy production and system generation efficiency of different type of PV systems. (5) Study environmental parameters of PV systems.	Meteonorm/ NASA
RETScreen	RETScreen is a software for renewable energy project analysis based on Excel. It is used to evaluate various energy efficiencies, energy generation for different renewable energy sources, energy saving benefits, cost of life cycle, emission reduction, and financial risk. It also contains database of components, cost, and meteorological data, and is capable to help users to decide the technology of clean energy and renewable energy, and the financial feasibility.	It is a powerful software that assesses the economy, greenhouse gases, finance, and risks for different applications of wind energy, PV, small hydropower, energy conservation, combined heat and power generation, biomass heating, solar heating, and geothermal heat pump. Calculation of energy generation of PV system is only one function of this software, therefore, the capability of design is slightly lower than PV system.	NASA

Tab. 8.4 (continued)

Design Software	Software Features	Major Function	Meteorological Database
SketchUp	SketchUp is a convenient 3D design software and compatible with GoogleEarth. It has a specialized plug-in "Skelion" for PV power station design.	This software is mainly used to plot the design sketch for the arrangement of PV system, analysis the shading and measurement of array spacing. The information such as tilt angle; orientation angle can be obtained after modeling. It also supports the one-click arrangement of PV modules.	None
PVSOL	PVSOL is a software used for simulation and design of PV system. It is excellent in database building and provides detailed meteorological data for most regions in Europe and America with interval of 1 h. The meteorological data contain solar irradiance, wind speed, and ambient temperature at elevation of 10 m for specific locations. All data can be visualized by table or curve with time interval of day, week, or month. In addition, the software also includes abundant load data, characteristic data of 150 types of PV modules, 70 types of accumulators, and 150 types of inverters for off-grid and grid-connected systems. It has high design flexibility due to the fact that data can be extended by customized by users.	The database of modules and inverters is provided by suppliers and updated periodically. The software is able to choose the optimal inverter configuration and generate Meteonorm 7.1 meteorological data automatically. It is compatible with Google Map for 3D simulation and shading simulation, the results of electrical diagram and module arrangement sketch can be exported into CAD. The simulation report contains information of solar irradiance, yearly energy generation, efficiency of PV modules, system efficiency, and system efficiency losses. Besides, this software is capable for economic benefit and environmental protection benefit analyzing. It separates the PV system into three groups: off-grid, grid-connected, and mixed system, and treats different groups differently.	Meteonorm
Summary	In summary, the PV system is more accurate and powerful in respect of functionality and meteorological data.		

Tab. 8.5: Comparison of meteorological database in common PV design software.

Meteorological Database	Meteonorm	NASA
Feature	Meteonorm is a commercial software with data from Global Energy Balance Archive, WMO/OMM, Swiss Meteorological Agency, and other authorities and institutions. It contains solar radiation data over 7,750 meteorological stations worldwide. Most of the 98 meteorological stations in China are archived in this database. In addition, the software also provides other multi-year average monthly irradiations obtained by interpolation methods at any other location without meteorological radiation observation data.	NASA database is the first choice of many users because it is free and convenient. Users can acquire weather and radiation data anywhere with this software. It is obtained by NASA through calculation of satellite observation data. The resolution of NASA data is 3–110 km. The NASA ground radiation database first obtains the radiation data at top of atmospheric through satellites or other means with high accuracy. Then through data such as cloud distribution maps, ozone layer distribution maps, and suspended particle distributions, the total surface radiation level data are obtained through complex modeling and calculations. The accuracy of this step is constrained by many factors.
Comparison	MN6.1 corresponds to radiation data from 1981 to 2000 time period, while MN7.1 corresponds to radiation data from 1991 to 2010 time period. MN7.1 data is generally smaller than MN6.1, and this difference is even more pronounced in large cities in China.	The NASA data includes the monthly average radiation in specific region from 1983 to 2005 time period. In most cases, NASA data is higher than Meteonorm data, with a maximum of more than 10%.
Summary	From the perspective of time, the Meteonorm data is closer to the actual situation in China than NASA data. Compared with the data of China's multi-year meteorological stations, NASA data is higher, and Meteonorm data is closer to China's actual statistical data. At present, there are 98 national ground radiation observation stations, 17 of which are primary stations (include total radiation, direct radiation, scattered radiation, reflected radiation, and net radiation), 33 secondary stations (total radiation and net radiation), and 48 tertiary stations (total radiation). However, the observation data of 17 direct radiation stations can only reflect the characteristics of time changes in their local areas, and cannot provide an overall distribution of solar radiation in the country, which is not able to meet needed details in engineering applications. Based on current development trends, the amount of solar radiation on the ground in China's central and eastern regions, industrial cities and large cities has decreased significantly. Therefore, no matter which source of solar radiation data is used, the amount of radiation data provided by the database needs to be reviewed and adjusted to ensure the accuracy of PV power generation estimation. In summary, when choosing a database, it is recommended to consider using actual measured data preferentially, followed by Meteonorm database or multiple data sources.	

Data source: Training documents of Share Solar Power Ltd.

In conclusion:
(1) The PVSystem or PVSOL software is more suitable for photovoltaic system design in China. The SketchUp software is useful for 3D design simulation.
(2) It is recommended to consider using actual measured data in project location preferentially, followed by Meteonorm database or multiple data sources when choosing a database.

8.2.2 Grid-connected PV power station and grid-connection

8.2.2.1 Requirement of grid connection for PV power station
The connection between the grid-connected PV system and the power grid is an important link. The design of large- and medium-sized grid-connected PV power stations in China should comply with following standards:
- GB/T 50865 Code for Design of PV Power Generation Connecting to Distribution network
- GB/T 50866 Design Code for PV Power Station Connecting to Power System
- GB/T 29319 Technical Requirements for Connecting PV Power System to Distribution Network
- GB 19964 Technical Requirements for Connecting PV Power System to Power System
- GB/T19939 Technical Requirements for Grid Connection of PV system;
- NB/T 32006 Technical Code for Power Quality Testing of PV Power Station, and so on.

The quality of electric energy transported from the grid-connect PV power station to local AC load and grid should follow requirements in respect of harmonics, voltage deviation, voltage unbalance, DC component, and voltage fluctuation and flicker.

8.2.2.1.1 Harmonic and waveform distortion
After the PV power station is connected to the power grid, the harmonic voltage of the PCC shall meet the requirements of GB/T 14549 Power Quality – Harmonics in public supply network, as shown in Tab. 8.6.

8.2.2.1.2 Voltage deviation
After the PV power station is connected to the power grid, the voltage deviation at the common connection point shall meet the requirements of GB/T 12325 Power Quality – Deviation of Supply Voltage, which stipulates that the sum of the absolute values of the positive and negative deviations of the voltage at the common connection point for PV power stations with capacity of 35 kV and above shall not exceed

Tab. 8.6: Limit of public power grid harmonic and voltage.

Grid nominal voltage (kV)	Total power grid distortion ratio (%)	Harmonic voltage ratio (%)	
		Odd order	Even order
0.38	5.0	4.0	2.0
6	4	3.2	1.6
10			
35	3	2.1	1.2
66			
110	2	1.6	0.8

10% of the nominal value, and for the three-phase common connection point of PV power stations with capacity of 20 kV and below, the limit of voltage deviation is ± 7% of nominal voltage.

8.2.2.1.3 Voltage fluctuation and flicker

After the PV power station is connected to the power grid, the voltage fluctuation and flicker at the common connection point shall meet the requirements of power quality–voltage fluctuation and flicker GB/T 12326. The voltage fluctuation limit induced by PV power station alone at the common connection point relates to the frequency of fluctuation and the voltage level, as shown in Tab. 8.7.

Tab. 8.7: Limit of voltage fluctuation.

Voltage fluctuation frequency: r (times/h)	Voltage fluctuation: d (%)	
	LV, MV	HV
$r \leq 1$	4	3
$1 < r \leq 10$	3	2.5
$10 < r \leq 100$	2	1.5
$100 < r \leq 1,000$	1.25	1

Note: Low voltage (LV), $V_N \leq 1$ kV; medium voltage (MV), 1 kV $< V_N \leq 35$ kV; high voltage (HV) 35 kV $< V_N \leq 220$ kV

The voltage flicker value induced by the PV power station alone at the common connection point shall be grouped and treated, respectively, based on the ratio of the installed capacity of the PV power station to the power supply capacity and the system voltage, and according to the provisions of GB/T 12326 Power Quality Voltage Fluctuation and Flicker. The limits are listed in Tab. 8.8.

Tab. 8.8: Flicker limit values at various voltage levels.

System voltage level	LV	MV	HV
Short-term flicker value: P_{st}	1.0	0.9（1.0）	0.8
Long-term flicker value: P_{lt}	0.8	0.7（0.8）	0.8

Note: (i) The measurement period for P_{st} and P_{lt} are 10 min and 2 h, respectively;
(ii) The values in parentheses of MV column can only be applied to the conditions where all users of the PPC connection are at the same voltage level.

8.2.2.1.4 Degree of voltage unbalance

After the PV power station is connected to the power grid, the degree of voltage unbalance for the three-phase voltage at the common connection point shall not exceed the limit specified in GB/T 15543 Power Quality – Three-Phase Voltage Unbalance. The degree of negative sequence voltage unbalance for the public connection point shall not be larger than 2%, and for short time, it shall not be larger than 4% in. Among it, PV-induced negative sequence voltage unbalance shall not exceed 1.3%, and for short time, it shall not exceed 2.6%.

8.2.2.1.5 DC component

When the PV power station is connected to the grid and operating, the DC current component injecting into the grid shall not exceed 0.5% of its AC-rated value. For PV power stations that are directly connected to the power grid without transformers, the DC component can be relaxed to 1% due to special factors such as inverter efficiency.

8.2.2.1.6 Power factor

The power factor of large- and medium-sized PV power stations shall be continuously adjustable in the range of 0.98 (leading) to 0.98 (lagging). In its reactive output range, large- and medium-sized PV power plants shall have the ability to adjust reactive power output based on grid-connected voltage levels, and to participate in grid voltage regulation. For the small PV power stations, when the output active power is greater than 50% of its rated power, the power factor shall not be less than 0.98 (leading or lagging); when the output active power is between 20% and 50%, the power factor should equal to or greater than 0.95 (leading or lagging).

8.2.2.2 Grid-connected type of PV system

8.2.2.2.1 Single-branch grid-connected type

For the grid-connected PV system with low power, PV modules can be directly connected to a single inverter after being connected in series or in parallel with each other, and the output of the inverter is connected to the power grid after passing through a watt-hour meter. At the same time, the operating parameters can be recorded and stored by connecting RS-485/232 communication interface to personal computers, as shown in Fig. 8.3.

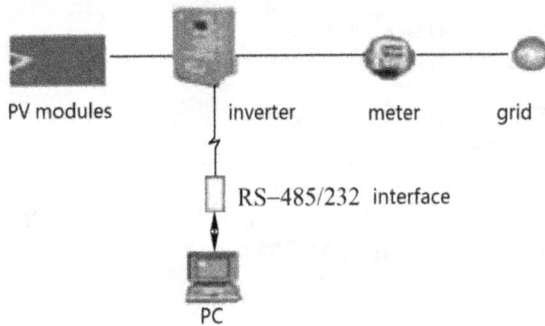

PV modules   inverter   meter   grid

RS–485/232 interface

PC

Fig. 8.3: Single-branch grid-connected PV system.

This grid-connected type is particularly suitable for small household PV systems with power between 1 kW and 5 kW, such as the small PV system installed on the roof, which can be tested according to GB/T 19064 Solar Home System Specifications and Test Procedures.

8.2.2.2.2 Multibranch grid-connected type

This type is suitable for applications where the power of the system is large, and the working conditions of the entire PV system are not the same. For the PV systems with shadow on sub-arrays, or with different tilt angle or azimuth angle, or there are a variety of models, and different voltage PV sub-arrays work at the same time, the recommended method is to equip an inverter for each sub-array (or multi-MPPT inverter), then connect the output into grid after a rate meter, as shown in Fig. 8.4.

The grid-connected inverters that are equipped can have different specifications. The operating parameters of each inverter, power generation amount, and fault records can be obtained by CAN bus, or connection to a PC through RS-485/232 communication interface. This type of grid connection is widely used, especially in the BIPV system. In order to meet the requirements of building structures, the operating conditions of various PV sub-arrays are often different. Therefore, this connection type is usually applied with string inverters.

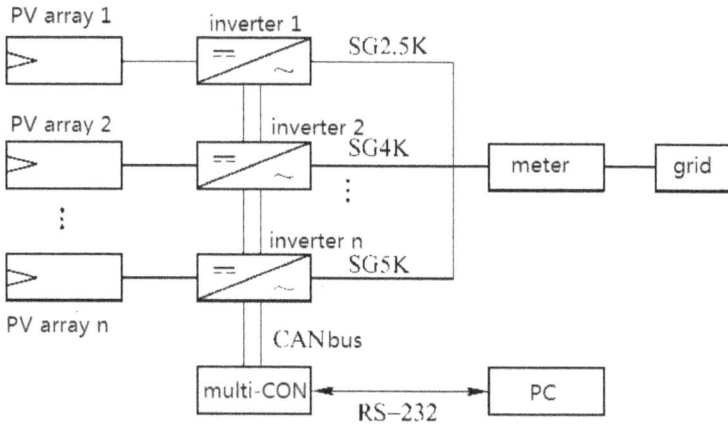

Fig. 8.4: Multibranch grid-connected PV system.

8.2.2.2.3 Parallel-connected grid-connected type

This type is suitable for larger-power grid-connected PV systems. Each sub-array is required to have modules with same power and voltage and are connected in series or parallel, and the installation angle of the PV sub-array is also the same. In this way, multiple inverters can be connected in parallel. When the solar irradiance in the morning is not strong, the data collector first randomly selects one inverter to be put into operation. When the irradiance on the array surface gradually increasing, another inverter is put into operation when the first inverter reaches its full load. At the same time, a controller divides the load equally to each inverter. As the solar irradiance continuously increase, other inverters are put into operation in turn. During the sunset, the controller turns off the inverter one by one, the input and output of the inverter are completely distributed by the controller according to the total power of the PV array, which can minimize the loss of inverters due to low load. Because inverters work in turn, it is not put into operation when it is not necessary, thus greatly prolonging the service life of the inverters. The parallel-connected grid-connected configuration is shown in Fig. 8.5.

PV power stations installed in desert or open spaces can use this connection method.

8.2.2.3 Grid connection method of PV system

8.2.2.3.1 Small-scale grid-connected PV system

For example, a residential roof PV system usually consists of PV arrays connected to a combiner box, and then a DC lightning protection switch, a grid-connected inverter, and an AC lightning protection switch, and finally is directly incorporated into the 220 V/380 V power grid. The data collection and recording device can be

PV array

inverter

meter — grid

CAN bus

RS–232

multi-CON ←—→ PC

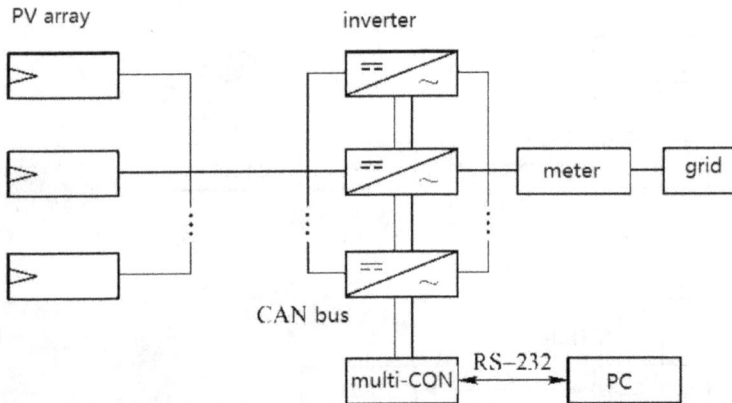

Fig. 8.5: Parallel-connected grid-connected PV system.

installed when needed. There are the following two ways of connecting grid-connected PV systems to the grid.

8.2.2.3.1.1 "Net metering" method

The output of the PV system is usually connected to the load side of residential meter, as shown in Fig. 8.6. The power generated by the PV system first meets household load power demand. The extra energy will be input into the power grid. In the rainy days and at night, the power supply to the household load will be provided by the power grid. In this case, only one bi-directional meter is needed. When the user uses the power from power grid, the power meter is rotating forward; when the PV system supplies power to the power grid, the power meter is rotating backward. In this way, when the user pays the electricity fee according to the number displayed in the power meter, the cost of the electricity generated by the PV system is already deducted.

There is another type under this method, which is called "connected but not feed-in," or "irreversible connection." PV system can generate electricity for household load, and when they are not enough, they can be supplemented by the power grid. However, if the power generated by the PV system is greater than the household demand, the extra energy cannot be allowed to feed into the grid. At this time, the PV system should be equipped with reverse power protection equipment. When it is detected that the reverse flow exceeds 5% of the rated output, the reverse power protection should disconnect the PV system from the power grid within 0.5 to 2 seconds. The irreversible grid connection does not require complex grid connection function.

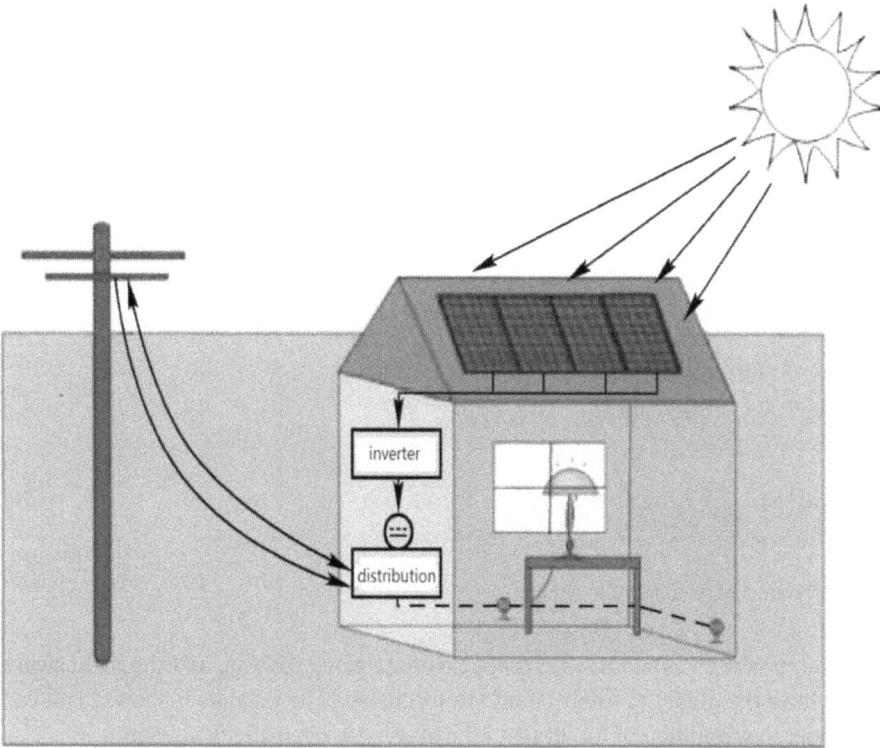

Fig. 8.6: Diagram of net metering method.

8.2.2.3.1.2 Feed-in tariff method

The output of PV system is usually connected to the grid side of residential electricity meter. The power generated by the PV system is totally feed into the power grid, and the household load power is supplied by the power grid. Therefore, two meters are needed to be equipped for "buy" and "sell" purposes. This method is shown in Fig. 8.7.

To promote the "solar roof" plan, many governments adopted a specific support policy to encourage private users to install residential PV systems. In this way, the green power generated by the user's rooftop PV system is purchased by power grid companies at a higher price, while the electricity used by the user is provided by the power grid and paid at a normal price.

8.2.2.3.2 Large- and medium-sized PV power stations

Same with conventional power stations, the large- and medium-sized PV power stations input all the generated power into the power grid. However, because of the large number of PV modules, it needs to be divided into many sub-arrays and equipped with many combiner boxes. Sometimes it also needs multiple DC distribution cabinets. PV

Fig. 8.7: Diagram of feed-in tariff method.

array output must contain a switch capable of arc extinguishing, and the main switch should have the ability to safely cutoff the 1.25 times of its maximum short circuit current under the condition of 1.25 times of the maximum open circuit voltage.

Power generated by the PV array is connected to the inverter through the DC combination cabinet, then the low-voltage AC output from the inverter is passed through the AC combination cabinet and finally incorporated into the high-voltage power grid through a step-up transformer. The step-up transformer should be connected with proper method to isolate the DC components and harmonic components generated by the inverter system, and a disconnect device with contact points shall be installed at the connection between PV power station and public power grid. The AC side of the PV system should also be equipped with grounding detection, overvoltage, overcurrent protection, monitoring instruments, and measurement meters. The lightning protection devices shall also be equipped at both the AC and DC terminals. In addition, main control and monitoring systems should also be installed. Their functions include digital signal sensing and acquisition, as well as necessary data processing, recording, transmission, and display. The diagram is shown in Fig. 8.8.

Currently, the ground-based PV power stations and distributed PV power stations with centralized access to the grid in China are connected to the power grid with "feed-in tariff" method, and government provides subsidies for PV power through benchmarking prices (including bid-to-price below the benchmark price).

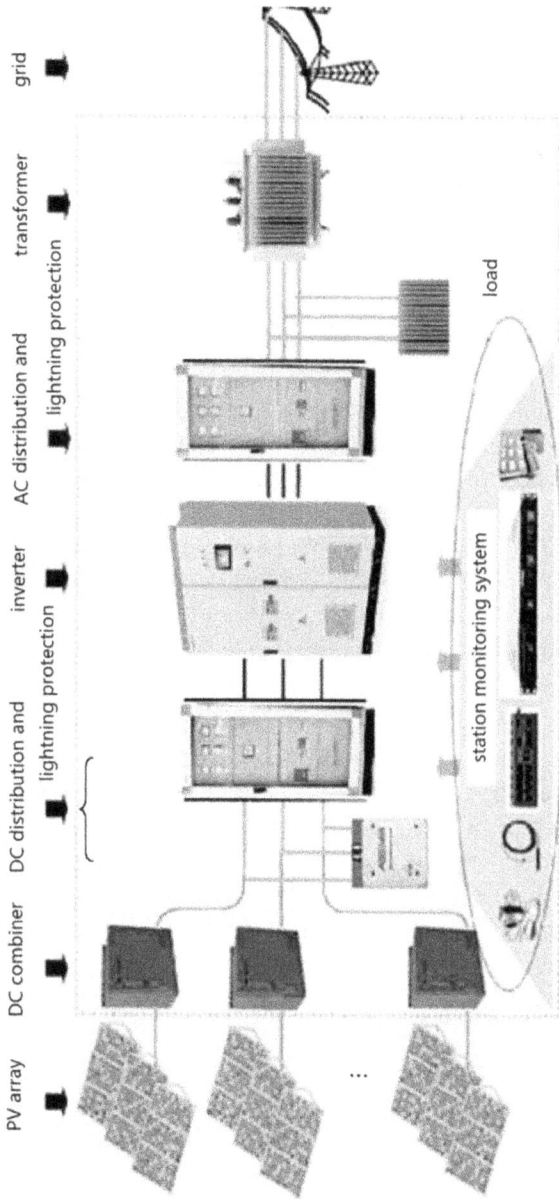

Fig. 8.8: Diagram of large- and medium-sized PV systems.

8.2.3 Basic flowchart of grid-connected PV system design

8.2.3.1 Collecting primary data

It requires to collect information about the installation location of PV system, local meteorological and geographical conditions, power grid status, and power consumption (reduction), and so on and to measure the capacity scale of PV system, and so on.

8.2.3.2 Site selection and investigation

The sites for ground-based PV power stations should be located in areas with flat terrain or low gradients from the north to the south, and avoid areas where air is often heavily polluted, and geology of dangerous rocks, debris flows, karst development, landslide sites, fault zones, key-protected cultural site, and open-pit mines or underground shallow mining areas with mining value. In the case of cultural relics and mineral buried under the ground floor of the site, in addition to get the permission of the cultural relics and mineral resources departments, the safety of the sites after excavation of the cultural relics and mineral deposits should be evaluated as well. Site selection should make full use of non-cultivable land and bad land, and should not damage the existing water system. It is also required to protect local vegetation, reduce earthwork excavation, save land, reduce house demolition, and population migration during site selection process. No PV power station can be built in flood control and flood discharge areas. The site selection should consider the access corridor of the power system when the power station reaches the planned capacity.

For the PV power stations located in the mountainous area, flood control design is required as well as the measurement to prevent mountain floods and flood discharges. These facilities should be designed for flash floods with a frequency of 2%.

The site investigation should obtain information of the terrain, topography of the installation site, as well as the orientation, area, and specific dimensions of the installation site of the PV system. Observation in the site should include whether there are obstacles such as tall buildings or trees to block the sunlight, and roughly determine whether it needs to be divided into several sub-array installations. In particular, the BIPV system requires a detailed understanding of the specific conditions of the site in order to determine the location of each sub-array and its influence on the orientation and inclination of the installation. It is also necessary to plan and arrange the specific settings of the auxiliary building, as well as the location of access to the grid, and so on.

8.2.3.3 General layout arrangement

With regard to the general arrangement of station sites for PV power stations, the principle of land use saving should be implemented. Through the system optimization, the area of production sites, living quarters, and construction land in the PV area of the whole station should be controlled. The scope of land use should be determined according to the planned capacity according to the needs of construction, and it should be phased out, requisitioned, and leased in batches. The general design of the station area should include the following contents: (1) PV array; (2) booster station (or switch station); (3) collector circuit within the station; (4) local inverter boost station; (5) station roads; (6) other protective function facilities (flood protection, lightning protection, fire prevention); and (7) support facilities for maintenance people.

The installation of PV systems on buildings should not reduce the sunshine standard of adjacent buildings. The installation of PV systems to existing buildings requires the safety check of buildings and electrical structures, and should meet the safety requirements of building structures and electrical installations.

8.2.3.4 Determine equipment configuration and model

Key points of selection for PV modules: (1) based on factors such as solar radiation, climate characteristics, site area, and others, and determined by technical and economic comparison; (2) For areas with high solar radiation and high direct light component, crystalline silicon PV modules or condenser PV modules are preferred; (3) Thin-film PV modules are recommended in areas with low solar radiation, high scattering light components, and high ambient temperatures; (4) For PV systems combined with buildings, with proper economic and technology situation, PV modules that are coordinated with the building structure should be selected. PV modules of building materials type should meet the technical requirements of the corresponding building materials or components.

The combiner box should be verified according to environmental conditions such as ambient temperature, relative humidity, altitude, pollution level, and seismic intensity. The combiner box should have the following protection functions: (1) lightning protection device (surge protector) should be installed; (2) the input circuit of the combiner box should have anti-backflow and overcurrent protection; for a multi-stage converging PV system, if the front stage has anti-backflow protection, the protection of against reverse flow at latter stage is not necessary; (3) the output circuit of the combine box should have function of isolation protection; and (4) monitoring devices are recommended.

The functions and models required for auxiliary equipment such as AC/DC power distribution cabinets, lightning protection switches, step-up transformers, and data acquisition systems that need to be configured in PV systems should be determined according to the relevant technical specifications.

8.2.3.5 Select the appropriate grid inverter

The following items are needed to be considered:
- On-grid or off-grid;
- Rated power and maximum current;
- Inverter conversion efficiency;

- On-site environmental assessment;
- Size and quality;
- Protection and security features;
- Warranty and reliability;
- Cost and availability;
- Additional functions (monitoring, charger, control system, maximum power point tracking, etc.);
- Nominal DC input and AC output voltage.

For grid-connected PV systems, special grid-connected inverters must be equipped, and their output waveforms, frequencies, and voltages are all strictly required. In addition, it must have a series of functions such as detection, grid connection, alarm, and automatic control and measurement. In particular, the function of preventing islanding is necessary to ensure the safety of PV systems and power grids.

The inverter performance for grid-connected PV systems are required to contain the continuously adjustable ability for active power and reactive power. The inverters for large- and medium-sized PV power plants should also have low-voltage (or zero voltage) traversal function. Inverters used in areas with humid tropics, severe industrial pollution, and coastal areas should consider the effects of moisture, pollution, and salt spray. Plateau-type products or derating method should be used for inverters used at altitudes above 2,000 m in plateau areas.

The configuration capacity of the inverter in the PV system should match with the installation capacity of the PV array, and the maximum allowable DC input power of the inverter should not be less than the actual maximum DC output power of the corresponding PV array. At the same time, the maximum power operating voltage range of the PV module string should be within the maximum power tracking voltage range of the inverter.

For large- and medium-sized PV power plants of a certain capacity, how to configure the inverter should be carefully considered comprehensively. Because inverters have different specifications and models, such as 1 MW PV power plants, whether to choose a 1 MW inverter with a relatively high power, or several small string inverters, such as two 500 kW or 20 sets of 50 kW, should be considered from different aspects. High-power inverters have high efficiency, relatively low cost and convenient maintenance. However, an inverter with too much power will have a greater impact on the power quality of the grid point when it is put into or out of service, and if it fails, it will cause downtime and the consequences will be more serious. The reliability and quality of product also need to be considered. In order to adapt to the development needs of large- and medium-sized PV power plants, the power of the grid-connected inverters provided by the production plants is getting higher and higher, and the device selection depends on whether the product has long-term use of mature experience, such as quality assurance.

In summary, the power of inverters and number of needed inverters should be considered with factors mentioned above as well as the cost in order to determine a safe, reliable, and economical solution. Finally, the specifications and models of grid-connected inverters can be determined.

8.2.3.6 Determine the grid-connection method of PV system

For PV systems such as large- and medium-sized terrestrial PV power stations, designs of multi-level convergence, decentralized inversion, and centralized grid-connected methods are generally used. After decentralized inversion, the voltage is generally boosted locally. Then the number of loops and the voltage level of the collector circuit should be determined after a technical and economic comparison.

The connection method should be decided according to the PV system capacity class and the relevant regulations, determine the grid connection mode of the PV system, specify the nodes connected to the grid, and implement the location and connection method of the proposed substation.

8.2.3.7 Determine the number of PV modules connected in series and parallel

The number of series connected PV modules in PV array can be calculated according to the following formula:

$$N \leq \frac{V_{dc\,max}}{V_{oc}[1 + (t - 25)K_V]} \tag{8.6}$$

where N is the number of PV modules connected in series (take an integer); V_{dcmax} is the maximum DC input voltage allowed by the inverter; V_{oc} is the open circuit voltage of PV modules; t is the limiting temperature under the PV module working conditions; and K_v is the open circuit voltage temperature coefficient of PV modules.

In reality, the number of PV modules in series can be calculated as the integer of ratio that the MPPT voltage range of the inverter to take the median value divided by the optimal output voltage of the PV module used. For example, with a 100 kW inverter, the MPPT input voltage range is 450 to 820 V with an intermediate value of 635 V. If the PV module power is 180 W, the optimal output voltage is 35 V. An array of 18 modules used in series has the string output voltage of 630 V and the power of each module string is 3,240 W. Then according to the capacity of the submatrix, determine the number of parallel PV modules.

8.2.3.8 Calculate the optimal PV array tilt angle

The inclination corresponding to the maximum solar radiation (optimal tilt angle) received on the tilted surface throughout the year can be calculated based on the data of the solar radiation on the long-term level in the local area. At the same time, the solar radiation on the surface of the tilted PV array can be obtained for the whole year and each month.

8.2.3.9 Conduct overall design of the project site to determine the layout of PV arrays

The installation plan of the PV modules, including the connection cable direction and the location of the combiner box, implement lightning protection and grounding measures, should be determined according to the size of the site, the size, and the tilt angle of the PV modules. In general, the distance between the lowest point of the PV module and the ground should not be less than 300 mm.

8.2.3.10 Analyze the influence of distributed access capacity on distribution network

When a distributed PV system is connected to a distribution network, if there is more than one PV system connected to the same public connection point, the impact on the power grid must be analyzed overall. When the total capacity of the PV system exceeds 25% of the maximum load in the power supply area of the upper transformer, special analysis of reactive power compensation and power quality is required.

8.2.3.11 Assess costs and benefits

Estimating the power generation of PV power plants and assessing their cost of generation, economic, and social benefits [6].

8.3 The design of off-grid PV system

The optimal design of the off-grid PV system is to make use of the configuration of the PV system, so as to ensure the long-term reliable operation of the PV system and fully meet the power demand of the load. At the same time, it can minimize the capacity of the installed PV arrays and batteries, save investment cost, and obtain the best economic benefits [7]. At this stage, the price of PV modules is still relatively high. The design of PV system should be based on the requirements of the load, the local meteorological and geographical conditions and comprehensive consideration of various factors, in accordance with the principle of energy balance. However, due to the many factors involved in the operation of off-grid PV systems, design calculations are difficult. Many documented PV power system design methods are not perfect or very complicated. Some even do not take into account the specific characteristics of PV power generation. Those non-reasonable methods cause inefficient PV products (such as some solar street lamps) or engineering, which cannot run stably for a long time, and some even cannot work properly.

Therefore, adopting scientific design and calculation methods to optimize the design are the key to build a reasonable and complete PV system.

8.3.1 Requirements of off-grid PV system design

The most important part of the construction of off-grid PV system is the capacity design, which includes determining the capacity of PV arrays and batteries, and determining the inclination of PV arrays. Under the premise of fully satisfying the user's load demand, the capacity of the PV array and the battery should be reduced as much as possible to achieve the best combination of reliability and economy, and the tendency to blindly pursue low cost or high reliability should be avoided. At present, it is particularly necessary to correct the practice of one-sided emphasis on low-cost investment, arbitrarily reducing system capacity, or selecting low-performance, low-cost products for market competition.

The design of PV systems and products should be optimized according to the requirements of the load and the meteorological and geographical conditions (such as latitude, solar irradiation and the longest continuous rainy days) of the place of use [8]. The two types of data should be fully mastered before design.

PV power system design is based on monthly energy balance [9].

8.3.2 Technical conditions

8.3.2.1 Load properties

The first step is to determine whether the load is DC or AC, whether it is an impact load or a non-impact load and whether it is an important load or a general load. DC load can be directly supplied by battery, while AC load must be equipped with inverter.

Different types of AC loads have different characteristics:
- Resistive loads, such as incandescent bulbs, electronic energy-saving lamps, and electric heater have a current and voltage in phase with no impulse current.
- Inductive loads, such as motors, refrigerators, and pumps whose voltage leads the current have an impulse current.
- Power electronic loads, such as fluorescent lamps with electronic ballasts, televisions, and computers have impulse current.

Inductive loads have surge current at start-up, their size varies according to the load, and their duration is not the same. For example, the surge current of motor is 5 to 8 times of the rated current and the time is 50 to 150 ms; the surge current of refrigerator is 5 to 10 times of the rated current and the time is 100 to 200 ms; for the color television, the surge current of the degaussing coil and are 2 to 5 of rated current times, time is 20–100 ms.

The actual load size and usage is changeable and can be distinguished by all-day use time. It can be roughly grouped into three kinds of load: daytime, evening, and whole daily time. For loads that are only used during the daytime, most of them can be directly supplied by the PV system, which reduces the losses caused by charging and discharging of the battery, so the capacity of the PV system equipped can be appropriately reduced. For loads used only at night, the capacity of the PV system will increase accordingly. The required PV system capacity for whole daily time load is between the range of first two types. In addition, the load can also be divided into balanced load, seasonal load, and random load based on the time of use throughout the year. A balanced load means a load that consumes the same amount of electricity every day. For simplification, the load for which the average monthly power consumption does not vary by more than 10% can also be regarded as a balanced load with the same average power consumption.

8.3.2.2 The concept of several kinds of sunshine
In different situations, different concept of sunshine will be used.

8.3.2.2.1 Possible sunshine duration
The possible sunshine duration refers to the total number of hours from sunrise to sunset every day without any obscuration. The calculation formula is:

$$Possible\ sunshine\ duration = \frac{2\omega}{15} = \frac{2}{15}\arccos(-\tan\varphi\tan\delta)$$

where ω is local sunrise, sunset angle, φ is local latitude, and δ is declination angle.

It is the possible longest sunshine time in the area. The sunshine hours at different locations are related to the local latitude and date. It can be calculated by formula (2–8) without measurement. For example, the time available for the winter solstice in Shanghai is 9.98 h; the possible sunshine duration on September 22 is 11.91 h.

8.3.2.2.2 Sunshine duration
Sunshine duration means the time period when the Sun reaches a certain irradiance (usually 120 W/m^2) at a certain location. Therefore, the sunshine duration measured by the meteorological observatory are less than the possible sunshine duration, and the sunshine duration measured at different locations in different years are different. For example, the average sunshine duration measured in 1971–1980 in Shanghai was 1963.4 h, with an average of 5.38 h per day. During the same period, the annual sunshine duration in the Lasa area was 3010.5 h, and averaged 8.24 h per day, while the annual sunshine duration in Chongqing was 1117.6 h, only 3.06 h per day.

8.3.2.2.3 Percentage of sunshine

The percentage of sunshine provided by the meteorological observatory refers to the ratio of sunshine duration to the possible sunshine duration:

Percentage of sunshine = sunshine duration / possible sunshine duration

8.3.2.2.4 Twilight duration

Before sunrise and after sunset, when the Sun is 0° to 6° below the horizon, light is scattered through the atmosphere to earth surface to produce a certain light intensity. This kind of light is called civil dawn twilight in the morning and civil dusk twilight in the evening. The average dawn and dusk time increases with latitude, especially in summer.

8.3.2.2.5 Peak sun hour

The peak sun hour is the number of hours under the standard test conditions STC (1,000 W/m², 25 °C, AM1.5) converted from the local solar irradiance. As shown in Fig. 8.9, the curve in the figure is the relationship between the actual local solar irradiance and the time. The solar irradiance changes all the time. The area under the curve is the solar irradiation of the day. Since the output power of the solar cell is measured under the standard test conditions, the area under the curve should be converted into a rectangle with a height of 1,000 W/m², then the width is a peak sun hour. Obviously, the peak sun hour should be used when calculating the output of PV arrays.

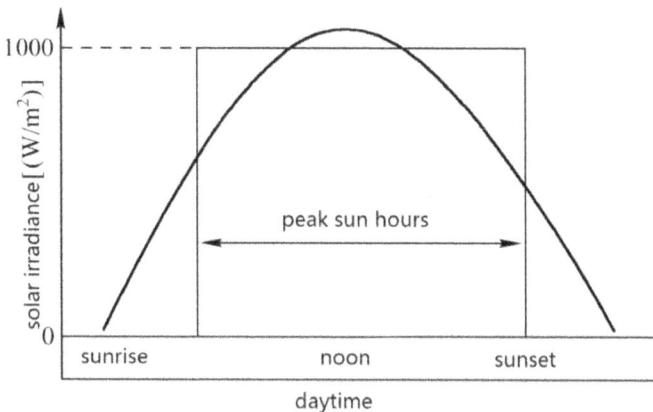

Fig. 8.9: Diagram of peak sun hours.

For example, the possible sunshine duration for winter solstice in Shanghai is 9.98 h, but the irradiance of the sun is 1,000 W/m² during this 9.98 h. It changes at any time, such as the cumulative solar radiation measured on this day. The amount is 2,300 Wh/m², then the peak sun duration on this day is 2.3 h.

From 1981 to 2000, the average annual peak sun duration in Lasa region was 5.33 h; during the same period, the average peak sun hours in the Shanghai region was 3.46 h; and the average peak sun hours in the Chongqing region was 2.44 h. These regions can roughly represent the high solar radiation, medium, and low levels of peak sun duration in different regions of China.

8.3.2.3 Temperature effect

It is well-known that when the temperature of the PV module rises, its open circuit voltage will decrease, and the output power will decrease too.

Some design methods consider the influence of the temperature coefficient of PV modules when determining PV array capacity, thereby increasing the capacity. As mentioned in the literature, since the output power of PV modules drops when the temperature rises, and the system is required to ensure normal operation even at the highest temperature, the output power of PV array at the standard test temperature (25 °C) is:

$$P = \frac{I_m V}{1 - \alpha(t_{max} - 25°)}$$

where P is the output power of PV array; I_m is the array output current; V is the array voltage; t_{max} is the highest temperature of the module, α is module power temperature coefficient.

Above equation is equivalent to treating the whole PV array as working at the highest temperature all year round, which is obviously a conservative method. In fact, the commonly used off-grid systems with standard component in series with 36 solar cells have a working voltage of around 17 V. When this PV array is used to charge a 12 V battery, in addition to meeting the requirements of the battery's float voltage, blocking diodes, and line voltage drop, the effect of lowering the voltage during the summer temperature increase is also considered, and usually the summer solar irradiance is higher and the array power generation is often excessive, which completely makes up for the reduced power due to temperature increase. Therefore, it is not necessary to additionally consider the influence of temperature when calculating the capacity of the PV array.

Under special circumstances, such as designing PV power generation systems for hot regions such as Africa, it is generally only necessary to increase the safety factor of the system. Only in rare circumstances, it is needed to consider increasing the number of series-connected PV cells in PV modules. However, when the temperature is low, the output capacity of batteries may be affected. In winter, thermal insulation should be properly considered when the working temperature is lower than 0 °C.

8.3.2.4 Battery autonomy days

The battery autonomy days refer to number of days that can be sustained by the power supply stored in the battery completely without the power supply of the PV array.

It is related to two factors: the degree of demand of the load on the power supply and the maximum number of rainy days for the place where the PV system is installed. In general, the maximum number of consecutive rainy days in the place where the PV system is installed can be used as a reference for maintaining the number of days used in the system design, but the requirements of the load on the power supply must also be considered comprehensively. For PV applications where the load is not critical for power supply, the number can be designed as 3 to 5 days. For those whose load is very critical to the power supply, a battery autonomy day of 7 – 12 days is needed. The so-called less demanding system usually means that the user can adjust the load demand slightly to adapt to the inconvenience caused by bad weather. A strict system refers to a system in which load is very important and will be affected if the power supply is broken, such as the systems used for military purpose, communication, navigation, or important health facilities such as hospitals, clinics, and so on. In addition, the installation site of the PV system must also be considered. If it is in a remote area, it is necessary to design a larger battery capacity because it takes a long time for the maintenance personnel to arrive at the site.

8.3.3 Selection of PV array tilt angle

8.3.3.1 PV array should be placed as tilt if possible

In order to receive more solar radiation energy on the surface of the PV array, according to the daily operation pattern, the surface of the array is preferably installed toward the equator (with an azimuth angle of 0°), that is, the Northern Hemisphere faces true south and the Southern Hemisphere faces true north. The modules also should be installed obliquely for the following reasons:
(1) Increase the amount of solar radiation received on the surface of the array throughout the year

In the Northern Hemisphere, the Sun is mainly operating in the southern half of the sky. If the surface of the PV is tilted to the south, it can obviously increase the amount of solar radiation received throughout the year.
(2) Change the distribution of solar radiation received on the array surface in each month

In the Northern Hemisphere, the Sun runs overhead during summer with large solar zenith angle, while during winter, it runs southwards with a small solar zenith angle. Therefore, if the PV array is tilted to the south, the amount of solar radiation received in the summer can be reduced, and the amount of solar radiation received in the winter can be increased. That is, the annual solar irradiation will be balanced. This is particularly important for off-grid PV systems, which are limited by battery capacity due to charging. During the summer, the solar radiation is large and after the battery is fully charged, the excess power from the PV array cannot be used. Therefore, it is hoped that the solar radiation received by the surface of the PV array in each month can be consistent as much as possible.

Any PV system should be installed with a tilt angle, except on the equator, or for the installation on vehicles (such as automobiles and ships) in which directions are constantly changing and should be installed horizontally.

8.3.3.2 Determine the optimal tilt angle

The optimum tilt angle for PV array should be determined according to different circumstances. For off-grid PV systems, the situation is relatively complex. Some of the earlier literature proposed that the angle corresponding to the maximum solar irradiation in the local design month (the month with the lowest solar radiation level on the horizontal plane, usually in December in the Northern Hemisphere) could be used as the array tilt angle, which is actually inappropriate, because it tends to decrease the amount of solar radiation received by the array in summer.

Some literature also suggested that the installation angle of the PV array is equal to the local latitude, or the local latitude plus 5° to 15°. There are other literatures suggesting that:

- when the latitude is 0° to 25°, the inclination angle is equal to the latitude;
- when the latitude is 26° to 40°, the tilt angle equals to the latitude plus 5° to 15°;
- when the latitude is 41° to 55°, the tilt angle equals latitude plus 10° to 15°;
- when latitude is greater than 55°, tilt angle equals latitude plus 15° to 20°, and so on.

These are not proper. In fact, even in two places with the same latitude, the solar radiation and its composition often have great difference. For example, the latitudes of Lasa and Chongqing in China are basically the same (only 0.05° difference), while the solar radiation on the horizontal plane of Lasa is more than double of that in Chongqing. The direct solar irradiation in Lasa accounted for 67.7% of the total irradiation, while direct irradiation in Chongqing accounted for only 33.8%. The distribution of solar radiation in different months on different slopes varies widely. It is obviously unreasonable to add the same degree as tilt angle. Similarly, it is also unreasonable that some literature ignores the actual conditions of the load and lists the specific degrees of array tilt angle for off-grid PV systems.

The first step to determine the best tilt angle of off-grid PV system is to distinguish between different types of load conditions.

Determination of optimal tilt angle for independent PV system with balanced load power supply must comprehensively consider the factors such as the balance and maximum of the solar radiation received on the array. Under the condition of satisfying the load power requirements, comparing the sizes of the PV arrays and the battery capacity required for various tilt angles. Finally, the tilt angle of the array corresponding to the required number of battery autonomy days and the minimum capacity of the configured PV array can be obtained through iterated calculation. It is found from the calculation that even if other conditions are the same, due to different tilt angles, the distribution of solar radiation on the array varies from month to month. For different battery autonomy days, the required cumulative system deficit is not the same, and its corresponding array optimal tilt angle is not necessarily the same.

For the seasonal load, typically the light-controlled solar lighting system, the load of such systems varies with the season, and its characteristics are determined by the intensity of natural light to determine the length of load working time. The load consumes a large amount of power during winter. Therefore, the design must focus on the winter, so that the irradiance obtained in the winter on the array is large. The corresponding optimal tilt angle should be larger than that of the power supply matrix with a balanced load.

In summary, the general rule of the tilt angle of the PV array is: for the same location, the tilt angle of the grid-connected PV system is the smallest, followed by the off-grid PV system that supplies the balanced load, for the off-grid PV system that supplies power for light control loads consumes large amounts of electricity in winter, and the best inclination angle of the array is also generally large.

The design steps are discussed below in accordance with different types of PV system [10].

8.3.4 Design of balanced load PV systems

8.3.4.1 Determine the load electricity consumption

List the power consumption, operating voltage, and average daily hours of use for various power loads, and record the information of auxiliary equipment of the system, such as power consumption of controllers, inverters, and so on. Select the battery operating voltage V, calculate the average daily power consumption of the load Q_L (Ah/d), and specify the number of battery autonomy days n (usually n take 3 to 7 days).

8.3.4.2 Calculate solar radiation on the PV array

There are many ways to calculate solar radiation on the PV array. One method of calculation is: according to local geographical and meteorological data, first set a tilt angle β, and calculate the average monthly solar monthly solar radiation H_t received on array surface according to the method of calculating the monthly average daily solar radiation published by Klein and Theilacker in Chapter 2. Then, calculate the total annual solar radiation $\overline{H}_t$. Where η1 is the efficiency of the input circuit from

$$T_t = \frac{H_t}{1000(W/m^2)} = H_t(h/d)$$

In this way, H_t is numerically equal to the average daily peak sun hours T_t, which is also replaced by H_t with unit of kWh/m²/d in the future.

8.3.4.3 Calculate monthly surplus or deficit of power generation

For a certain inclination, the minimum current output from the PV array is:

$$I_{min} = \frac{Q_L}{\overline{H}_t \eta_1 \eta_2} \tag{8.7}$$

where η_1 is the efficiency of the input circuit from the PV array to the battery, including dust shading loss on array surface, component mismatch loss, component attenuation loss, blocking diode and line loss, battery charging efficiency, and so on; and η_2 is the efficiency of the output circuit from the battery to the load, including the discharge efficiency of the battery, the efficiency of the controller and the inverter, and the line loss.

The idea of formula (8.7) is: in this situation, annual power output of the PV system is exactly equal to the annual power consumption of the load. However, the actual condition is that after the summer battery is full, some energy cannot be used, so the output current of the PV array cannot be lower than this value.

Similarly, the maximum current of the output required by the PV array can also be obtained from the minimum value $H_{t\text{-}min}$ of the average solar radiation within the 12 months.

$$I_{max} = \frac{Q_L}{H_{t\cdot min}\eta_1\eta_2} \tag{8.8}$$

The idea for the eq. (8.8) is: the modules are assumed to be working under the minimum solar irradiance throughout the year, so the PV system power generation will be greater than the load power consumption in any month. Since the battery is used as an energy storage device, it is allowed to charge and store energy when the PV array power generation amount is greater than the load power consumption in the summer, and can be used when the PV array power generation is insufficient in the winter. It does not need to have a surplus every month, so the calculated current is the largest output current of the PV array.

The actual working current of the PV array should be between I_{min} and I_{max}. The monthly power generation Q_g of the PV array can be calculated with a random current value between I_{min} and I_{max} and is

$$Q_g = NIH_t\eta_1\eta_2 \tag{8.9}$$

where N is the number of days of the month; H_t is the amount of solar radiation on the tilted plane of the month.

And the monthly power consumption is:

$$Q_c = NQ_L \tag{8.10}$$

Therefore, the monthly surplus or deficit of power generation is:

$$\Delta Q = Q_g - Q_c \tag{8.11}$$

ΔQ is the surplus amount of the month if $\Delta Q > 0$, which represents that the amount of power generated by the PV system is greater than the amount of power consumed during the month. In addition to satisfying the load, the power generated by the PV system has excess energy to charge the battery. If the battery is fully charged at this time, the excess power is usually wasted and becomes useless energy. ΔQ is the deficit amount of the month if $\Delta Q < 0$. It indicates that the electricity generated by the PV system is insufficient during the month, which requires the battery to provide part of the stored energy.

8.3.4.4 Determine the cumulative deficit amount $\sum |-\Delta Q_i|$

List the monthly surplus or deficit in 2 years, if there is only one continuous deficit period with $\Delta Q < 0$, the cumulative amount of deficit is the sum of the monthly deficit amount during this period. If there are two or more discontinuous deficit period with $\Delta Q < 0$, then the amount of surplus ΔQ_i should be deducted between consecutive two indebted periods and the accumulated deficit amount $\sum |-\Delta Q_i|$ is obtained.

8.3.4.5 Determine output current of the PV array

Substitute the cumulative deficit amount $\sum |-\Delta Q_i|$ into equation below:

$$n_1 = \frac{\sum |-\Delta Q_i|}{Q_L} \tag{8.12}$$

Compare the number of accumulated days of deficit "n_1" with the number of battery autonomy days "n" of the specified battery, if $n_1 > n$, it means that the current under consideration is too small, so that the deficit is too large. Therefore, the current I should be increased and recalculated and vice versa. The PV array output current I_m can be obtained when n_1 is similar with n.

8.3.4.6 Calculate the optimal tilt angle

The array output current I_m obtained above is the current that can satisfy battery autonomy days of n at a certain specified tilt angle, and may not be the optimal tilt angle. Then people should change the tilt angle, repeat the above calculations, and repeat the comparison to find the smallest array output current I_{min}. The corresponding tilt angle is the optimal tilt angle β_{opt}.

8.3.4.7 Obtain capacity of battery and PV array

The battery capacity "B" is:

$$B = \frac{\sum |-\Delta Q_i|}{\text{DOD} \cdot n_2} \qquad (8.13\text{a})$$

where DOD is the depth of discharge of the battery, usually is in the range between 0.3 and 0.8.

Then, based on eqs. (8.12) and (8.13a),

$$B = \frac{nQ_L}{\text{DOD} \cdot n_2} \qquad (8.13\text{b})$$

In fact, according to the known conditions, the required battery capacity can be obtained. Above complicated calculation process is mainly to determine the best operating current of the PV array and the array capacity as:

$$P = kI_m(V_b + V_{d)} \qquad (8.14)$$

where k is the safety factor, usually has value of 1.05 to 1.3, and can be determined by the importance of the load, the uncertainty of the parameters, the load in the day or night work, the impact of temperature and other necessary factors; V_b is the battery charging voltage; V_d is the voltage drop of blocking diodes and cables.

8.3.4.8 Decide the best configuration

A series of B–P combinations can be obtained with different value of battery autonomy days. Then, the best combination of battery and PV array capacity can be determined according to factors such as product model and unit price. Finally, the number of PV modules and batteries to be used should be checked to determine that the series voltage is in line with the original design requirements. Otherwise, the capacity of each PV module and battery should be re-selected. Therefore, the final capacity of PV array is often not an integer.

In summary, there can be various combinations of PV arrays and battery capacities that meet the requirements for load power consumption and the number of battery autonomy days. Therefore, it is necessary to find out the array output current that satisfies the above-mentioned requirements for different tilt angles, and to compare them repeatedly. The obtained tilt angle corresponding to the minimum output current of I_m is the best tilt angle. The capacity of the PV array can be determined from the minimum output current I_m value, and the battery capacity can be obtained from the battery autonomy days. A series of B–P combinations can be obtained by changing different battery autonomy days, from which the optimal configuration of battery and PV array capacity can be decided.

These calculations are quite complicated and require special computer software for calculation. Figure 8.10 shows the optimized design of off-grid PV system.

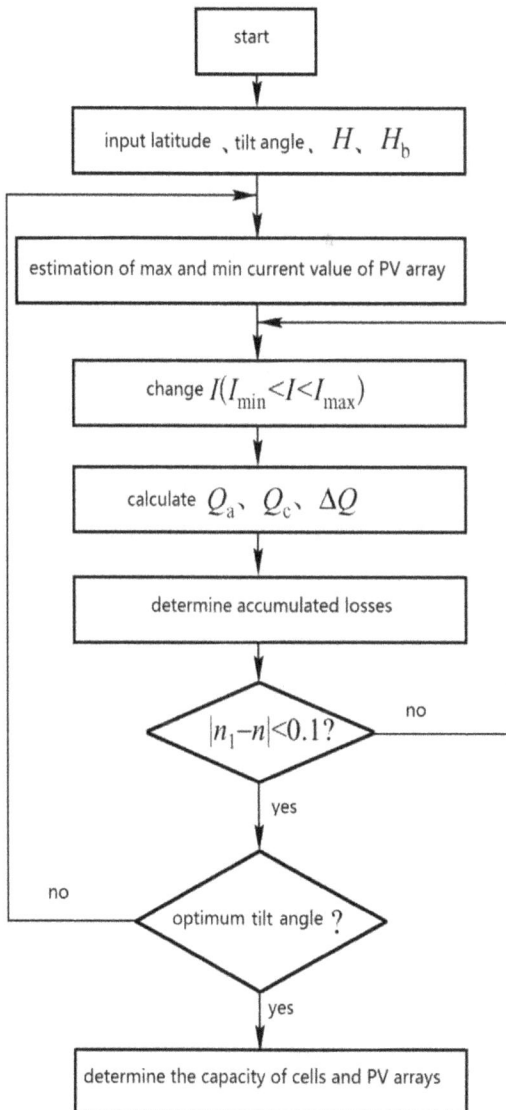

Fig. 8.10: Optimization design block diagram for off-grid PV system.

8.3.4.9 Case Analysis

Example 8.3 Design a set of solar street lights for the Shenyang area. The power of lamps is 30 W. These lamps work 6 h every day with the working voltage is 12 V, and the number of battery autonomy day is 5 days. What is the capacity of the PV array and battery and the tilt angle of the array?

Solution First calculate the daily power consumption of the load:

$$Q_L = \frac{30W \times 6h/d}{12V} = 15Ah/d$$

The latitude of Shenyang is 41.44°, and arbitrary angle of 60° is used to calculate the monthly average daily solar irradiation H_t exposure on the PV arrays in each month. Then the annual average daily solar irradiation can be obtained as $\bar{H}_t = 3.809$ kWh/m^2/d, and find out the minimum daily solar radiation in December as $H_{tmin} = 2.935$ kWh/m^2/d.

Set the parameters of $\eta_1 = \eta_2 = 0.9$ and substitute them into eqs. (8.7) and (8.8):

$$I_{min} = \frac{Q_L}{\bar{H}_t \eta_1 \eta_2} = \frac{15}{3.809 \times 0.9 \times 0.9} A = 4.86A$$

$$I_{max} = \frac{Q_L}{H_{t.\ min} \eta_1 \eta_2} = \frac{15}{2.935 \times 0.9 \times 0.9} A = 6.31A$$

Randomly choose the I between I_{min} and I_{max} as 5.2 A,

Calculate the power generation quantity Q_g for each month with eq. (8.9), and list the power consumption Q_c for each month, so as to obtain the power generation surplus and deficit for each month ΔQ. The specific values are shown in Tab. 8.9.

Tab. 8.9: Monthly energy balance with condition of $\beta = 60°$ and $I = 5.2$ A.

Month	H_t (kWh/m²/d)	Q_g (Ah)	Q_c (Ah)	ΔQ (Ah)
January	3.347	437.0	465	−28.0
February	4.162	490.8	420	70.8
March	4.436	579.3	465	114.3
April	4.209	531.1	450	81.1
May	4.105	536.0	465	71.0
June	3.812	481.7	450	31.7
July	3.489	455.6	465	−9.4
August	3.660	477.9	465	12.9
September	4.206	531.4	450	81.4
October	4.040	527.5	465	62.5
November	3.317	419.1	450	−30.9
December	2.935	383.2	465	−81.8
January of next year	3.347	437.0	465	−28.0

According to Tab. 8.9, there were deficits in July and November to December of the same year and January of next year, so there were two default periods, of which the deficit was −9.4006 Ah in July. However, there was a surplus of 12.916 Ah in August, which can make up all deficit. Therefore, it is not necessary to add the deficit in July as 9.4006 Ah. The accumulated deficit for the whole year is the sum of the deficits from November to January, that is,

$$\sum^{|-\Delta Q_i|} = |-30.971 - 81.808 - 28.016| \,(A \cdot h) = 140.69 \, A \cdot h$$

Substitute this number to eq. (8.12):

$$n_1 = \frac{\sum^{|-\Delta Q_i|}}{Q_L} = 9.38$$

It can be seen that the result is much larger than the required battery for 5 days, which indicates that the PV array current taken is too small, so the PV array current should be increased and the calculation should be performed again. The above steps are repeated continuously, and the final result is $I = 5.476$ A. The energy balance of each month is shown in Tab. 8.10.

Tab. 8.10: Monthly energy balance with condition of $\beta = 60°$ and $I = 5.476$ A.

Month	H_t (kWh/m^2/d)	Q_g (Ah)	Q_c (Ah)	ΔQ (Ah)
January	3.347	460.2	465	−4.9
February	4.162	516.8	420	96.8
March	4.436	610.0	465	145.0
April	4.209	559.3	450	109.3
May	4.105	564.4	465	99.4
June	3.812	507.3	450	57.3
July	3.489	479.8	465	14.8
August	3.660	503.3	465	38.3
September	4.206	559.6	450	109.6
October	4.040	555.5	465	90.5
November	3.317	441.4	450	−8.7
December	2.935	403.5	465	−61.5
January of next year	3.347	460.2	465	−4.9

As can be seen from Tab. 8.10, there are also deficits in November–December of the year and January of the following year. But the total deficit amount is $\sum |-\Delta Q_i| = 74.5$ Ah.

It can be found that n_1 is 5 days, which is basically consistent with the required $n = 5$ days. Therefore, the current can be determined as $I_m = 5.5$ A.

However, this is only the array current that satisfies the battery autonomy days when the tilt angle = 60°, and is not necessarily the minimum current of the PV array. Then, the tilt angle should be changed again, and the same current should be used to compare the accumulated deficit (or n_1) until the minimum current that basically corresponds to the number of days of autonomy $n = 5$ days is obtained. Then, this angle is the best tilt angle. Finally, $I_m = 5.47351$ A is obtained, and the corresponding tilt angle is 62°. The energy surplus and deficit of each month is shown in Tab. 8.11.

Tab. 8.11: Monthly energy balance with condition of $\beta = 62°$、$I = 5.474$ A.

Month	H_t (kWh/m²/d)	Q_g (Ah)	Q_c (Ah)	ΔQ (Ah)
January	3.348	460.2	465	−4.8
February	4.147	514.8	420	94.8
March	4.392	603.7	465	138.7
April	4.132	549.6	450	99.6
May	4.014	551.7	465	86.7
June	3.722	495.1	450	45.1
July	3.413	469.1	465	4.1
August	3.592	493.6	465	28.6
September	4.153	552.3	450	102.3
October	4.017	552.2	465	87.2
November	3.315	441.0	450	−9.1
December	2.939	403.9	465	−61.1
January of next year	3.348	460.2	465	−4.8

Finally, the working current I_m of the PV array is 5.474 A; the best tilt angle β_{opt} of the array is 62°. Substituting I_m and β_{opt} into (8.14) to obtain the PV array capacity P = 107.3 W.

Set DOD as 0.8 and substitute into (8.13b) and the obtained capacity B of battery is 104.2 Ah.

Actual configurable PV module capacity is 110 W, battery capacity is 105 Ah at 12 V.

8.3.4.10 Off-grid PV system capacity discussion

In some articles when calculating the capacity P of an off-grid PV array, the formula applied is

$$P = k \frac{Q_L V \times 365}{H \eta_1 \eta_2}$$

where P is the capacity of the PV array; Q_L is the daily electric energy consumption of the load; V is the floating charge voltage of the battery; H is the local average annual peak sun hours; k is the safety factor; η_1 is input circuit efficiency; η_2 is output circuit efficiency.

Actually this method is unreasonable due to the reasons below:

(1) V should be the working voltage of the array, in addition to the floating charge voltage of the storage battery, the voltage drop of the circuit to the storage battery (including the voltage drop of the blocking diode) needs to be added.

(2) H should be the annual total solar irradiation on PV array with unit of kWh/m²/ d, instead of being the annual mean peak sun hours of local area.

(3) The equation above indicates that the annual PV array power generation is equal to the load power consumption, which is not the case in reality. Because in the off-grid PV system, part of the power is usually wasted in the summer, and the power generated by the PV array cannot be fully utilized. For example, the $\sum \Delta Q_i$ is calculated as 617.24 Ah in Tab. 8.11 during the whole year and is wasted.

Other method of calculating the number of PV modules connected in series and in parallel in PV array is

Number of parallel PV modules = daily average load (Ah)/Coulomb efficiency × Module daily output (Ah) × Degradation factor

Number of Series PV modules = System Voltage (V)/ModuleVoltage (V)

Then multiply the two numbers to get the capacity of the PV array. This method only meets the requirement of energy balance in the whole year, but does not take into account the fact that excessive energy cannot be used after the storage battery is full in the off-grid PV system in the summer. Therefore, this configuration is too small to guarantee normal operation in winter.

In *Example 8.3.*, if the number of storage battery autonomy day is changed, the optimal tilt angles of the different PV arrays and the required PV arrays and storage battery capacity can be obtained, as shown in Tab. 8.12.

It can be seen that when the battery autonomy day increases, the battery capacity needs to be increased, and the capacity of the PV array can be decreased accordingly. However, as the number of battery autonomy day increases, the capacity of the matrix decreases less and the battery capacity increases faster. Therefore, the number of battery autonomy day should be determined according to the needs of

Tab. 8.12: The system configuration with different battery autonomy days.

Battery autonomy day n (d)	2	3	4	5	6	7	8
Optimal tilt angle β_{opt} (°)	64	64	62	62	62	62	62
PV array capacity P (W)	115.5	111.5	108.5	107.3	106.0	104.8	103.6
Battery capacity B (Ah)	41.7	62.6	83.3	104.2	125.0	145.8	166.7

the load and the condition of the local rainy days, and the total investment of the PV system.

In addition, due to the difference in the number of battery autonomy day, the cumulative amount of deficits is different, and the distribution of power generation varies from month to month, so the optimal tilt angle of the PV array is not necessarily the same.

It also need to be noticed that if the load is an inductive type, large inrush current will be generated during startup. If the equipped battery has a small capacity, when the voltage drops greatly, it may cause the load to fail to start properly and affect the work. Therefore, when working with inductive loads, the capacity of the batteries and inverters to be installed should be appropriately increased.

8.3.5 Design PV systems with seasonal load

For PV systems with seasonal load, the load power consumption varies with the season and cannot be treated as a balanced load. Such as the power supply for solar refrigerators, the summer power consumption is relatively large.

At present, most of the applications are light-controlled solar PV lighting systems. Light-controlled lighting system has a feature that the intensity of natural light determines the length of load working time. The daily working hours of this type are not the same, so the load power consumption is not the same, and it is the contrary to the rules of the sun's sunshine time. In the summer when sunshine time is long and solar irradiance is high, the needed time of lamp lighting is short, which is exactly opposite in the winter. Therefore, the operating conditions of such light-controlled lighting systems in PV power applications are the most demanding and must be designed with special care.

The first step is to evaluate the time needed for lighting. By formula (2–8), the number of hours without sunshine between sunset and sunrise is

$$t = 24 - \frac{2}{15} \text{arc} \cos(-\tan\varphi\tan\delta) \tag{8.15}$$

where φ is local latitude, and δ is sun declination angle. The value of t is calculated in eq. (8.15). t varies with δ as well as different region. For example, in Haikou area ($\varphi = 20.03°$), on summer solstice (June 21st, $\delta = 23.45°$), t is 10.1 h; on winter solstice (December 21, $\delta = -23.45°$), t is 13.2 h. However, in Harbin area ($\varphi = 45.45°$), on the summer solstice day, $t = 8.5$ h, and on winter solstice, $t = 15.5$ h. It indicates that the higher the local latitude, the greater the difference in the time needed for lighting in the evening.

Under normal circumstances, the sky still has light within half an hour before sunrise and after sunset. For reasons of economy, it is possible to not turn on the lights. If the operating current of the load is i, the daily electric energy consumption of the load should be

$$Q_L = (t-1)i \tag{8.16}$$

The monthly electricity consumption is

$$Q_c = NQ_L \tag{8.17}$$

where N is number of days in that month.

Obviously, the electric energy consumption varies from month to month, with less summer and more winter, which is the characteristic of seasonal load.

The optimization design steps of the light-controlled solar lighting system are basically the same as those of the off-grid PV system, except that the daily power consumption Q_L is not the same. Therefore, the first design step is not to determine the daily electric energy consumption, but to obtain the working current i. Then, according to formula (8.15), the daily working time t is determined, and the power consumption of each day can be obtained from eq. (8.16).

8.3.6 PV system design for load with special requirements

The loss of load probability (LOLP) is usually used to evaluate the reliability of power supply system, which is defined as [11]

$$\text{LOLP} = \text{Annual power outage time} \,/\, \text{year} - \text{round time}$$

The value of LOLP is between 0 and 1. The smaller the value, the higher the reliability of the power supply. If LOLP = 0, it means that the power supply can be guaranteed at any time. The power outage time in the year is zero. In fact, even if a conventional power grid is applied to supplies power to a big city, it will also lose power for an average of several hours each year due to faults or overhauls. The LOLP can only reach magnitude order of 10^{-3}. Due to the current high price of PV energy, for general-purpose systems, the rate of load shortage can be 10^{-2} to 10^{-3} order.

However, in some special occasions, such as off-grid PV systems that provide power for important communication equipment, disaster monitoring and reporting equipment, and military equipment, it is necessary to meet the requirement of zero-minute power-off. For this kind of off-grid PV system, the design must be especially careful. Otherwise, the result may affect the long-term reliable operation of the PV system and cause serious consequences. Of cause, it cannot also blindly increase the safety factor of the system and cause a lot of waste.

For an off-grid independent PV system with a LOLP = 0, the same optimization design steps mentioned above can also be used, except that the number of battery autonomy days is first substituted with $n = 0$, so that the power output of PV array in each month is greater than the load power consumption, and the capacity of the PV array can be determined. However, it should be noted that when calculating the capacity of PV modules, $n = 0$ is taken into account, which does not mean that PV systems do not require batteries. Obviously, the batteries must be used for power supply at night and on rainy days. When calculating the battery capacity, people can refer to the local maximum number of consecutive rainy days, determine a reasonable number of battery autonomy days n, and finally get the battery capacity.

8.4 PV power system hardware design

The previously discussed contents mainly concern the software design of PV system, which is the core part of the entire PV system design. However, in order to build an efficient, safe, and reliable PV system, it also requires a series of supporting hardware design [12].

8.4.1 Site layout

According to the site conditions, determine the location of the installation of PV arrays. The PV site requires a reasonable layout with overall nice appearance and easy connection. The PV module surface should not be shaded by buildings or trees. Otherwise, in the shaded areas, there will not only be no electricity output, but also result a "hot spot" effect, which can damage PV module. Due to the large number of PV modules in a PV power station, the module arrays need to be installed in rows, one behind the other. In order to prevent the front row blocking the sunlight on the rear row, sufficient distance must be maintained between the front and rear rows (south and north). Therefore, when designing the overall layout of the site, it is necessary to determine the minimum distance between rows. In order to do so, the length of the shadows of the obstructers should be obtained.

8.4.1.1 Length of obstructer shadows

When installing PV arrays, if there are shaded trees or buildings in front of the PV array, its shadow will block the sunlight in the modules. Figure 8.11 is a real example where the distance between the front and rear arrays is too small and the front array blocks the sunlight of rear row. Therefore, the length of the obstructer shadows should be calculated at first in order to determine the minimum distance between two rows. Figure 8.12 is a diagram of calculating the minimum distance between front and rear rows.

Fig. 8.11: Example of improper distance between front and back PV rows.
Source from: NABCEP PV Installer Resource Guide

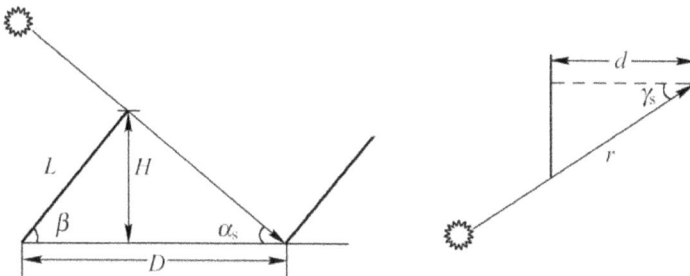

Fig. 8.12: Minimum distance calculation between front and back rows.
L – heights of PV array; D – distance between two rows; β – tilt angle; γ_s – solar zenith angle; α_s – solar azimuth angle; r – length of the projection of the incident sunlight onto the horizontal plane between the front and rear rows

It can be seen from Fig. 8.12, if the obstructer height is H, the length of its shadow is d, then according to geometric relationship:

$$\frac{H}{r} = \tan \alpha_s, \ r = \frac{H}{\tan \alpha_s}$$

From the top view, it can be obtained:

$$\frac{d}{r} = \cos \gamma_s, \ r = \frac{d}{\cos \gamma_s}$$

Both equations are equal, that is,

$$\frac{H}{\tan \alpha_s} = \frac{d}{\cos \gamma_s}$$

Therefore,

$$d = \frac{H \cos \gamma_s}{\tan \alpha_s}$$

The sine of the solar zenith angle is

$$\sin \alpha_s = \sin \varphi \sin \delta + \cos \varphi \cos \delta \cos \omega$$

Substitute that into the cosine of the solar azimuth angle:

$$\cos \gamma_s = \frac{\sin \alpha_s \sin \varphi - \sin \delta}{\cos \alpha_s \cos \varphi} = \frac{(\sin \varphi \sin \delta + \cos \varphi \cos \delta \cos \omega) \sin \varphi - \sin \delta}{\cos \alpha_s \cos \varphi}$$

$$= \frac{(\sin^2 \varphi - 1) \sin \delta + \cos \varphi \cos \delta \cos \omega \sin \varphi}{\cos \alpha_s \cos \phi} = \frac{\sin \varphi \cos \delta \cos \omega - \cos \varphi \sin \delta}{\cos \alpha_s}$$

Therefore,

$$d = \frac{H \cos \gamma_s}{\tan \alpha_s} = H \frac{\sin \varphi \cos \delta \cos \omega - \cos \varphi \sin \delta}{\cos \alpha_s \tan \alpha_s} = H \frac{\sin \varphi \cos \delta \cos \omega - \cos \varphi \sin \delta}{\sin \alpha_s}$$

$$= H \frac{\sin \varphi \cos \delta \cos \omega - \cos \varphi \sin \delta}{\sin \varphi \sin \delta + \cos \varphi \cos \delta \cos \omega} = H \frac{\cos \omega \tan \varphi - \tan \delta}{\tan \delta \tan \varphi - \cos \omega}$$

For the length of the shadow of the obstruction, the generally used principle is: the rear PV row is not shaded between 9:00 am and 3:00 pm on the winter solstice day. Therefore, substitute the declination angle $\delta = -23.45°$ and hour angle at 9:00 am and 3:00 pm on the winter solstice day $\omega = 45°$:

$$d = H \frac{0.707 \tan \varphi + 0.4338}{0.707 - 0.4338 \tan \varphi} \tag{8.18}$$

Let $d = Hs$

$$s = \frac{0.707 \tan \varphi + 0.4338}{0.707 - 0.4338 \tan \varphi} \tag{8.19}$$

where s is shading coefficient and only depends on local latitude. When the latitude gradually increases from 0, the beginning of the shadow coefficient s increases slowly. When the latitude increases to 50° and above, s increases rapidly. When latitude reaches 58.46°, s becomes infinity and becomes negative afterwards. Since the China is in the south of 58°N, it is not difficult to determine the value of the shadow coefficient s.

8.4.1.2 Minimum distance between two rows

Based on Fig. 8.12,

$$D = L \cos \beta + d, \text{ and } H = L \sin \beta$$

Therefore:

$$D = L \cos \beta + L \sin \beta \frac{0.707 \tan \varphi + 0.4338}{0.707 - 0.4338 \tan \varphi}$$

$$= L \cos \beta + L \sin \beta \cdot s \tag{8.20a}$$

As long as the local latitude is known, and the array height and tilt angle are determined, the minimum distance between two rows of PV arrays can be calculated.

Example 8.4 Calculate the minimum distance between two rows of PV arrays installed in Shanghai, with an array height of 1.5 m and a tilt angle of 22°.
Solution The latitude of Shanghai is $\varphi = 31.12°$, and $\tan\varphi = 0.6037$.
   The tilt angle $\beta = 22°$, then substitute the $\cos\beta = 0.927$; $\sin\beta = 0.375$
   Plug these into eq. (8.20a), D can be calculated as $D = 2.477$ m.
   Some literatures describe the formula for calculating the distance between the front and rear rows of a PV array (actually, the length of the shadow of an object whose height is H) is

$$d = \frac{0.707H}{\tan[\arcsin(0.648 \cos \varphi - 0.399 \sin \varphi)]} \tag{8.20b}$$

where H is height difference between the highest point of the front row and the lowest position of the rear row, and φ is local latitude.
   When the installation site is determined, the local latitude is fixed, and the shadow length of the obstacle with height H is also determined. For Shanghai area, $d = 1.904 H$ is calculated with eq. (8.20b).
   However, this formula is incorrect because the concept of azimuth angle is confused with hour angle during calculation. The value of d for Shanghai should be calculated as $d = 1.938 H$ with eq. (8.20a).
   It can be seen that if the formula (8.20b) is applied to the calculation, the distance between the two rows obtained is too small, and during operation, the front row will block the sunlight in the rear row.

8.4.1.3 Layout of PV arrays

After determining the minimum distance between the front and rear rows of PV arrays, the optimal tilt angle of the array can be decided according to the actual size of the site and the size of the PV modules used. At the same time, factors such as string connection and parallel connection of PV modules should also be considered. The final reasonable site layout should be obtained after repeat arrangement and comparison.

8.4.1.3.1 The minimum height between the ground and the lowest point of PV module is 0.3 m. The actual height that affected by the parapet wall is

$$H = 1.5m - 0.3m = 1.2m$$

which is equivalent to raising the roof by 0.3 m.

Therefore, the shadow length of the parapet wall can be calculated with eq. (8.18):

$$d = H\frac{0.707\tan\varphi + 0.4338}{0.707 - 0.4338\tan\varphi} = 1.2 \times \frac{0.707\tan 31.17 + 0.4338}{0.707 - 0.4338\tan 31.17}m = 1.93m$$

In this way, the length that is available for installation after considering the shadow length caused by the parapet wall in the north-south direction is 60 m − 1.93 m = 58.07 m.

8.4.1.3.2 Calculate the distance between two rows

In order to avoid the front row blocking the sunlight in the rear row in north-south direction, it is necessary to calculate the minimum distance between two rows.

Substitute the array height of $L = 1.50$ m and tilt angle $\beta = 22°$ into eq. (8.20a):

$$D = L\cos\beta + L\sin\beta \times \frac{0.707\tan\varphi + 0.4338}{0.707 - 0.4338\tan\varphi} = 2.477m$$

Therefore, the distance between two rows is selected as 2.5 m.

The projection of the height of the PV array in the north-south direction is

$$L\cos\beta = 1.5\cos 22° = 1.39 \text{ m}$$

In this way, the aisle width between two rows is

$$D = 2.5m - 1.39m = 1.11m$$

This distance is basically suitable for installation, maintenance, and overhaul.

8.4.1.3.3 Calculate the number of PV rows

The number of PV rows that can be installed in the north-south direction is

$$N_1 = 58.07/2.5 = 23.23$$

Take the integer of result as 23 rows.

8.4.1.3.4 Calculate the number of PV modules in each row

The width of each PV module is 0.808 m, taking into account the need to leave gaps and borders, the width is determined as 0.83 m. The roof of the building is 100 m long and has parapet walls on both sides of the east and west, so there is a gap of 2 m on each side.

In addition, it is necessary to take the length of site in the east-west direction into consideration. If it is full, it will cause inconvenience to installation and maintenance, and at least 2 maintenance walk ways with a width of 1 m should be left in the middle.

In this way, the length of available PV array the east-west direction is

$$100 - 2 \times 2 - 2 \times 1 = 94\text{m}$$

And the number of modules that can be installed in the east-west direction is

$$N_2 = 94/0.83 = 113.25$$

Take the integer of result as 113.

8.4.1.3.5 Calculate the total power of installed modules

The total number of PV modules that can be installed is

$$N = N_1 N_2 = 23 \times 113 = 2599$$

And the total power is

$$P = 2599 \times 180 = 467.82\text{kW}$$

Note that this is the maximum capacity that can be installed, not necessarily the actual amount of installation. When calculating the number of specific installations, it is also necessary to determine the number of series connected PV modules N_c and the number of parallel connections based on the input voltage and rated power requirements of the inverter. $N_c \times N_b$ is not necessarily equal to N. In this case, it is recommended to change the number of N_c and make the $N_c \times N_b$ to be as close as possible to N but less than N to obtain the number of parallel module strings N_b, so as to finally determine the installed capacity.

Example 8.5 There is an east-west warehouse in the Shanghai area, and a grid-connected PV array is required to be installed on the roof. There is a parapet wall around the roof, with the height of 1.5 m, east-west length of 100 m, and north-south of 60 m. The PV module is composed of 72 pieces of crystalline Si cells in series. The module has a power of 180 W and a size of 1.50 m x 0.81 m. What is the total power of the PV array that can be installed on the roof? What is the layout of this site?

Solution The latitude of the Shanghai area is 31.17°. The optimal tilt angle of the grid-connected PV can be calculated form discussion above as 22°.

Consider a single module vertically installations in the height direction.

First calculate the number of rows can be installed in north-south direction.

In order to avoid mud splashing on the surface of the array during rain, the effect of snow on the ground, and for the convenience of installation, the PV array should not be installed close to the ground.

The discussion above is the case where only one module is mounted in the height direction for each row. It is also possible to install two or more modules in the height direction. But this requires recalculating the distance between the two rows and repeating the arrangement. At the same time, it also required to consider the line connection and other issues, especially the design of the array racks, and to check whether its mechanical strength and stiffness are enough to avoid accidents such as strong wind. Therefore, the total capacity of a PV module that can be installed on a site cannot be calculated simply by the number of PV modules that can be accommodated. It is also necessary to fully consider the requirements of the input voltage of the inverter, the installation of modules, and the arrangement of electrical wire connection. Usually the capacity of a large- and medium-sized PV power plant is not exactly an integer.

8.4.1.4 Design of array racks

After the layout of the PV array is determined, the supporting structure, such as a rack and a base can be designed according to the dimensions of the selected modules, the number of series and parallels connection, and the tilt angle.

8.4.2 Distribution room and electrical design

Reasonably arrange the electrical room (including power distribution room, switch station, booster station and so on) on the site. Set the location for devices in the order of: DC power combiner boxes (including lightning protection), controllers, or grid-connected inverters, AC power combiner boxes (including lightning protection), step-up transformers, and equipment for measurement, data recording, storage, display, and communication. Try to design the site with reasonable layout, reliable wiring, and easy operation. Also consider where and how to connect to the grid. For off-grid PV system, it is necessary to make these devices as close to the battery as possible, but they also can be separated from each other.

The size of storage battery room is based on the number and size of the batteries. Rationally design the battery racks and its structures. The connection wires should be short and tidy, and the room should be dry and well ventilated.

Determine the connection method of the modules according to the string and parallel requirements in optimization design. When the number of serial and parallel modules is large, the hybrid connection method is preferred, and bypass diodes should be connected in parallel at both ends of the module (in case modules do not have bypass diodes), and at the same time, the position of the blocking diode and the connection method must be decided.

Reasonably arrange electric wires, and try to adopt the shortest connection method. Determine the position and connection of the combiner boxes, switches, and connectors.

For relatively important projects, electrical principles (including main wiring diagrams) and structural drawings should be plotted to facilitate maintenance and inspection.

8.4.3 Selection of auxiliary equipment

8.4.3.1 Storage battery
According to the optimal design results, determine the battery voltage and capacity, select the appropriate battery type and model, specifications, and then determine the numbers and their connection method.

In general, lead-acid batteries can be used. In severe operating conditions, sealed valve-regulated lead-acid batteries are used in some cases.

8.4.3.2 Controller
According to the requirements of the load and the degree of importance of the system, determine the adequate and necessary functions of the controller for the PV system, and install the corresponding controllers.

It is not the case that the more controller functions the better. On the contrary, increasing functions of controller will not only increase the cost, but also increase the possibility of failures.

8.4.3.3 Inverter
PV systems must be equipped with appropriate inverters for AC load. Off-grid PV systems usually use integrated controllers and inverters. In some cases, the electric grid can supply power to the load when the PV generation capacity is insufficient, but it is not allowed to send PV power back to the grid when there is excessive PV power generation. In this case, the controller and the inverter must have irreversible function to prevent reverse power transmission, so as to ensure that the excess PV power does not feed into the grid.

8.4.3.4 Lightning protection device

The solar PV power station is a third-level lightning protection structure. The set up lightning arrester, deflector and properly grounded should follow the requirements listed below, and other related standards:
- GB 50057 Code for Design Protection of Structures Against Lightning;
- GB/T 32512 Technical Requirements for Protection of PV Power Station Against Lightning;
- DL/T 1364 Technical Code for Protection of PV Power Station Against Lightning;
- JGJ 203 Technical Code for Application of Solar PV Systems of Civil Buildings.

8.4.3.5 Fire Safety

The solar energy PV power station should be equipped with portable fire extinguisher. The configuration of the fire extinguisher should comply with the relevant provisions and requirements in
- GB 50140 Code for Design of Extinguisher Distribution in Buildings;
- GB 50229 Code for Design of Fire Protection for Fossil Fuel Power Plants and Substations.

When the capacity of a single transformer in a solar PV power station is 5,000 kVA or above, an automatic fire alarm system should be set up and fire signal remote transmission function should be provided. The form of the automatic fire alarm system for solar PV power stations is categorized as a regional alarm system. The devices of various detectors and fire alarm devices should follow the relevant provisions and requirements, such as GB 50116 Code for Design of Automatic Fire Alarm Systems.

The selection of fire detectors in rooms of various types of equipment should use different types of smoke or temperature detectors according to the characteristics of the installation site, and the layout and selection requirements should comply with the relevant provisions and requirements of GB 50229 Code for Design of Fire Protection for Fossil Fuel Power Plants and Substations.

8.5 Other designs

When designing and installing the PV system, it is necessary to consider the effect of temperature. PV module operating temperature should be reduced as much as possible, especially in hot regions, and appropriate cooling methods should be applied, such as maintaining a certain interval between modules, especially BIPV which is recommended that there should be some gaps for ventilation and cooling.

The output capacity of the storage battery will be affected when the ambient temperature decreases. The capacity decreases by 1% for every 1 °C drop in temperature when ambient temperature is lower than 20 °C. Especially in northern regions, the low temperature in winter have a serious impact on battery capacity. Certain measures must be taken to keep the ambient temperature of the battery within a certain range, such as heating, insulation, or burying underground. It should also be noted that not the higher the temperature, the better the battery; if the temperature is too high, the self-discharge of the battery will increase, and the depletion of the electric plate will accelerate.

In addition to the above design, a standardized design, which includes various procedures such as spare parts, packaging, transportation, construction, acceptance, and commissioning, are required. Sometimes it also needs to design some standby power.

After the completion of the design, it is necessary to train stuff and provide them with documents, materials, drawings, and other information. These documents include design materials, installation manuals, personnel training manuals, operation and maintenance manuals, operation records, and warranty guarantee documents. In addition, spare parts are also needed.

Finally, technical and economic analysis (including cost accounting and analysis of economic and social benefits) are also required.

Overall, the design and construction of large- and medium-sized PV systems is a comprehensive system engineering project with many influence factors [13]. In order to ensure the reliable, rational, safe, and economical operation of PV system, it is required to get sufficient data and information of the site at the time of design, adopt advanced optimization design methods, and pay attention to every aspect to build a complete PV system.

References

[1]   R. Haas. The value of photovoltaic electricity for utilities, technical digest of the international PVSEC-7, 1993, Nagoya, Japan 365–366.
[2]   A Mellit. Sizing of photovoltaic systems: A review. Revue des Energies Renouvelables, 2007, 10(4), 463–472.
[3]   R.N. Chapman. Sizing handbook for stand-alone photovoltaic/storage systems[J]. Sandia Report SAND87-1087·UC-63, 1995.
[4]   B Marion. et al., Performance parameters for grid-connected PV systems. NREL/CP-520-37358. February 2005.
[5]   S. A Klien and J.C. Thcilacker. An algorithm for calculating monthly-average radiation on inclined surfaces. Journal of Solar Energy Engineering, 1981, 103, 29–33.
[6]   Luzi Clavadetscher, et al. Cost and performance trends in grid-connected photovoltaic systems and case studies. Report IEA PVPS T2-06: 2007. December 2007.
[7]   Jinhuan Yang, Xiaolu Huang and Jun Lu. A new method for designing independent photovoltaic power generation system. Acta Energiae Solaris Sinica, 1995, 16(4), 407.
[8]   J. M. Gordon. Optimal sizing of stand-alone PV solar power systems. Solar Cells, 1989, 43(2), 71.

[9]   Martin A. Green (translator: Xiuwen Li, et al.), Solar cells operating principles, technology and
       system applications. Beijing: Publishing House of Electronics Industry, 1987.
[10]   Sandia National Laboratories, Stand-alone photovoltaic systems: A handbook of
       recommended design practices. SAND87-7023 Updated July 2003: 15.
[11]   P. H Tsalides and A. Thanailakis. A loss-of-load-probability and related parameters in
       optimum computer-aided design of stand-alone photovoltaic systems. Solar Cells, 1986, 18,
       115–127.
[12]   G Xiangyang and K. Manohar. Design optimization of a large scale rooftop photovoltaic
       system. Solar Energy, 2005, 78, 362.
[13]   Florida Solar Energy Centre. Photovoltaic system design course manual. FSEC-GP-31-86,
       1986, Cape Canaveral. Florida

Exercises

8.1   What are the main factors related to the power generation of grid-connected
      PV power stations? When the construction site and capacity are determined,
      how to increase the power generation of PV power stations?

8.2   Why should PV array be tilted towards the equator?

8.3   How to determine the optimum tilt angle of grid-connected and off-grid PV
      Systems? Which angle is usually larger at the same location?

8.4   Outline the overall goals and general processes of PV system design.

8.5   There is an off-grid PV system for communication. The load power is 150 W
      and the operation is 8 h a day. The battery pack depth of discharge DOD is
      designed to be 60% and the safety factor is 1.2. In order to ensure that the
      load can still work normally for 5 consecutive rainy days, how much capacity
      is required for storage battery?

8.6   There is a solar street light with lamp load power of 50 W, working voltage of
      12 V, night light 8 h, battery discharge depth of 50%, output circuit efficiency
      of $\eta_2 = 0.9$. The longest rainy days in this region is 3 days. Assuming that the
      battery is fully charged before rainy days, how much capacity (Ah) is required
      at least?

8.7   There is a PV power station with a capacity of 1 MW and performance ratio of
      0.8. The average solar irradiation received on the PV array is 4 kWh/m²/d.
      What is the maximum annual power generation of this PV power station?

8.8   To build a 10 MW crystalline silicon PV power station in Xining, if the perfor-
      mance degradation rate r = 0.8%, the performance ratio PR = 0.8. The lifetime
      is 25 years, and the array is installed at the optimum tilt angle, calculate its
      annual power generation and total power generation.

8.9   In 2016, the expected China's electricity demand is 6.5 trillion kWh. If a PV
      power station is planned to install on the Gobi Desert with an area of
      1.3 million square kilometers to complete this target, how much area is the
      PV power station? (Assume the performance ratio of photovoltaic system is

80%, and the average solar irradiation on the array is 8150 MJ/m²/y, the system land occupation rate is 15 m²/kW.)

8.10 The slopes of some seawalls in Shanghai are facing south. The length of the seawall is about 60 km, and the length of the slope is 10 m. If the anti-seawater corrosion, plastic frame PV module with efficiency of 15% will be used to lay the seawall. What is the estimated annual total electricity generation? Assume the performance ratio is 75%, the average solar radiation amount on the seawall slope is 4800 MJ/m²/y, and the area utilization ratio of the seawall slope is 95%.

8.11 A user purchased 20 pieces of 150 W PV modules and a 3 kW grid-connected inverter to build a PV system. The optimal operating voltage of the module is 19.2 V, the open circuit voltage is 23 V, the inverter withstand voltage is 400 V, and the MPPT operating range is 170–300 V. How can the components be serially and parallelly connected to achieve safe and efficient design goals?

8.12 The customer requires that the PV system output power be 250 W to charge the 12 V battery. The module is consisted with several 125 mm x 125 mm cells, the optimal operating voltage for each cell is 0.48 V, the optimum operating current is 5.45 A, and the package and connection line losses are 10%.

(1) How many modules are needed? How to connect them? How many cells dose each modules consists of?

(2) Draw a simple electrical connection diagram, and marked positive and negative electrode.

8.13 The latitude of the Guangzhou region is N23.10°. There is a large tree with a height of 10 m in front of the PV array. What is the length of the shadow?

8.14 The latitude of the Lanzhou region is N36.03°. The PV array height is 1.6 m, and the area is 3 m × 3 m. It is installed at a tilt angle of 25° to the south. What is the minimum distance between two arrays?

Chapter 9
PV projects construction, acceptance, and maintenance

Photovoltaic power generation system (also simply called PV system) is a modern power supply system involving multiple fields of expertise. To build a reasonable, reliable, and economical PV system, in addition to optimizing the design and selecting high-quality modules and BOS, careful installation and commissioning are also very important. Careless operation may affect the power generation efficiency of PV systems, cause failures, or even induce severe safety issues and threaten human beings and devices. The quality of PV systems depends to a certain extent on the technical level and professionalism of the installation personnel. It requires workers and stuff possessing necessary scientific knowledge and professional skills. Only people who pass a rigorous review process and obtain a PV system installer certificate can carry out installation work [1].

The construction organization design of PV project should follow the guidance of GB/T 50795 Code for Construction Organization Planning of Photovoltaic Power Project, including the determination of construction organization, personnel allocation, construction schedule, construction preparation, and general layout of construction. The construction and acceptance of PV projects should be carried out in accordance with the provisions of the GB 50794 Code for Construction of PV Power Station and GB 50796 Code for Acceptance of Photovoltaic Power Project.

9.1 Installation of PV power generation system

Large- and medium-sized PV station construction is a complex system project that requires the preparation of a comprehensive feasibility study and project implementation plan. After set up of the project, all administrative examination and approval procedures, including land use, environmental protection analysis, construction, and fire control plan will be processed, and the conditions for grid connection will be obtained. Then, it is needed to get a power connection permit from the local electrical authority. The installation of PV stations can be started only when the site, capital, technology, manpower, equipment, materials, and design data are all fully prepared and evaluated [2].

The construction of PV system includes civil engineering and installation [3]. Civil works include earthworks, foundations for mounting rack, site and underground facilities, buildings and structures, and so on. Installation works include the installation of mounting racks, PV modules, combiner boxes, inverters, and other

https://doi.org/10.1515/9783110524833-009

equipment, as well as secondary electrical systems, lightning protection and grounding, overhead lines, and cable installations.

9.1.1 Preparation before installation

Large- and medium-sized PV system should meet the following requirements before installation [4]:

Qualifications of construction organizations, special operation stuff, construction machinery, construction materials, and measuring instruments. They should have obtained relevant certifications and permits.

- Design documents and construction blueprints are ready and have been reviewed and approved.
- Construction process design and construction plan have been approved.
- Sites, electricity, roads, communications and other conditions have met the normal construction needs.
- Preinstalled foundations, mounting holes, embedded parts, concrete pouring products, preembedded pipes, and facilities have been completed. Those preliminary works should meet the requirements of the design drawings, and have been inspected and accepted.
- All necessary equipment and materials have been delivered to the site and have been kept in order.
- Accessory buildings and facilities such as control rooms and power distribution rooms have been completed.

9.1.2 Installation of PV arrays

9.1.2.1 Preparation before PV array installation
(1) Site survey check
In general, the following items about the site should be checked:
- If it is appropriate to install a PV array on site, then the size and orientation of the installation site should match the design;
- If there are any building or tree blocking in front of the PV arrays;
- How to install the PV arrays in the field;
- Whether the distance between the front and rear rows of arrays meets the design requirements;
- Whether the location of prepared foundation, mounting holes, embedded parts, and others are correct;

- Whether the placement of the balance of system (BOS) components are appropriate;
- How to connect PV systems with existing grids.

(2) Determine the construction plans and prepare documents such as design blueprints.

(3) Leveling the site, pouring foundations, and embedded parts. Foundations and embedded parts must be securely fixed to the ground. Check and recalculate the bearing capacity, strength, and stability of foundation.

If a PV array is installed on the roof, the structure design, materials, durability, and strength of the building should be reviewed and checked beforehand to confirm the supporting capacity of the roof can indeed withstand the mass of the PV array and the additional loads such as wind pressure and snow. For PV array mounting racks, they should use galvanized steel connectors extending from the reinforced concrete foundation, or use stainless steel fasteners with the foundation. The main steel bar of the concrete foundation should be anchored to the main structure. When the structural conditions are limited and the anchoring cannot be performed, it should take other methods to increase the adhesion between the foundation and the main structure. Preembedded parts of steel foundation and top of concrete foundation should be protected with anticorrosion coating to meet the designed anticorrosion level. After the foundation casting is completed, waterproof treatment should also be carried out to prohibit water leakage. It should meet the requirements of the standard GB 50207 Code for Acceptance of Construction Quality of Roof.

The cable passing through the roof, floor, or wall must have a waterproof seal between the waterproof casing and the main body of the building. The surface of the building should be smooth and clean.

Application technologies of PV system on civil structures involve planning, architecture, structure, electrical, and other specialized fields. The construction should be complied with relevant standards and regulations, such as:
- GB 50352 Code for Design of Civil Buildings
- GB 50368 Residential Building Code
- GB 50055 Code for Design of Electric Distribution of General-Purpose Utilization Equipment
- GB 50052 Code for Design Electric Power Supply Systems
- GB/T 16895.6 Low-voltage Electrical Installations – Part 5–52: Selection and Erection of Electrical Equipment – Wiring Systems
- JGJ/T 16 Code for Electrical Design of Civil Buildings
- JGJ 203 Technical Code for Application of Solar Photovoltaic System of Civil Buildings
- GB 50797 Code for Design of Photovoltaic Power Station

(4) During the construction of the PV systems, the structure and ancillary facilities of the building should not be damaged, and the capacity of the building to withstand various loads within the design service life should not be affected. If it is unavoidable to cause partial damage during the PV installation, it should be repaired at the end of the construction.

(5) Based on number of PV arrays, installation dimensions and optimal design, array tilt angle can be calculated, and the mounting rack can be made. Mounting rack material and dimension should meet the design requirements. The overall structure of the PV array should be determined according to the quality of the PV modules, the size of the mounting rack, the local wind load, and the amount of snow and other factors, so that the array has sufficient strength, stiffness, and stability.

In general, the mounting rack is made of galvanized steel or angle iron with protective paint. As for the PV array installed on the coast or on the island, stainless steel may also be used in consideration of the prevention of salt corrosion.

(6) PV modules and frames, mounting racks and fixing bolts, connecting cables and bushings, combiner boxes and other accessories must be shipped to the site before installation.

(7) Tools, equipment, and spare parts needed for installation must be fully prepared, especially when the site is in a remote area. Otherwise the lack of a tool or a bolt can cause great trouble and even affect the progress of the entire project.

9.1.2.2 Site installation

The installation of ground-mounted PV systems generally follows the following steps [5]:

(1) Check all supporting foundations are installed in place with correct interval distance according to design requirements.

(2) Take out PV modules from the shipping box and check the status. Measure and examine the V_{oc}, I_{sc}, and other technical parameters of each module under the sun. It will be really difficult to find the faulted module after installation and operation if a module with defects is installed without inspection. Therefore, it is necessary to perform this simple and effective test for each component before installation, or at least a reasonable portion of total modules should be checked.

(3) Prior to installation, PV modules should be grouped according to their technical parameters. Those modules with similar current at maximum power point should be connected in series and modules with similar voltage at maximum power point should be connected in parallel. Since large- and medium-sized PV systems often have tens of thousands of PV modules, it is not easy to classify all modules, but at least it is necessary to ensure that the same string consists of modules with the same type and power.

(4) During installation, PV modules are usually placed face down on a clean, non-rigid platform side by side. The module shipping box can be used as a work bench if needed. The position of module junction box should be placed according to the series and parallel requirements to make it easy to operate when connecting the wires.

(5) Place the mounting rack over the modules so that the mounting holes of the rack are facing down and aligned with the mounting holes of modules.

(6) Firmly fix the mounting rack to all modules with stainless steel screw bolts, spring washers, and screw nuts.

(7) According to the design requirement for serial and parallel connection of modules, connect the positive and negative poles of the module with wires, if there is no preinstalled output wires on the module. The wire connection principle is: as thick and short as possible to reduce line loss. However, it should also be noted that when installing in the summer, the wire connection should not be too tight, and there should be a margin to avoid poor contact or even broken of wire caused by the decrease temperature in winter. The positive and negative leads of the PV array output and the grounding wire should be distinguished with cables of different colors or labels so as not to confuse the polarity and cause accidents.

Note: the definition of wire (cable) color is usually as follows:
- DC cable: positive (+) – brown; negative (-) – blue; grounded line (neutral) – light blue;
- Three-phase electricity (three-phase four-wire system): A phase line – yellow; B phase line – green; C phase line – red; N line (neutral, neutral) – light blue; ground line – yellow green;
- Single-phase electricity: phase line – red; neutral line – blue; ground line – yellow-green;
- Internal wiring of devices and equipment – black
- AC circuit with double-core conductors or twisted pair cables – Red and black in parallel, Red is live wire, black is neutral wire.

The connection between the wires must be reliable, and two wires cannot be twisted together carelessly. A rubber sleeve should be use as out-layer for connection cables, rather than ordinary electric tape. It is also recommended to put insulation sleeve on the outside of the wire. The connection of wires should comply with the requirements of GB 4706.1 Household and Similar Electrical Appliances – Safety Part 1: General Requirements, and other standards.

Moreover, it is necessary to use stainless steel clips, straps, saddle clips with a protective cover, or antiaging plastic clips to fix the wires to the protective tube or array mounting rack to prevent poor contact due to long-term wind shaking.

After wiring, close the combiner box cover.

(8) The mounting rack with modules can be fixed on the foundation by stainless steel bolts, spring washers, and screw nuts. If necessary, it can also be welded, but be careful to avoid distortion of the modules.
   - The welding of steel structures should meet the requirements of the standard GB 50205 Code for Acceptance of Construction Quality of Steel Structure.
   - The anticorrosion treatment should be carried out for the welding of PV arrays.
   - The anticorrosion construction should meet the requirements of the standard GB 50212 Code for Construction of Building Anticorrosive Engineering, and
   - GB 50224 Code for Acceptance of Construction Quality of Anticorrosive Engineering of Buildings.
(9) After checking the tilt angle, fix the four bases and the mounting racks. Make sure the mounting racks are firm, tidy and reliable.
(10) Based on different situation on site, it is also possible to install the mounting racks, then install the modules to the mounting rack, and connect the wires.
(11) If there is more than one subarray, the wiring can be connected to distribution box or combiner box first and then connected to grid.
(12) Lightning protection devices must be installed in the mountain top, lightning-prone areas, or near important PV systems. The PV arrays should be within the protection scope. The lightning protection and grounding of PV systems and grid-connected interface equipment should follow the guidance in:
   - GB/T 32512 Technical Requirements for Protection of Photovoltaic Power Station Against Lightning
   - DL/T 1364 Photovoltaic Power Station Lightning Technical Regulations
   - SJ/T 11127 Overvoltage Protection for Photovoltaic Power Generating Systems – Guide
(13) For PV array installed on a roof, it also needs to pay attention to the following items:
   - For PV modules integrated with building materials installed on a roof, the rainproof connection structure between the upper and lower and the left and right sides must be strictly constructed. Water leakage is strictly prohibited. The appearance must be neat and smooth as well.
   - The ventilated layer on the back of the PV array should not be filled with debris and should maintain well ventilation.
   - The conductive frame and mounting rack should be reliably connected to the building grounding system. The grounding of the electrical system should meet the requirements of the standard GB 50169 Code for Construction and Acceptance of Grounding Connection Electric Equipment Installation Engineering.

- The components of the PV system should comply with the corresponding requirements in GB 50016 Code for Fire Protection Design of Buildings. The minimum requirement for components installed on the roof is fire rating C (basic fire rating).
- The installation of PV systems on buildings should not reduce the lightning protection rating of buildings and should comply with the requirements of the GB 50057 Code for Design Protection of Structures Against Lightning.

9.1.2.3 Safety precautions [6]

(1) The installer must pass safety education before construction. The construction site should be equipped with necessary safety equipment and ensure safety of construction workers.
(2) No construction work during rainy days.
(3) When the roof slope is more than 10°, a footboard should be provided to prevent people or objects from falling off.
(4) It is strictly forbidden to stand on the glass surface of PV modules to avoid glass breakage or damage, or slip off from the glass.
(5) The output of the PV array cannot be short-circuited. When installing in sunlight, it is best to cover the PV modules with opaque materials such as black plastic films.
(6) PV system products and components must not be impacted or damaged during storage, handling and installation. Special attention should be paid to prevent components from being impacted by hard objects.
(7) When lifting PV modules, the bottom should be lined with wooden boards or packaging cartons to prevent the lifting cable from damaging the components. Before the lifting operation, safety measures around the working locations should be taken. Pay attention to hoisting machinery and objects when hoisting, do not hit surrounding buildings and other facilities.
(8) Install fences around PV arrays to prevent animal intrusion or vandalism.

9.1.3 Installation of electrical equipment

9.1.3.1 Technical requirement

Large- and medium-sized PV systems should be equipped with independent power distribution rooms or control equipment rooms. Distribution cabinets, instrument cabinets, grid-connected inverters, monitors, and batteries (if the system has energy storage devices) should be placed in the equipment room.

(1) The wiring and equipment deployment of the PV system should meet the design specifications of the low-voltage power system and the design specifications of the PV system. Installation of electrical equipment of PV systems should comply with the requirements of the standard GB 50303 Code of Acceptance of Construction Quality of Electrical Installation in Building.

The withstand voltage level of the DC circuits should be higher than 1.25 times the maximum output voltage of the PV array; the rated carrying capacity should be higher than the setting value of the short-circuit protection device, and the setting value of the short-circuit protection device should be higher than 1.25 times the nominal short-circuit current of the PV array. In addition, the line loss should be controlled within 2%.

The connection between the PV system and the power grid should be marked with visible break points, which should be electrically isolated by transformers.

The construction of cables should comply with the requirements of the standard GB 50168 Code for Construction and Acceptance of Cable System Electric Equipment Installation Engineering.

(2) PV systems on buildings are rated as application class A, and their design should meet the requirements of class A.

Application level A refers to a system, which has a DC voltage above 50 V or power above 240 W, and is likely to be in contact with or approaching by the public. Equipment suitable for application class A shall be equipment that meets the requirements of safety Class II equipment.

Class II equipment refers to: equipment that protects against electric shock not only depends on basic insulation, but also on additional safety protection methods (such as equipment with double insulation or reinforced insulation). The protection against electric shock of this type of equipment is neither dependent on the protective earth nor on the protective measures of the installation conditions.

(3) Insulated isolation devices should be provided on both DC side and AC side of the inverter and the system.

The connection of PV system into the public power grid should comply with the standards:

- GB/T 50865 Code for Design of Photovoltaic Generation Connecting to Distribution Network
- GB/T 50866 Design code for PV power station connecting to Power System
- GB/T 29319 Technical Requirements for Connecting Photovoltaic Power System to Distribution Network
- GB 19964 Technical Requirements for Connecting Photovoltaic Power Station to Power System
- GB/T 19939 Technical Requirements for Grid Connection of PV System

The necessary electric shock warning and safety measures for preventing electric shock should be taken into consideration at DC side of the PV system. Isolation devices should be installed between the PV system and the public power grid. The isolation devices shall have obvious disconnection point indication and complete disconnection function.

PV system should be equipped with a dedicated low-voltage switchgear for grid connection at the grid-connected location, with special signs and warning texts and symbols such as "warning" and "dual power supply."

All junction boxes (including the connection boxes of the system, PV arrays, and PV strings) should be provided with a warning label indicating that the devices in the junction box may still be charged when the junction box is disconnected from the PV inverter.

(4) Insulation performance

The insulation resistance between the main circuit of the PV array, the combiner box, the inverter, and the protection device and the conductive equipment surfaces shall not be less than 1 MΩ. The insulation performance should withstand the voltage of 2,000 Vac for 1 min and avoid flashover or breakdown.

(5) PV systems connected to the public power grid should have functions such as polarity reversal protection, short-circuit protection, grounding protection, overheating and overloading protection, alarm function, and antiislanding protection.

9.1.3.2 Installation location

If the controller and inverter are installed indoors, the power distribution room must be prepared beforehand. The location selection of storage area should avoid high corrosive, high dust, high temperature, and high humidity environments. In particular, it should be avoided to a possible metal objects drop in. The location of power distribution should be as close as possible between PV array and users to reduce the line loss. The controller cannot be placed directly on the battery because the corrosive gas generated by the battery can affect the electronic components inside the controller.

The small- and medium-sized controllers and inverters can be fixed on the wall or placed on the workbench as required. The large-scale controllers and inverters are generally placed directly on the ground with a certain distance from the wall for installation, maintenance, and ventilation, to avoid direct sunlight on controllers and inverters.

If controllers and inverters are installed outdoors, controllers and inverters must have protective functions to against moisture.

9.1.3.3 Wiring connection

Select the appropriate insulated wire and cable. The selection of appropriate conductor diameter should be based on the rated current, and according to the relevant specifications or data provided by the manufacturer. The wires with too small diameter will increase line loss or overheating, resulting in wasted energy and decreased efficiency. In severe cases, the insulating layer may melt, resulting in a short circuit or even a fire. If the diameter is too large, it will cause higher initial cost. Refer to Tab. 9.1 for the current carrying capacity of different cables.

Tab. 9.1: Carrying capacity of different cables.

Cross-section area (mm²)	Number of cable cores	Multicore cable diameter (mm)	Current carrying capability (A)
1	1	1.13	12
1.5	7	0.5	14
2.5	7	0.67	17
2.5	50	0.25	20
4	7	0.85	29
6	7	1.04	37
10	7	1.35	51
16	7	1.70	66

Usually, a series connection of modules can use the wires from the junction box on the module back directly, because the modules are designed with the diameter of the lead wire sufficient capacity to carry the maximum current output from the module. However, in a parallel connection, the current passing through the wire will increase accordingly, so the corresponding wire diameter will increase.

For the low-voltage DC part of the off-grid PV system, special consideration should be given to line loss on both power and voltage. For example, a 200 W load is powered by a 220 Vac and a 12 Vdc power supply, the current flowing through the 220 Vac line is 0.91 A, and the voltage drop across a 10 m long cable with a cross-sectional area of 2.5 mm² is small with 0.13 V. However, the current flowing through the 12 Vdc line is 16.7 A. The voltage drop across the same cross-sectional area and the same length of AC wire is 2.44 V. After passing through this wire, the voltage is only 9.56 V, and the electrical equipment can no longer work properly. Therefore, it is necessary to increase the cross-sectional area of the wire or other methods to solve this problem.

9.1.3.4 Wiring sequence

When the controller and inverter are unpacked, first check if there are warranty cards, factory inspection certificates, and product specifications; whether the appearance is damaged, and whether the internal connections and screws are loose. If there are problems, contact with the manufacturer in time.

When the controller is wired, the switch should normally be placed in the "off" position. Connect the battery first, followed by the connection with the PV arrays. Close the switch when there is sunlight and observe if normal charging current flows. Finally, connect the controller to the load.

Place the inverter's input switch in the "off" position and wire it. Pay attention to the positive and negative polarity when wiring and ensure the quality and safety of the wiring. After the line is connected, check whether the DC voltage input from the controller is normal. If it is normal, the inverter's output switch can be turned on under no load to make the inverter start to work.

A switch without arc suppression function is not allowed to connect or disconnect under load, to prevent accidents caused by pulling an arc.

9.1.4 Installation of storage battery packs

The commonly used lead-acid battery in PV systems has sulfuric acid as the electrolyte. The battery is potentially dangerous with improper transportation, installation, or maintenance. Therefore, people working on the battery must be familiar with the installation procedure.

9.1.4.1 Installation precautions
Before working with the battery, metal decorations on the hands and neck should be removed. Wear a nonmetallic hard hat and protective clothing, including acid-proof gloves, apron, and protective eyepieces.

Don't stand on battery.

There should be a clean water source beside the battery, in the case of acid splashing on the skin or eyes, rinse immediately.

9.1.4.2 Set-up for battery room
The battery room should be located as close as possible to the PV array. For medium-to-large size of PV systems, the battery room must be separated from the electricity distribution room where electrical equipment such as controllers and inverters are installed in.

The battery room should be dry, clean, well-ventilated, and free from direct sunlight. The distance between battery room and heat source must not be less than 2 m. The indoor temperature should be kept in 10–25 °C.

Insulation between the battery and the ground should be provided. Generally, wood or other insulation materials should be placed under the battery to avoid discharge due to a short circuit between the battery and the ground. A large amount of batteries can be placed on a dedicated battery rack.

The battery cannot be turned upside down or subjected to any mechanical shock and heavy load. The location should be convenient for wiring and maintenance.

9.1.4.3 Connection wire

Check whether the voltage of each battery is normal. After confirming that the electrolyte is at the normal level and the battery is full, connect the battery to the load.

According to the design requirements, the batteries should be connected in series or in parallel. Note that the positive and negative electrodes cannot be connected incorrectly.

Connect battery poles carefully to prevent short circuits. If a metal tool or object falls between the battery poles, an electric discharge will be formed, which will generate a large current and spark, and may damage the equipment or cause accidents.

If there are more than one battery connected in series, in order to avoid accidental electric shock and short-circuit, usually it can leave the connection terminal to the controller open, and after all connections are completed and a normal voltage is confirmed, connect batteries to the controller.

The battery pole and the clamp must be in close contact with each other. Otherwise, due to poor contact, it will increase the resistance, or even cause a broken circuit. A thin layer of suitable oil film may also be applied at each connection point to prevent corrosion.

9.1.4.4 Preparation of electrolyte

Some batteries need to be charged before use, while dry-shipped lead-acid battery can only be used after adding electrolyte.

When preparing the sulfuric acid electrolyte, slowly injected sulfuric acid into the distilled water, and constantly stirred with a glass rod. Injection of water into the sulfuric acid is strictly forbidden, to prevent the spattering of sulfuric acid and hurting people.

After adding the electrolyte to the battery, tighten the lid of the filling hole to prevent debris from falling into the battery.

At the end of the installation, measure the voltage, positive and negative polarity of the battery, and check the connection quality and safety. For the open valve type of battery, measure and record the electrolyte density.

After the battery is installed, all data must be recorded and archived.

9.2 Commissioning of PV systems

9.2.1 Preparation before commissioning

PV system commissioning consists of equipment commissioning and system commissioning. Before commissioning, the installation work should have been completed and accepted. The construction project should meet the following conditions.

– All decoration work should be completed and cleaned.
– Installation of special facilities such as air-conditioning or ventilation equipment has been completed and put into operation.
– Tasks which cannot be performed or could affect operator's safety after system connection shall be completed in advance.

The commissioning of PV system should be under the responsibility of a qualified engineer and may work together with relevant organizations and equipment suppliers, and start with checking whether the installation and conditions are in accordance with equipment instructions and related standards and procedures.

PV system commissioning should be done on a sunny day and should be performed when the sunshine and wind condition reach a stable level. It is best to test the system in 2 hours before and after solar noon.

Prior to commissioning the PV system, the surface of the PV array should be cleaned and all switches should be off.

Prepare instruments, meters, tools, and notebooks for testing.

9.2.2 Commissioning of PV array

9.2.2.1 General review
First, carefully inspect the PV area, to check if the appearance of the PV array is normal and smooth, the module surface is clean, press the assembly with hand to check if the connection is loose, and whether the wiring is fixed and the connections are good.

Check whether the materials and components used in the PV system meet the design requirements and whether the installation quality meets the relevant standards and specifications.

Check whether the corresponding insulation protection meets the requirements of Class II safety equipment.

Check the grounding wire and measure the grounding resistance. The grounding resistance value should be less than 10 Ω.

9.2.2.2 Inspection and measurement for PV arrays
Usually, the PV modules in the PV array are of the same model and specification. Based on the technical parameters provided by the manufacturer, the open circuit voltage of a single module can be found. The product of single module voltage with the number of series connections should be roughly equal to the open-circuit voltage at two ends of the PV strings. If the difference between actual measurement and calculated value is too large, it is likely that there are problems such as module

damage, polarity reversal, or poor contact at the connection. People can check the open circuit voltage and connection status of each module to eliminate the fault.

Be caution when conducting the measurement with too many modules in array.

Measuring the short-circuit current at two of the PV module string, the obtained value should basically meet the design requirements. If the difference is too large, some modules may have poor performance and should be replaced.

After all PV modules are inspected and recorded, it is ready for the next stage of commissioning.

The parallel connection can only be done when all measured open-circuit voltages of all parallel-connected PV modules are basically the same. After the parallel connection, the voltage is almost unchanged, and the total short-circuit current should be substantially equal to the sum of the short-circuit currents of the individual PV strings. Be caution when measuring the short-circuit current. If the current is too high, it may jump sparks and cause equipment or personal accidents.

9.2.2.3 PV string test

(1) The following conditions should be met before testing:
  - All PV modules should be connected correctly according to the number and model listed in the design documents.
  - Each circuit cable inside the combiner box should be connected and marked clearly and accurately.
  - The fuse or switch in the combiner box should be off.
  - The grounding of the combiner box and internal lightning protection module should be fixed, reliable, and have good connection.
  - The test should be conducted under the condition of irradiance not less than 700 W/m^2.
(2) Testing should include the following requirements:
  - The polarities of the tested PV modules in the combine box should be correct.
  - The open circuit voltage deviation between strings of the same PV module under the same test condition should not exceed 2%, and the maximum deviation should not exceed 5 V.
  - In the case of power generation, use a clamp multimeter to check the current in the PV module string in the combiner box. Under the same test conditions and when the irradiance is not less than 700 W/m^2, the current deviation between the strings of the same PV module should not exceed 5%.
  - The temperature of the wire of the PV module string should be free of abnormal conditions such as abnormally hot.

(3) Before the inverter is put into operation, all the combiner boxes connected to the inverter unit should be tested. After the inverter is put into operation, the sequence of in and out service of PV strings in the combiner box should meet the following requirements:

   (i) If the main switch of the combiner box has the arc suppression function, the on and off of service sequence of the combiner box should be performed according to the following steps:

      – Turn-on small series switches or fuses of the PV strings and then turn-on the main switch of the combiner box.

      – Turn-off the main switch of the combiner box and then turn-off small series switches or fuses of the PV strings.

  (ii) If a fuse box is used for the total output of the combiner box and the switches of the branch circuit PV strings have arc suppression function, the sequence is:

      – Turn-on the total output fuse of the combiner box first and then turn-on small series switches or fuses of the PV strings.

      – Turn-off all small series switches or fuses of the PV strings and then turn-off the total output fuse of the combiner box.

 (iii) When the fuse box is used for both the total output of the combiner box and all branch circuits, disconnect the inverters before turn-on and turn-off the fuse box.

(4) Measure PV array technical parameters.

If there are more than one subarray, check the quality of each one according to the above method, then connect the positive and negative output from the arrays to the combiner box or controller, measure the total working current and voltage of the square array, and record the parameters.

9.2.3 Commissioning of controller

9.2.3.1 Documentation check

Check whether the product specification and factory inspection certificate of controllers are complete.

9.2.3.2 Performance testing

If possible, the performance of the controller should be fully tested to verify whether it meets the specific requirements of:

– GB/T 19064 Solar Home System Specifications and Test Procedure
– NB/T 32006 Test Code for Power Quality of Photovoltaic Power Station

For a general off-grid PV system, the controller's main function is to prevent the battery from overcharging and overdischarging. The controller should be tested separately before connecting it to the PV system. A suitable DC voltage regulator can be used to provide a stable operating voltage for the controller. Through adjusting the voltage, it is able to check whether the voltage meets the requirements at conditions of disconnection when fully charged, restoring the connection when charging is needed, and disconnect at low voltage.

Some controllers have an output voltage regulation function that can change the input voltage within certain range. Check this function and measure whether the output voltage is stable.

In some off-grid PV systems, a public power grid is used as a backup power source. When the battery is not fully charged during a long period of continuous rainy days, the controller switches to the grid for charging. After it is fully charged, it is disconnected and powered by the PV system again. When commissioning, it is necessary to confirm whether this function is working as expected.

Check whether the maximum self-consumption of the controller exceeds 1% of its rated operating current. If the controller also has the functions such as smart control, equipment protection, data acquisition, status display, fault alarm, and so on, it can also perform appropriate verifications about these functions.

For small PV systems or certified controllers that have been commissioned before delivery and no damage during transport and installation, these tests may not be performed on site.

9.2.3.3 Connecting wires

After the controller is tested individually, it can be connected to the battery according to the design requirements. Some controllers do not have antireverse function, so it needs to make sure the connection has correct polarity. Then, the positive and negative outputs of the PV array are connected to the corresponding inputs of the controller. Finally, check whether the output voltage of the PV array is normal and whether there is charging current.

After the record is made, the commissioning of the DC input is completed.

9.2.4 Commissioning of off-grid system inverter

For off-grid PV systems, it must be equipped with off-grid inverters for AC load. Before commissioning, it is necessary to check whether the product specifications and factory inspection certificates of inverters are complete.

The inverter should be subject to comprehensive inspection when conditions permit, and its main technical analysis indicators should comply with the requirements of GB/T 19064 Solar Home System Specifications and Test Procedure.

Measure the output voltage of the inverter and check whether the output waveform, frequency, efficiency, and load power factor meet the design requirements.

Test the inverter protection, alarm, and other functions, and take records.

9.2.5 Commissioning of grid-connected inverter

The grid-connected inverter is the core component of the grid-connected PV system. It is related to the normal operation of the PV system and the safe operation of the grid. The commissioning work is very important [7]. The grid-connected inverter must have a product specification and a factory inspection certificate.

9.2.5.1 Performance testing

Before the grid-connected inverter is connected to the PV system, the output AC power quality and protection function should be tested separately.

(1) Power quality test

PV power stations that are connected to main public grid side should be equipped with a power quality online monitoring device at the grid connection point. PV power stations that are connected to grid distribution side should be equipped with a power quality monitoring device at the user metering point. The power quality data of large- and medium-sized PV stations should be able to be transmitted remotely to the utility dispatching department. For small-size PV stations, it should be able to store more than one year of power quality data, which can be used by utility companies when necessary.

Note: The definitions of the grid connection point and common connection point are as following:

- Grid connection point refers to the connection point between the PV system and the power grid (which can be either the public power grid or the user power grid). For a PV station with a booster station, it refers to a busbar or node on the high-voltage side of the booster station; for a PV station without a booster station, it refers to the output point of the PV station.
- Common connection point refers to the connection point where more than one user connects to the grid.

The power quality can refer to the power quality test circuit diagram shown in Fig. 9.1.

If the deviation of the grid voltage and frequency can be kept within 50% of the maximum allowable deviation, the AC power simulator with adjustable voltage and frequency can be omitted, and the system can be directly connected to the grid for testing.

After making the connections, measure the following parameters:

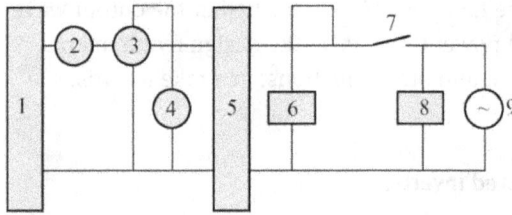

Fig. 9.1: Power quality test circuit diagram.
1 – PV array; 2 – DC current meter; 3 – DC power meter; 4 – DC voltmeter; 5 – grid-connected inverter; 6 – power quality analyzer; 7 – disconnector of grid connection point; 8 – AC load; 9 – AC power supply with adjustable voltage and frequency (grid simulator) (current capacity is at least five times the current supplied by the PV system)

(i) Working voltage and frequency

After the PV system is connected into the power grid (or grid simulator), measure and record the voltage and frequency at the grid connection point three times, and determine whether the PV system affect the power grid based on the requirements in GB/T 12325 Power Quality – Deviation of Supply Voltage.

(ii) Voltage fluctuations and flicker

According to GB 12326 Power Quality – Voltage Fluctuation and Flicker, measure and record the fluctuation and flicker of the grid voltage at the grid connection point, and judge whether the influence of the PV system on the grid meets the requirements of the standard.

(iii) Harmonic and waveform distortions

Before the PV system is connected to the grid, use a power quality analyzer to measure and record the grid (or grid simulator) harmonics.

After the PV system is connected into the power grid, measure harmonic current and voltage of the PV system with a power quality analyzer to determine whether the harmonics meets the requirement of GB/T 14549 Quality of Electric Energy Supply– Harmonics in Public Supply Network.

(iv) Power factor

Before and after the grid-connected operation, when the PV system power output is greater than 50% of inverter-rated power, adjust the output and phase of the inverter, and measure and record the output power efficiency and power factor of the PV system at the grid connection point continuously with a power quality analyzer. Then, determine whether the requirements of the equipment and system technical conditions are met.

(v) Unbalance of output voltages (applicable to three-phase output at the common connection point)

Before and after the PV system is connected into the power grid, measure and record the three-phase voltage unbalances at the grid connection point with a power quality analyzer. Then, evaluate whether it meets the requirements of the GB/T 15543 Power Quality – Three-Phase Voltage Unbalance, and the contract requirement set by the local utility company. In general, the allowable voltage unbalance is 2%, and it must not exceed 4% in short time.

(vi) Output DC component

When the PV system is connected to the grid and working, it detects the DC current component under different output power (33%, 66%, 100%) using a power quality analyzer. The DC current component inject to the grid should not exceed 0.5% of its AC rating. When the DC component is greater than the limit, the PV system should be automatically disconnected from the grid.

(2) Protection function test

The protection function test can be conducted with an AC power regulator, and the test consists of items below:

(i) Over/under voltage protection

Based on the indexes in Tab. 9.3 of GB/T 19939 Technical Requirements for Grid Connection of PV System, change the voltage value of the AC power regulator item by item and measure the reaction value and time of the protection system of the grid-connected system, which should meet the requirements of related standards.

According to the GB 50797 Code for Design of Photovoltaic Power Station, the response when the grid voltage is abnormal should meet the following requirements:
- The output voltage of the grid-connected PV system should match the grid voltage.
- Large- and medium-sized PV stations should have a low-voltage ride-through capability (Fig. 9.2). When the voltage of the grid connection point is on or above the voltage curve in Fig. 9.2, the PV system should maintain the grid-connection operation. When the voltage at the grid connection point is higher than 110% of the grid-rated voltage, the PV system operation status is determined by the PV station designed functions. The low-voltage ride-through requirements for large- and medium-sized PV power stations which connected to the grid user side are determined by the local utility company. In the Fig. 9.2, V_{L2} is the minimum voltage limit for normal operation and can be taken as 0.9 times of rated voltage. V_{L1} can be taken as 0.2 times of rated voltage. T_1 is the time needed to maintain the grid connection when the voltage drops to 0, T_2 is the time needed to maintain the grid connection when the voltage drops to V_{L1}, and T_3 is the time that the PV station can connect into the grid. The values of T_1, T_2, and T_3 shall be determined

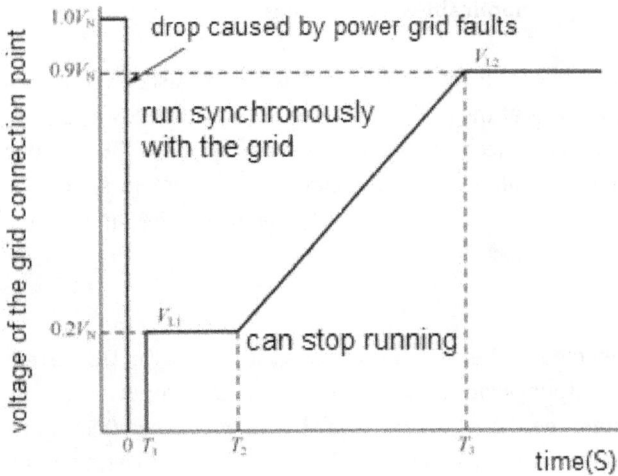

Fig. 9.2: Low-voltage ride-through capability requirements for large- and medium-sized PV power stations.

according to the actual conditions of protection and reconnection response time, usually $T_1 = 0.15$ s, $T_2 = 0.625$ s, and $T_3 = 2$ s.

– When the voltage of a small-size PV station at grid connection point is in different operating ranges, the response time of the PV station to abnormal grid voltage should meet the requirements on Tab. 9.2.

Tab. 9.2: PV power station response requirement for abnormal grid voltage.

Voltage at on-grid point	Maximum opening time
$V < 50\%V_N$	0.1 s
$50\%V_N \leq V < 85\%V_N$	2.0 s
$85\%V_N \leq V \leq 110\%V_N$	Continue working
$110\%V_N < V < 135\%V_N$	2.0 s
$135\%V_N \leq V$	0.05 s

(ii) Over/under frequency protection

According to the requirements of GB/T 19939 Technical Requirements for Grid Connection of PV System, change the frequency of AC power regulator (the changing rate should not be faster than 0.5 Hz/s), measure the protection system response frequency and time, and they should meet the requirements of related standard.

Based on GB 50797 Code for Design of Photovoltaic Power Station, the response to abnormal grid frequency should meet the following requirements:

– Grid-connected PV systems should run synchronously with the grid.
– Large- and medium-sized PV plants should have a certain ability to withstand abnormal frequency of the power grid. The response time requirements of large- and medium-sized PV stations when the grid frequency is abnormal shall meet the requirements of Tab. 9.3. When the grid frequency exceeds the range of 49.5 to 50.2 Hz, the small-size PV station should stop feeding power into the grid within 0.2 s.
– When the system frequency restored to the normal grid continuous operation state within the specified time, the PV power station should not stop the power transmission.

Tab. 9.3: Response time to abnormal frequency for large- and medium-sized PV stations.

Grid frequency	Response time requirements
$f < 48$ Hz	Determined by the minimum frequency allowed by the PV station inverters or grid requirement.
48 Hz $\leq$ $f < 49.5$ Hz	At least 10 min is required.
49.5 Hz $\leq$ $f \leq 50.2$ Hz	Continue working
50.2 Hz $<$ $f < 50.5$ Hz	PV station should have the capability of continuous operation for 2 minutes, but at the same time have the ability to stop sending power to the grid within 0.2 s. The actual operation time is decided by local utility company; without permission, the PV station in the off status may not be reconnected to the grid.
$f \geq 50.5$ Hz	Stop sending power to the grid within 0.2 s. The PV station in the off status may not be reconnected to the grid.

(iii) Antiislanding Effect

A large- and medium-sized PV stations should have protection device to make sure that the PV station can be disconnected in a fault event, and the PV station may not be provided with antiislanding protection [8]. Small-size PV plants should have the ability to quickly detect islands and immediately disconnect from the grid, and their antiisland protection should be co-operated with the grid-side line protection. In case that the grid-connected line is connected to other power loads with T type at the same time, the response time of the antiislanding protection of the PV station should be less than the line protection reclosing time at the grid-side.

When the PV system is connected to the grid and working, make the voltage of the simulated power grid loss voltage, the antiislanding protection should react within 2 seconds to disconnect the PV system from the power grid.

(iv) Grid restoration to normal condition

During the protection function tests of over/under voltage, over/under frequency, and antiisland effect, the PV system is shut down or disconnected from the power grid due to the adjustment of AC power regulator or simulator. After recovering to the normal working range of the grid, the grid-connected PV system with the automatic restoration and grid-connected function should connect to the grid within the specified time period.

(v) Short circuit protection

Simulate the short circuit condition at the grid connection point, measure the output current and the disconnection time of the PV system, which should meet the requirements of GB/T 19939 Technical Requirements for Grid Connection of PV System. When the network is short-circuited, the over-current of the inverter should not exceed 150% of the rated current, and the PV system should be disconnected from the power grid within 0.1 s.

(vi) Feedback current protection

When the PV system is designed to operate in a nonfeedback mode (also called irreversible grid connection mode), it cannot feedback current into the grid, even if the PV system generate extra electricity more than the local usage. In this case, the nonfeedback protection function should be tested.

The nonfeedback current protection test circuit is shown in Fig. 9.3.

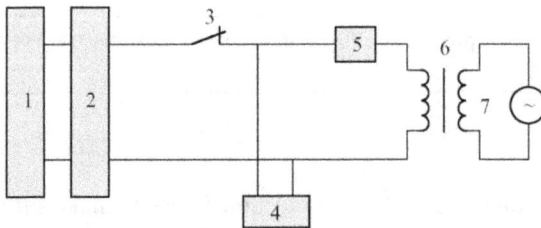

Fig. 9.3: Reverse current protection reference test circuit.
1 – PV array; 2 – grid-connecting inverter; 3 – disconnector at grid connection point, 4 – variable AC load; 5 – Reverse current detection device; 6 – isolation transformer; 7 – AC power regulator or simulator

When the PV system is connected to the grid through an isolation transformer and in operation, adjust the AC load from large to small, or the output power of the PV system, until a feedback current occurs in the PV system transformer to the grid. Record the response value and time of the PV system feedback current protection, or the disconnection to the power grid. These values should meet the design requirements. When it is detected that the feedback current is higher than 5% of the rated output of the inverter, the PV system should stop the feedback to the grid within 0.5 to 2 seconds.

If it is guaranteed that the grid-connected inverter used has been tested and qualified before leaving the factory, and there is no damage during transportation, these tests may not be performed on site.

9.2.5.2 Connecting wires

First connect the grid-connected inverter to the PV array and measure the working current, voltage and output power of the DC side. If it meets the requirements, connect the grid-connected inverter to the grid and measure the voltage, power, and other technical data of the AC side. At the same time, record the parameters such as solar irradiation intensity, ambient temperature, and wind speed, to determine whether they meet the design requirements.

If all components of the PV system work normally, they can be put into trial operation. Record various operating data at regular intervals during trial operation.

After a certain period of trail operation, if there is no abnormal condition, it can be completed and accepted.

9.3 Acceptance of PV project

The acceptance of a PV project consists of four phases: subengineering projects, project start-up, project trial run and transfer, and project completion.

This section is applicable to the acceptance of newly built, reconstructed, and expanded projects for ground and rooftop PV generation connected to electric grid at 380 V or higher voltage levels. It does not apply to BIPV and residential PV systems.

Acceptance of PV systems should comply with the requirements of the standard GB 50796 Code for Acceptance of PV Project, and GB 50794 Code for Construction of PV Power Station.

9.3.1 Acceptance of subengineering projects

The subengineering projects of PV system consist five parts: civil engineering, installation engineering, plant engineering, security and protection engineering, and fire protection engineering. Acceptance process of a subengineering project is organized by the construction organizations (composed of persons in charge of construction, design, supervision, commissioning, and others). Each subproject consists of several areas. The acceptance of the subengineering project should be organized by the supervising engineer and based on the acceptance of these areas. These areas should pass a self-checking by each responding organizations.

9.3.1.1 Civil engineering

Civil engineering includes PV module support foundation, site and underground facilities and building (structure), and other subengineering projects: (a) PV module support foundation includes individual concrete (bar) foundation and pile foundation; (b) sites and underground facilities including site leveling, roads, cable trench, water supply, and drainage facilities; (c) inverter rooms, power distribution rooms, comprehensive buildings, main control buildings, booster stations, and fences (walls) of buildings (structures) should comply with the standards of GB 50300 Unified Standard for Constructional Quality Acceptance of Building Engineering, GB 50205 Code for Acceptance of Construction Quality of Steel Structure, and related regulations for design.

9.3.1.2 Installation engineering

Installation works include mounting rack, PV modules, combiner box, inverter, electrical equipment, lightning protection, grounding, wire and cable installations, and so on.

(1) The mounting racks include the fixed mounting rack and the tracking mounting rack. The installation of fixed racks should comply with the standards:
  - GB 50794 Code for Construction of PV Power Station
  - GB 50205 Code for Acceptance of Construction Quality of Steel Structure

The installation of tracking racks should comply with the standards:
- GB/T 29320 Technical Requirement for Tracking System of PV Power Station
- The relevant provisions of GB 50205 Code for Acceptance of Construction Quality of Steel Structure

(2) The installation of lightning protection and grounding should comply with the standards:
  - GB/T 32512 Technical Requirements for Protection of Photovoltaic Power Station Against Lightning

- DL/T 1364 Photovoltaic Power Station Lightning Technical Regulations
- SJ/T 11127 Overvoltage Protection for Photovoltaic Power Generating Systems – Guide
- GB 50169 Code for Construction and Acceptance of Grounding Connection Electric Equipment Installation Engineering
- GB 50057 Code for Design Protection of Structures Against Lightning

(3) Installation of wires and cables includes PV array DC cables, AC cables, and overhead cables.

9.3.1.3 Other subengineering projects

The requirements for plant engineering, security engineering and fire protection engineering are as follows:

(1) The plant and vegetation restoration of the construction area meet the design requirements.

(2) The security engineering includes alarm systems, video security monitoring systems, and entrance and exit control systems. The acceptance of security engineering should follow the requirements of the standard GB 50348 Technical Code for Engineering of Security and Protection System.

(3) The design drawings of the fire protection engineering should be examined and approved by local fire department. The fire performance and fire resistance of the building (structure) components shall comply with the relevant requirements of the standard GB 50016 Code for Fire Protection Design of Buildings. Installation of safety exit sign and fire emergency lighting fixtures should comply with the relevant provisions of the standard GB 13495 Fire Safety Signs, and GB 17945 Fire Emergency Lighting and Evacuate Indicating System.

9.3.2 Acceptance of project start-up

After the project start-up achieves the acceptance conditions, the construction organization should submit an application for acceptance to the development organization. Multiple similar PV systems may submit an application for acceptance at the same time. After receiving the acceptance application, the development organization should establish an acceptance committee to carry out the project start-up acceptance according to the actual conditions of the project.

The project start-up acceptance committee is composed of construction, supervision, commissioning, production, design, relevant government departments, and utility companies. The construction organization, equipment manufacturer, and other participating organizations should also attend the project start-up acceptance inspection.

9.3.3 Acceptance of project commissioning and transfer

After the project start-up acceptance is completed and the project has been achieved the commissioning and transfer conditions, the construction organization should submit an application for project commissioning and transfer acceptance process to the development organization. After receiving the application for acceptance, the development organization should establish an acceptance committee to carry out the acceptance process according to the actual conditions of the project.

This committee is formed by the development organization and consists of construction, supervision, commissioning, production operation, design, and other relevant organizations.

During the acceptance process, equipment and system commissioning should be conducted under conditions of fine weather with solar radiation of not less than 400 W/m². After commissioning of the PV generation project, the continuous grid-connecting operating time with no fault should be less than the time needed to achieve total accumulated solar radiation received by PV arrays of 60 kWh/m².

9.3.4 Acceptance of project completion

The acceptance of the project completion will be carried out when the commissioning and transfer production acceptance is completed. After receiving the acceptance application, the development organization establishes a project completion acceptance committee based on the actual conditions of the project.

The project completion acceptance committee is composed of the relevant departments in charge of environmental protection, water conservancy, fire control, and quality supervision. Development organizations, design, supervision, construction and major equipment manufacturers or suppliers, and other organizations should also send representatives to participate in the acceptance process.

9.4 Maintenance and management for PV systems

Before the grid-connected PV system is formally put into operation, it must go through grid-connected procedures with the power grid regulator. A PV system can only formally operate after getting a permit and the "networking agreement" from the power grid regulator. During the operation, it should also be subject to regular inspections by the power grid regulator, to ensure the safety and stability of the power grid. The power grid regulator should periodically check and verify the power metering devices of the PV system.

9.4.1 Routine maintenance

PV system operation and management staff should have the necessary professional knowledge and high responsibility to carry out maintenance and management work with a serious and responsive attitude.

Carry out daily inspection to observe the operation status of the PV system, to understand whether the power meter display is working in the normal range, whether the measurement is correct and effective, and to make a good record of the operation [9].

9.4.1.1 Array inspection
Observe whether the surface of the PV array is clean and remove dust and dirt in time. Rinse with clean water or wipe with a clean rag, but do not use chemicals for cleaning. Check if there is any disconnection in the array.

9.4.1.2 Equipment inspection
Check whether there is defect appearance such as corrosion, damage for all devices, and whether there are temperature anomalies by touching the device shell with the back of the hand. Check whether the exposed conductors have any insulation aging, mechanical damage, or whether there is water in the cabinet. Check for other abnormal conditions, such as small animals penetrate inside the equipment. Whether there is abnormal noise in the equipment or odor in the operating environment. If any, causes should be found out and be resolved immediately.

9.4.1.3 Battery maintenance
Observe the charge/discharge status of the battery. Maintenance staff should wear protective glasses and protective equipment during the maintenance work, and use insulated devices to prevent personal accidents or battery short circuit. It is also recommended to wipe the dirt and dust outside the battery frequently and keep the room clean. If the battery has a seal cap or vent plug, the vent must always be checked and kept open. Pay attention to the electrolyte level of the battery and do not expose the plates and separators above the liquid surface.

9.4.2 Periodical inspection

In addition to routine inspections, it is also needed for professionals to conduct periodical inspections as follows.

Check and understand the operation records, analyze and judge the operational status of the PV system. If problems are found, professional maintenance instructions should be provided immediately.

Equipment appearance inspection and interior inspection mainly involve connection parts, wires and cables, especially high-current carrying wires, power devices, and places where corrosion is likely to occur.

For the inverter, clean the cooling fan regularly and check whether it is normal. Clean the dust inside the enclosure regularly (if accessible). Check whether the screws of each terminal are tight. Check whether there are any traces left after overheating and damaged devices. Check whether the wires are aging.

Periodically check and maintain the density of the battery electrolyte and replace the damaged battery in time.

If conditions permit, the infrared (IR) detection method can be used to inspect PV arrays, wires, and electrical equipment to identify abnormal heating and fault points and fix them in time.

For the whole PV system, check the system's insulation resistance and grounding resistance every year according to the system drawing.

Perform inspection and test of the power quality and protection function of the whole project once a year for inverters and control devices.

All records, especially professional inspection records, should be filed and properly kept.

The grid-connected PV system does not have battery and load problems. The key is grid-connected inverters, which generally have the function of controls. With the products from modern reputable manufacturers with strict inspection and proper certification, the quality can be guaranteed. If the grid-connected inverter does not work properly, the manufacturer should be notified immediately for repairing.

Due to the large number of complex factors involved in off-grid PV system, the questions of whether the system design is reasonable, whether the configuration of the PV array and battery capacity is appropriate, whether the equipment and parts and material quality are good or bad, and whether the processing and installation are in compliance with requirements, the operation and maintenance of the off-grid PV system should be more meticulous. Faults should be discovered in a timely manner and properly handled to ensure long-term normal operation of the PV system.

In summary, the management and maintenance of PV system is the key to ensuring the normal operation of the system. The PV system must be properly managed, carefully maintained, standardized operations carefully checked, and problems solved in time. Only in this way can the PV system be in a stable state of normal operation for a long period of time, and the social and economic benefits of PV generation can be fully obtained.

References

[1]   Bill Brooks, Jim Dunlop. PV Installation professional resource guide. 2016 NABCEP v. 7.
      http://www.nabcep.org.
[2]   Bill Brooks, Jim Dunlop. NABCEP PV installer resource guide 12.11. 2011.12/v. 5.1. http://
      www.nabcep.org/wp-content/uploads/2011/10/Web-PV-Installer-Resource-Guide-10-17-11.
      pdf
[3]   Ward Bower., et al., NABCEP PV installer job task analysis. 2011. 10. http://www.nabcep.org.
[4]   NABCEP Study guide for photovoltaic installer certification. Version 4.2-2009.4. http://www.
      nabcep.org/_Study_Guide-Revised_Version_3_-_08_05-FINAL.pdf
[5]   M. Hubbuch 1992. Advanced technology for mounting and connecting PV modules.
      Proceedings of the 11th E.C. Photovoltaic Solar Energy Conference, Montreux, Switzerland
      1338–1339.
[6]   Gray Davis. A guide to photovoltaic system design and installation. California Energy
      Commission Consultant Report 500-01-020, June 2001 http://www.energy.ca.gov/reports/
      2001-09-04_500-01-020.PDF.
[7]   Takigawa K. et al., Test and evaluation facility for photovoltaic inverters, 22nd IEEE
      Photovoltaic Specialists Conference, 1991, Las Vegas, 689–684.
[8]   Matsuda H. et al., Testing and evaluation of measures for preventing islanding of grid-
      connected residential-scale pv systems. Technical Digest of the International PVSEC-7,
      Nagoya, Japan, 1993, 385–386.
[9]   Garyl Smith, Roch Ducey, Lee Humble and Larry Strother. Maintenance and operation of
      stand-alone photovoltaic systems. Sandia national Laboratories Architectural Energy
      Corporation, December 1991. U.S. Government Printing Office 1997–573-127-42154.

Exercises

9.1   What conditions should a large-size PV system possess to be installed?

9.2   Briefly describe the installation sequence of PV systems.

9.3   What are the safety precautions when installing a PV system?

9.4   What are the general steps for commissioning a PV array?

9.5   What does the PV system debug mainly include?

9.6   What are the contents of the performance test during the commissioning of
      the grid-connected inverter?

9.7   What are the main aspects of daily maintenance of PV systems?

Chapter 10
Applications of photovoltaic systems

Since solar energy is a safe, reliable, convenient, and flexible renewable clean energy source, the scale and scope of application of PV power generation systems are rapidly expanding. It is transitioning from supplemental energy source to alternative energy source and will play an increasingly important role in the energy consumption field.

10.1 Classification of ground PV systems

The applications of PV system are gradually developing. At the beginning, they were classified mainly into two categories: space and ground applications. Later, ground systems were divided into three types: stand-alone systems, grid-connected systems, and hybrid systems. In recent years, it has been further divided into six types: pico systems, off-grid residential systems, off-grid nonresidential systems, grid-connected distributed systems, grid-connected centralized systems, hybrid systems, and space PV power stations.

10.1.1 Pico PV system

Pico PV systems have undergone significant development in the past few years, with the use of highly luminous sources (mainly LEDs), smart charge controllers, and highly efficient PV modules [1]. Some small PV modules have a power of only a few watts, but they can provide basic service needs, such as the solar portable lantern (Fig. 10.1) that combines a battery and a handheld lamp can be carried around, which is suitable for daytime charging, nighttime lighting, used to replace kerosene lamps and candles. It can also be applied to field activities, disaster relief, emergency lighting, and other places. According to the Renewables 2016 Global Status Report released by the Renewable Energy Policy Network for the twenty-first century (REN21), 15% to 20% of households in the world's none-electricity regions use solar lighting [2]. Globally, in mid-2015, 44 million sets of pico PV products were sold, with an annual output value of US$ 300 million. The largest market was in sub-Saharan Africa, with 1.37 million sets sold, and followed by South Asia, with 1.28 million sets sold [3]. In 2015, about 2 million sets of portable PV lanterns were sold worldwide. By the end of 2014, India had promoted 960,000 sets, Tanzania had 790,000 sets, Kenya had 765,000 sets, and Ethiopia had 662,000 sets. In the past 4 years in sub-Saharan Africa, the annual growth rate of solar lantern sales has exceeded 90%.

https://doi.org/10.1515/9783110524833-010

In addition to solar lighting, this type of pico PV system has also developed a variety of products, such as solar lawn lights, solar traffic lights, solar road studs, and solar chargers (Figs. 10.2–10.5).

Fig. 10.1: Solar lantern.

Fig. 10.2: Solar lawn light.

Fig. 10.3: Solar traffic light.

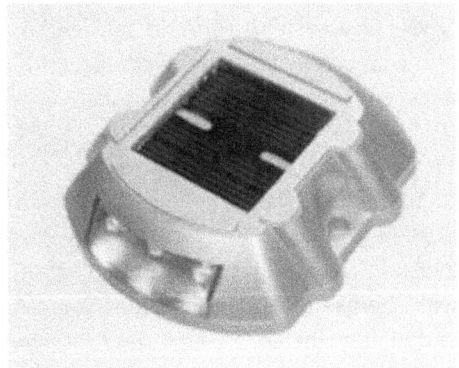

Fig. 10.4: Solar road stud.

Fig. 10.5: Solar charger.

As solar energy is widely available and easy to use, some products with low electricity consumption using PV power generation has its outstanding advantages. For example, when working or traveling in the field, solar energy backpacks (Fig. 10.6) can be used as emergency power supply for electrical appliances. Some items used indoors, such as calculators (Fig. 10.7), watches (Fig. 10.8), and toys (Fig. 10.9), can be applied amorphous silicon (a-Si) PV cells, which not only with the low cost, but also with better low light responses than crystalline silicon solar cells and lower temperature coefficients. In the case of the same rated power, the a-Si PV cells generate more actual power, so solar calculators and watches began to promote and use in very early days and achieved good results.

Fig. 10.6: Solar backpack.

Fig. 10.7: Solar calculator.

Fig. 10.8: Solar watch.

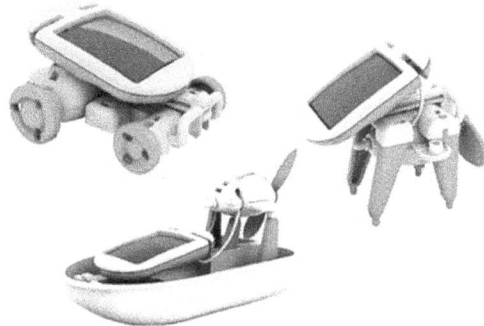

Fig. 10.9: Solar toys.

10.1.2 Off-grid residential system

There are still nearly 1.2 billion people in the world who are not able to use electricity. Most of them live in remote areas with underdeveloped economy. It is difficult to rely on extending the conventional power grid to solve the electricity problem. These areas are often rich in solar energy resources, and the application of solar power is very promising.

Off-grid PV systems were originally called stand-alone PV systems and were PV systems powered entirely by solar cells. PV arrays were the only source of energy in the system. Off-grid systems have always been the main application area for PV systems, and they are still playing an important role in many areas that do not have grid coverage. Until 2000, the installed capacity of global grid-connected PV systems exceeded that of off-grid systems.

Initially, off-grid small household PV systems were 10 to 500 W, which provided basic power for households far away from the public grid, meeting the needs of low-power appliances such as lighting, listening to radio, and watching TV. The most basic small-scale residential PV system, as shown in Fig. 10.10, is mainly composed of PV modules, batteries and controllers, and connecting wires.

Fig. 10.10: PV module, battery, and controller.

The system is simple, and some do not even use controllers. Later, it gradually developed and could support two or three energy-saving lamps and a 14-inch black-and-white TV. In the early days, due to the unreasonable design of the system, the unqualified components, improper maintenance, and other reasons, more failures occurred. Later, after gradual improvement, it has become a mature PV product. With the economic growth and improvement of living standards, users demand that the residential PV system be able to supply more home appliances (such as color

TVs and refrigerators), and accordingly the power of solar panels also develop to 500–1,000 W.

Solar power is not subject to regional restrictions and is an ideal power source for residential electricity in areas where there is no electricity [4]. For a long time, Asia has been the largest market for such off-grid systems. The average annual growth rate of Bangladesh has reached 60% in the past 10 years. By the end of 2014, 3.6 million units have been installed, 1.1 million units have been installed in India, there are 500,000 sets for Nepal and China and 320,000 sets for Kenya, and Latin America and Guyana installed 13,600 sets (884 kW). M-KOPA has sold 300,000 units in Kenya, Uganda, and Tanzania in East Africa since 2012. SunEdison announced that it will provide 20 million lighting systems by 2020. By the beginning of 2015, there were 6 million sets of small-scale residential PV power sources of 10–500 W in operation. The promotion and application of these off-grid residential PV systems have achieved good economic and social benefits (Fig. 10.11).

Fig. 10.11: Rural residential PV system.

In some developed countries, some places such as country houses and holiday huts built in mountainous areas or in remote areas have no access to electricity. According to calculations, it is economical to use an off-grid PV residential system even if it is 1–2 km away from the grid; so many users use PV power, as shown in Fig. 10.12 [5].

Fig. 10.12: Country house residential PV system.

10.1.3 Off-grid nonresidential system

The simplest off-grid PV system is a direct-coupled system. When PV arrays are generating power under sunlight, they directly supply power to the load. There is no energy storage device, so the load can only work when there is sunshine. Such systems include solar water pumps and solar wind caps.

The direct-connected PV system is very simple, and the block diagram is shown in Fig. 10.13.

Fig. 10.13: Direct-connected PV system block diagram.

10.1.3.1 PV pump

PV pumps are ideal for pumping in dry areas without grid electricity. They have the advantages of no need to connect to the grid, do not consume fuel, easy to move, easy to install, easy to maintain, have a long working life, and have no pollution. The more arid, the stronger the sunlight is, the more water pumped by the PV pump, it just meets the requirements, so it is particularly suitable for solving the problem of drinking water and a small amount of irrigation for people and animals in areas without electricity [6].

10.1.3.1.1 PV pump main components

PV pumps are electric motors driven by PV arrays under sunlight to pump water. Due to the special working conditions of the PV pump, in most cases, the pumping can be stopped at night or on rainy days. Therefore, it is not necessary to provide batteries that are expensive and difficult to maintain. Water storage tanks can generally be used to replace energy storage. Therefore, most PV pumps are used in the way that PV arrays are directly coupled to the motor. However, some PV pump systems are also equipped with batteries for reasons of efficiency.

The motor is a device that converts electrical energy into mechanical energy. According to the type of electrical energy used, it can be divided into two major categories: AC motor and DC motor. The AC motor is divided into synchronous and asynchronous. The DC motor can be divided into two types, based on mechanical or electrical commutators. Since the PV system generates DC power, most of the small- and medium-sized PV pumps use DC motors. Among them, the permanent magnet brushless DC motor has the advantages of high efficiency, easy maintenance, and working in a wide voltage range; therefore, it is widely used in small- and medium-sized PV pumps. Only large-scale PV pumps convert DC power to AC power through an inverter and use AC water pumps.

The water pump is a kind of fluid machinery that converts mechanical energy into water energy. Due to different principles and applications, it can be divided into centrifugal type, volumetric type, and other types. Among them, the centrifugal type relies on the rotation of the impeller to obtain the pressure and the speed of the water by centrifugal action. The advantage is that the simple structure, the stable performance, and the small starting torque. Therefore, a centrifugal pump is often used for a medium-volume PV pump.

As the cost of PV power generation is relatively high, it is required that all components such as motors, pumps, and controllers must have high efficiency. However, the conventionally used AC water pumps are often less efficient in order to reduce costs. Therefore, PV pumps cannot directly use a commercially available conventional AC water pump. PV water pumps usually use high-quality materials and are specially designed and manufactured to achieve better performance.

10.1.3.1.2 PV water pump design

When designing a PV pump, first select the type of motor and pump used according to the user's requirements for lift head and flowrate. Floating pump and surface suction pump can be used in low lift head and large flowrate applications. Volumetric pumps are commonly used in high lift head and low flowrate applications. Manual pumps are commonly used because of the low power required for low lift head and low flowrate applications, as shown in Fig. 10.14.

Next, we need to determine the effective power required by the PV pump and then calculate the required shaft power based on the efficiency of the motor and

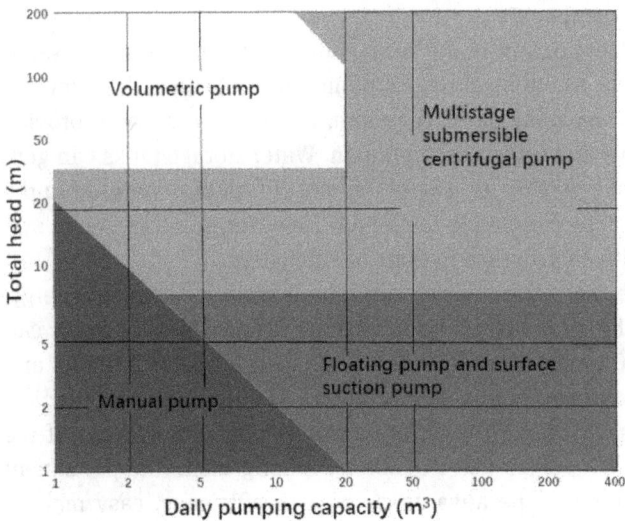

Fig. 10.14: Selection of pump types.
Source: Barlow et al.1993

pump unit. From the local solar radiation data, determine the required PV array capacity and also consider whether to use tracking or fixed angle and other factors, as well as working conditions, maintenance, and other aspects.

As the PV pump is working, the solar irradiance is constantly changing during the day, and the output power and voltage of the PV array will also change, which will affect the overall efficiency of the system. The controllers of modern PV pumps are mostly equipped with a MPPT function, which can automatically adjust the load curve so that the working point of the PV array is often near the maximum power point, so that solar radiation energy can be fully utilized to increase the utilization efficiency of the entire PV power and water pump system.

In short, the performance of a solar water pump is not only related to the characteristics of the motor-pump unit itself, but also related to the operating characteristics of the PV modules. There are many factors involved and the relationship between them is complex. Each component must have a relatively high efficiency and must reach the best match of performance between components. Therefore, in order to obtain better performance, according to the local meteorological conditions, consider the influence of various factors and optimize the system design as much as possible.

In recent years, due to the development of technology, the performance of PV water pump systems has been continuously improved and the application range is gradually expanding. As shown in Fig. 10.15, with the continuous reduction of the cost of PV power generation, PV water pumps have a broad development prospect.

Fig. 10.15: PV water pumping.

10.1.3.2 Solar fan hat

The solar fan hat is another type of direct-connected system. The solar module is installed on the top of the cap (Fig. 10.16) and connected to a small fan through a cable. Under the sun, the power from the PV module can drive the small fan to run and bring cooling air to people working outdoors during the hot summer.

Fig. 10.16: Solar fan hat.

Since solar power generation is subject to climatic conditions, the power generation generally does not match the load demand. Therefore, for off-grid PV systems, energy storage devices are usually required to store the excess power generated by the PV array to meet the load requirements in other times.

An off-grid PV system that provides power to DC load is mainly composed of a PV array, controller, battery, and blocking diode. Its DC power supply block diagram is shown in Fig. 10.17.

In addition to the above components, an off-grid PV system that supplies power for AC load must also be equipped with an inverter. The block diagram of its power supply is shown in Fig. 10.18.

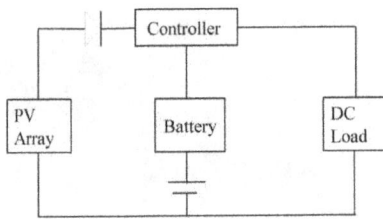

Fig. 10.17: Off-grid PV system DC power supply.

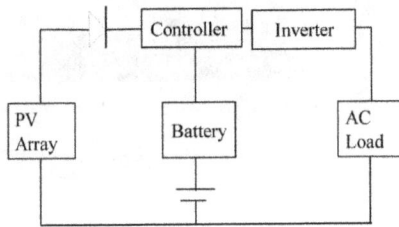

Fig. 10.18: Off-grid PV system AC power supply.

Off-grid nonresidential PV systems supply electricity to equipment in places without electricity and can provide on-site power supply, without extending the power grid, saving investment, reducing line losses, and being easy to use, safe and reliable, and easy to maintain. Its' application fields varies a lot.

10.1.3.3 Communication PV power

In terrestrial long-distance communication or signal transmission, repeaters or relay stations are usually set up at regular intervals, and a reliable power supply is required for their operation. However, conventional power grid may not be available in some station locations. Especially in high mountains or desert areas, it is impossible to rely on extending the power grid to supply electricity. Other types of independent power sources may be used, such as gasoline or diesel generators, wind turbines, closed cycle vapor turbo (CCVT) power generation, and thermoelectric generator (TEG), all with certain limitations. PV power generation is available everywhere and can be unattended. The use of PV power generation is an ideal choice. Therefore, in the early days of PV system applications, communications power sources are accounted for a considerable share.

For a large number of radio, TV, and telephone relay stations, even the power consumption for each of them is not large, it is difficult to rely on other power sources for safe and reliable power supply, and PV power generation has unique advantages. In areas where the AC grid cannot reliably supply power, PV communication power sources have been widely used, such as shown in Figs. 10.19 and 10.20.

At present, emergency telephone, such as on highways, has widely used PV power (Fig. 10.21). In some special places, such as military communications in the field, flexible thin film PV modules are very convenient to use and carry (Fig. 10.22).

Fig. 10.19: PV power supply for a microwave relay station.

Fig. 10.20: Communication relay station with PV power.

Fig. 10.21: Solar telephone station.

Fig. 10.22: Flexible PV modules for military communications.

10.1.3.4 Off-grid power source

10.1.3.4.1 Solar beacon light
In order to ensure ship navigation safety, it is often necessary to set beacon lights in ocean coastal lines or rivers. The kerosene lamp was used in early days. Buoy workers had to row a boat every day to turn on the lights and turn off the lights, no matter it was severe cold or heat, wind, or rain. Later, the use of dry battery as a one-time power supply has improved working conditions, but the consumption of dry battery is very large, and the cost is very high, and the brightness of the light is gradually darkened. The importance of beacon lights and the harshness of working conditions are exactly suitable for PV power generations, due to its safety, reliability, and capability for unattended operation. Since the 1970s, solar beacon lights started to be used with PV modules to charge the battery (Fig. 10.23). As a beacon light power, it can also be automatically switched on and off according to surrounding brightness. It also greatly reduces the labor intensity of workers and has significant social and economic benefits. In 1973, China installed the first set of solar beacon lights in Tianjin Port, which is the earliest terrestrial application area for PV systems in China. Some large coastal lighthouses also use solar energy (Fig. 10.24). Because of their large power, strong light intensity, and long range visibility, they significantly improved the navigation safety in the ocean.

Fig. 10.23: Solar navigation lights.

Fig. 10.24: Solar lighthouse.

10.1.3.4.2 Cathodic protection power supply for oil and gas pipelines

Petroleum and natural gas pipelines are used for long-distance transportation, but the corrosion of metal pipelines is widespread. If there is no protection, about 10–20% of metal materials could be lost every year. In order to effectively protect the metal pipes, the protected metal pipe can be subjected to external cathode polarization to reduce metal corrosion. This method is called cathodic protection. The impressed current method that is mainly used now consists of a DC power supply, a potentiostat, an auxiliary anode, and a reference electrode. DC power supply is mainly based on the requirements of the cathodic protection to provide the required DC voltage and current, usually through a step-down rectification of AC power to obtain DC power, which requires a set of step-down rectifier devices, but also consume some of the power. In regions where there is no reliable AC power source, diesel engines and other conventional power supplies were used. This is not only expensive, but also requires transportation of fuel. Operation and maintenance are also complex and require professional management. PV power generation has the advantages of safety, reliability, convenient operation, and low maintenance, and is especially suitable for unattended operation. It also provides direct current and does not require rectification equipment. PV systems are now widely used not only in oil and gas pipelines (Fig. 10.25), but also in piers, bridges, water locks, and other metal constructions. Many of them are used as cathodic protection power sources, which are important for preventing metal corrosion, and have achieved significant economic benefits.

Fig. 10.25: Cathodic protection PV power supply.

10.1.3.5 Emergency management and disaster relief power supply

In order to predict and prepare for natural disasters, it is often necessary to conduct measurements and observations in the field, such as to set up observation stations for earthquake, flood, wildfire, and so on. In the observation station, the measuring instrument requires power, and the manually managed stations need living electricity. Many locations are sited in high mountains or remote areas, often lacking of grid power supply. PV power generation is often the best choice for such unattended or low electric consumption sites [7]. PV system can fully meet the electricity needs of observation stations as long as properly designed and constructed. Figure 10.26 shows a PV power supply for field monitoring equipment. Figure 10.27 shows a PV power supply for the forest fire observation station.

Small-scale meteorological observing stations generally use PV power because their use electricity consumption is low (Fig. 10.28). In addition to observation equipment, some meteorological stations need extra power for communication. For example, in Huashan Weather Station, there is no power supply in the early days; the original transceiver report relied on a hand-cranked generator, which requires very high labor intensity. In 1976, Xi'an Jiaotong University developed a small-scale PV system for Huashan Weather Station (Fig. 10.29). The PV module capacity was 20 W. Since then, the transceivers have been freed from arduous physical labor. The PV power source has provided significant benefits.

10.1.3.6 Transportation power

10.1.3.6.1 Solar aircraft

With the advancement of science and technology, the efficiency of solar cells has increased, and prices have been declining. This has created many opportunities for new PV applications. The use of solar energy as a power source for aircraft is a long-held

Fig. 10.26: PV power for a field monitoring station.

Fig. 10.27: PV power for a forest fire observation station.

Fig. 10.28: PV powered weather station.

dream. Theoretically, as long as it can catch up the speed of the earth's rotation, so be exposed to sunlight continuously, the solar-powered aircraft can continue to fly, and the duration only limits to the life limit of the components. At the end of the twentieth century, people began to explore and develop several solar-powered aircraft.
(1) Apollo is shown in Fig. 10.30.
(2) Zephyr is shown in Fig. 10.31.
(3) Solar Impulse is shown Fig. 10.32.

As early as 2003, the Swiss explorer Bertrand Piccard proposed the idea of the First Round-The-World Solar Flight, powered only by the sun, with no fuel or polluting emissions. In 2006, the research team began to develop the first proto-type "Sunshine Impulse". In November 2009, the first time manned flight has been carried out. On July 7, 2010, the entire day and night flight was successfully completed. The flight time was 26 h 10 min 19 s which is a new record for solar aircraft.

Fig. 10.29: Huashan weather station with 20 W PV power supply.

Fig. 10.30: Apollo solar aircraft.

Fig. 10.31: Zephyr solar plane takeoff.

On the basis of "Sunshine Impulse" experience, Bertrand Picard's team developed "Sunshine Impulse 2" and installed 17,248 new high-efficiency ultra-thin solar cells with an energy conversion efficiency of 22.7%, these cells can produce up to 70 kW of power. The electricity generated is sent to four electric motor-driven propellers. Each propeller has 13.5 kW and 4 m diameter twin blades. Propeller stable

Fig. 10.32: Solar Impulse 2 solar aircraft.

Fig. 10.32 (continued)

speed can be increased to 525 rpm and cruising speed is 90 km/h. In order to meet the requirements of day and night flight, a 633 kg lithium-ion battery was installed in the aircraft. "Sunshine Impulse 2" has a wing width of 72 m, which is 3.5 m wider than the Boeing 747–8 wing span. The length of the aircraft is 22.4 m, which is close to the length of a basketball court. The machine height is 6.37 m, which is equivalent to three-floor building. But the weight is only 2.3 t, which is equivalent to the weight of a small truck. The surface of this solar aircraft is as thin as cicada's wings. Thanks to the carbon fiber honeycomb sandwich material used in the fuse-lage skeleton, the mass density of this material is only 25 g/m^2, which is three times lighter than paper, and its strength can fully meet the mechanical requirements of the aircraft.

On March 9, 2015, the "Sunshine Impulse 2" solar-powered aircraft departed from the UAE capital, Abu Dhabi, and piloted alternately by Bertrand Piccard and Andre Borschberg. The global flight has 17 stops, as Muscat of Oman, Ahmedabad, and Varanasi of India, Mandalay of Myanmar, and stayed in Chongqing and Nanjing of China, and then landed due to weather conditions in Nagoya of Japan, after a con-tinuous 118 h flight, the 8,200 km trans-Pacific journey from Nagoya to Hawaii was completed. On July 15 that year, "Sunshine Impulse 2" suspended global flights due to battery failure.

On June 23, 2016, the "Sunshine Impulse 2" solar aircraft landed in Seville of Spanish, after a 70-h flight. After stopping at Cairo, it returned to Abu Dhabi on July 26, 2016, and completed a global flight of 40,000 km. Although "Sunshine Impulse 2" can only carry one person, and its flying speed is also very slow, solar-powered aircraft will not become real vehicles in the near future. This flight is a new milestone in the history of human aviation and highlights the future of energy technology. It has a great significance for the development and use of clean energy.

10.1.3.6.2 Solar yacht

According to the statistics of the United Nations, the greenhouse gas emissions from commercial ships account for about 4.5% of the world's total emissions, and the use of clean energy as a power for water transport meets the requirements of energy conservation and environmental protection. The speed of tourist yacht is not fast, and its roof area is relatively large so more PV modules can be installed. All these have created favorable potentials for PV systems on yachts. As early as in 1982 on the World Expo held Knoxville of USA, China exhibited the "Golden Dragon" solar boat, which has attracted widespread attention. In recent years, with the concept of environmental protection becoming more and more popular, solar yacht has gained more development.

At present, the world's largest luxury yacht exclusively powered by solar energy is the white catamaran Tûranor (Fig. 10.33) with the Swiss flag. It is built in Germany and costs 18 million euros, and has 31 m long, 25 m wide and 95 t weight. More than 800 PV modules have been installed, with a total area of 512 m^2, driving two 60 kW motors. The energy storage device is a lithium-ion battery with a mass of 8.5 tons in two hulls and can be used for 72 h without sunlight. Its speed can reach 25 km/h and can carry up to 50 passengers. A voyage around the equator began on Monaco on September 27, 2010. Its long journey including the Trans-Atlantic Panama Canal, the Pacific Ocean, the Indian Ocean, the Suez Canal, and the Mediterranean Sea, it sailed as close to the equator as possible to get as much sunlight. The global travel was completed on May 4, 2012, and covered 52 cities in 28 countries.

There are already many solar yachts on the international market. In November 2011, Xiamen, China, also built a yacht with a length of 15 m and a width of 6 m. All of its power came from the PV array installed at the top, under bright sunlight. It can produce 200 W power per square meter, which can drive two 7.5 kW motors, the maximum speed of sailing is 9 km. The yacht can comfortably carry 40 persons (Fig. 10.34).

It can be expected that with the advancement of science and technology, the efficiency of solar cells will gradually increase, and the cost will continue to drop, solar yachts may be the earliest possible application for promotion in solar transportations.

10.1.3.6.3 Solar-powered car

The automobile is the most common mode of transportation in the modern era. It consumes a large amount of liquid fuel and causes serious air pollution. With the gradual depletion of oil reserves, people began to explore the use of clean renewable energy as a driving force, so solar cars came into being. For more than

Fig. 10.33: World's largest solar yacht Tûranor (2012).

20 years, various countries have produced a variety of solar cars. The first World Solar Car Challenge was held in 1987 and was held every 3 years in Australia since then. After 1999, it was held once every year. This event is currently the worldwide largest and longest running solar car competition, which reflects the current highest level of development in the solar car field. Its successful holding has also brought more and more attention to clean energy issues.

The 2015 World Solar Car Challenge was held from October 17 to 25 (Fig. 10.35). There were 45 solar cars from 25 countries State participate in the competition, including Colombia, South Africa, Iran, China, Sweden, Japan, and the United.

Fig. 10.34: Xiamen solar yacht.

Starting from Darwin, the capital of Australia's Northern Territory, it heads south to Adelaide, South Australia, and is 3,000 km long in total. The competition is from 8 am to 5 pm every day, the temperature sometimes exceeds 37.8 °C. After 5 days of hard work, the "Nuna 8" solar car from Delft University of Technology in the Netherlands has won the championship (Fig. 10.36). The car has a mass of only 150 kg and an average speed of 95–100 km/h. This is the sixth time the team has won this tournament. Another Netherlands "Twente" solar car was runner-up with only 8 min behind.

Recently, some organizations also announced that they have developed several solar cars.

As the area of the top of the car is small, and the direction will change constantly, solar cells can only be installed in line with the roof curvature. Even during the daytime, some solar cells often do not receive sunlight. Therefore, the provided electricity

Fig. 10.35: Solar car challenge is about to begin.

Fig. 10.36: Nuna 8 solar car arrives destination.

is limited, and the current high-efficiency solar cell price is very expensive. In addition, the car is a convenient means of transportation, and the use of solar cars is related to the weather. If it encounters rainy weather for a long time, use will cause problems. Therefore, it will take a certain time for solar car development to enter the actual market.

10.1.3.7 Lighting power

PV power generation can be used everywhere, and for small-sized power equipment, power can be supplied nearby and it is not necessary to connect the power grid remotely. Used as a lighting power source, it is very convenient, and it has begun to be widely used, and currently is the largest number of applications in PV systems [8].

There are many types of solar lighting systems, and the most widely used solar lighting types are as follows:
- Small-sized lighting fixtures, such as solar street lights (Fig. 10.37), projection lights, patio lights, lawn lights, and handhold flashlights.

Fig. 10.37: Solar street lights.

- Large-scale lighting systems, such as airport runway lights, hotel outdoor lights, billboard lights, and highway tunnel lights (Fig. 10.38).
- Traffic lights such as signal lights and warning lights (Figs. 10.3 and 10.4)

Solar lighting system is characterized by the daytime load basically does not consume electrical energy, when the sun is shining, the power generated by PV modules is stored in the battery, which is used at night.

At present, the most widely used are solar street lights, garden lights, and solar traffic lights. Such systems generally consist of PV modules, batteries, controllers, and lamps. Usually, all components are integrated together because of the overall

Fig. 10.38: Highway tunnel lights.

transportation and installation, there is no need to set up transmission lines or trenches to lay cables, no damage to the environment, installation, and use is very convenient, especially suitable for use in parks, sports fields, museums, and other places.

However, at present, solar lights is not widely used in main roads, due to the need to meet high illumination requirements, and the required power of light sources is relatively large, so the required capacity and area of PV modules are large, but too large modules are difficult to install on the top of solar street lights. Even if they can be installed, they are difficult to withstand loads such as wind, and their appearance is not attractive. Some solar lighting fixtures are currently having low reputations, mainly due to improper design, insufficient capacity for the configuration of solar cells and batteries, poor quality of supporting components, and improper installation locations. For example, some PV modules do not face the equator, or have buildings or trees block the sun, so severely affect the actual use.

10.1.4 Grid-connected distributed system

The distributed PV refers specifically to a PV system that is built near the user's site, operated by the user, and for user's self-consumption. It may use power grid as a backup if PV-generated power is not sufficient, or feedback to the grid if there are excessive power generated by the PV system.

PV modules are usually installed on the roof of a building, or nearby ground and structures. The capacity of a single project usually does not exceed 20 MW. PV power is generally connected to a distribution network below 10 kV, which can make up for the lack of stability of large power grids. It can continue to supply power in the event of an accident and become an important supplement to centralized power supply.

This type of PV system has many advantages:

- Solar arrays can be installed on buildings without taking up valuable land resources.
- The power supply in a short distance without long transmission line loss. In general, there is no need for additional distribution stations, which reduces the additional transmission/distribution costs.
- Because each system is independent of each other, it can control itself and avoid large-scale blackouts and has high safety. At the same time, when the electricity demand is high in summer, the PV power generation is also high. The operation is simple, the start–stop is fast, and the automatic operation is easy to realize.
- Since electricity can be feedback to the grid or supplied by the grid at any time, the PV system does not have to be equipped with energy storage, which not only save investment, but also avoid battery maintenance and periodical replacement.
- Because it is not limited by the capacity of the battery, the electricity generated by the PV system can be fully utilized.

The distributed grid-connected PV system was initially developed under the "Solar Roof Project" by some developed countries. Generally, 3–5 kW PV systems are installed on residential roofs, as shown in Figs. 10.39 and 10.40, connected with the grid. PV power mainly meets the needs of household electricity; excess power can enter the grid. Due to favorable financial incentives, distributed grid-connected PV systems have developed very rapidly and the types of applications have gradually increased. Currently, there are probably some of the following types.

10.1.4.1 Integration of PV and buildings

According to statistics, residential and commercial buildings now consume about 20.1% of total energy. The US proposed goal is to reduce the energy consumption of new buildings by 50%, and gradually renovate the existing 15 million buildings to reduce energy consumption by 30%. One of the important measures is to promote rooftop grid-connected PV systems. In order to realize the target of "zero-energy buildings," building integrated PV (BIPV) is a valuable option. In the "BIPV Position Paper" published by the European Union in December 2016, it was pointed out that CO_2 emissions from buildings and construction sectors accounted for 30%

Fig. 10.39: Residential rooftop PV systems.

Fig. 10.40: A 3 kW residential PV system in Shanghai.

of global emissions [9]. BIPV should be promoted to achieve the target of reducing greenhouse gas emissions by 40% compared to 1990 levels by 2030.

There are mainly two ways to combine PVs and buildings [10].

10.1.4.1.1 Building-attached PV (BAPV) (Fig. 10.41)

Fig. 10.41: BAPV.

Common PV modules are installed on the roof or balcony of a building, and the output of the inverter is connected to public power grid to supply power to the building. It can also be made an off-grid system and completely powered by the PV system. BAPV does not add any additional value besides generating electrical energy. It is usually installed after the completion of the building construction. This is the primary form of integration of PV systems and buildings.

10.1.4.1.2 Building-integrated PV (Fig. 10.42)

PV modules are integrated with building components. BIPV not only generates electricity, but also provides one or more functions of the building envelope, so it can reduce the overall cost of power generation and the building.

Unlike ordinary flat plate PV modules, BIPV modules must meet the performance requirements of building materials, such as thermal insulation, electrical insulation, wind resistance, rain protection, light transmission, and aesthetics, to provide both functions of power generation and building components. In addition, the BIPV module must have enough strength and rigidity to prevent damage, easy installation, and transportation, and also consider whether the service life is match to other building components. According to the needs of the construction project, a variety of PV modules have been produced that meet the performance requirements of roof tile, curtain wall, facade, canopy, and window. The shape of BIPV modules does

Fig. 10.42: BIPV.

Fig. 10.43: 12.5 MW rooftop PV power station in Italy.

not contain only a standard rectangle, but also a triangle, a diamond, a trapezoid, or even an irregular shape. The BIPV modules can also be customized based on different requirements, for example, frameless modules, transparent modules, and the junction box can be mounted on the side instead of being mounted on the back. In order to meet the requirements of construction, BIPV modules of various colors and different degrees of transparency have been developed, which can be selected by architects to match the color of the building more harmonious with the surrounding environment.

Fig. 10.44: General Motors rooftop 11.8 MW PV power station in Spain.

Fig. 10.45: Shanghai Hongqiao Hub Station roof 6.68 MW PV power station.

Because of the huge market potential of the combination of PV and architecture, it has started research and development very early in various countries. As early as 1979, the Solar Design Associates (SDA) of USA developed a large-scale PV module with an area of 0.9 m × 1.8 m under the support of Department of Energy, and built a household roof PV test system. In 1980, the famous "Carlisle House" was built at MIT. A 7.5 kW PV array was installed on the roof, combined with a

Fig. 10.46: Shanghai World Expo Theme Pavilion 2.825 MW PV power station.

Fig. 10.47: 11.5 MW rooftop solar PV plant in India.

passive solar house and a solar collector. In addition to supplying electricity, it also provided hot water and cooling.

More than 20 years ago, Sanyo Electric Co., Ltd. of Japan developed a tile-shaped a-Si PV module, with an individual output of 2.7 W. However, due to its high price and poor performance, it has not been widely applied. Later, after continuous development and improvement, various countries have successively introduced various forms of BIPV products.

In recent years, a large number of BIPV PV power stations have been built in various countries. According to PVRESOURCES statistics, as of September 27, 2016, the top 10 rooftop PV power plants in the world are shown in Tab. 10.1. Figures 10. 43–10.47 are photos of some roof PV power plants.

In 2012, Renault Samsung Motors in Busan, South Korea, built the world's largest rooftop PV system with a capacity of 20 MW and a total investment of 56 billion won (US$ 49.3 million). It can generate 25 M kWh of electricity per year and provide power to 8,300 households. The electricity demand can reduce CO_2 emissions by 10,600 tons per year.

Shanghai Automotive Asset Management Co., Ltd. invested 500 million yuan to build a PV integrated power station with a total capacity of 55 MW. In December 2015, the Ningbo Branch of Shanghai Volkswagen Automotive Co., Ltd. announced that after six months of installation and construction, the first phase of 20 MW has been built and connected to the grid for power generation. It is estimated that about 22 M kWh of electricity can be produced each year, which can save about 6,600 tons of standard coal and reduce CO_2 emissions by 17,000 tons.

In May 2016, India announced that it had invested 1,350 million Rupees and built 11.5 MW rooftop solar PV power plant (shown in Fig. 10.47) in Jammu Dhula Baba covering more than 42 acres of roof area. The electricity generated can be used by 8,000 households and will reduce CO_2 emissions by 400,000 tons in the next 25 years.

PV power generation and building are originally two different technical fields that are completely unrelated to each other. There are many problems that need to be solved to integrate these two fields. However, with the advancement of science and technology, BIPV new products will continue to emerge, and the large-scale application of BIPV will decrease its price further. The integration of PV and buildings will become one of the most important areas of PV applications and will also be accepted by more and more architects. The BIPV, which combines the huge construction market with the flourishing development of PV, is an important symbol of the application of PV from remote rural areas into cities and has a very broad prospect for development.

10.1.4.2 PV noise barrier system

Another type of PV application with the same two functions, which does not occupy land and has great economic value, is the PV noise barrier system. By installing PV modules on both sides of roads, such as railways, highways, urban light rail, and elevated roads, both noise reduction and power generation can be achieved, which can further reduce the cost of PV systems [11].

The USA and some countries in Europe began to research and build PV noise barrier systems in the late 1980s. In 1989, TNC Consulting AG built the world's first PV noise barrier system along the A13 highway near Chur in Switzerland (Fig. 10.48). The PV noise barrier system next to the Autobahn 92 in Germany during 2003 uses 1080 I-50 modules and 6,750 I-50 CER modules with a total power of 499 kW. The

Fig. 10.48: World's first PV noise barrier.

inverters use 12 SP40000 models from Sunpower and can generate 475 K kWh every year.

In 2010, Italy built a PV noise barrier system along the SS434 Transpolesana expressway near Oppeano (Fig. 10.49) with a length of 1,700 m, an average height of 4.8 m, and an area of 8,150 m². This system has a capacity of 833.28 kW and an annual power generation of ~793K kWh.

Fig. 10.49: 833 kW PV noise barrier in Italy.

According to PVRESOURCES statistics, as of December 29, 2015, there are 10 PV noise barrier systems with a capacity of more than 180 kW in the world (as shown in Tab. 10.2), the largest of which is on the top of the German A3 highway tunnel. It has a capacity of 2.65 MW and uses 16,000 solar modules, covering a length of 2.7 km (as shown in Fig. 10.50). In 2008, Shanghai's Rail Transit Line 3 also built China's first PV noise barrier system (Fig. 10.51) with a power of 10 kW and a length of about 360 m (Tabs. 10.1–10.4).

Fig. 10.50: 2.65 MW PV noise barrier in Germany.

Of course, noise barriers and PV power generation have their own characteristics; application conditions and requirements are not the same. There are some obstacles to use noise barriers with general flat plate PV modules. For example:

- The cost is higher than the normal noise barrier.
- Noise reduction is not as effective as special sound absorbing materials.
- According to the requirements of PV power generation, the PV array should face towards the equator, and there must be certain requirements for the installation tilt angle. However, the installation of noise barriers along both sides of roads tends to change directions, and it must be limited by road width and cannot be installed at the best inclination angle. These will affect the power generation of modules.
- As PV arrays are installed individually, they are arranged along a long distance along the highway, and the connection line is very long, which will cause a large line loss.

Fig. 10.51: Shanghai Rail Line 3 PV noise barrier.

Tab. 10.1: World's top 10 rooftop PV power plants.

Rank	Capacity (MW)	Location	Comments	Commission time
1	20	Busan, South Korea	Renault, Samsung facility	2012
2	13	Kallo, Belgium	Loghidden City, Katoen Natie	2010
3	12.5	Padova, Italy	Interporto Padova	2010–2011
4	12	Amritsar, India	RSSB-EES PV Roof System	2016
5	11.9	Maubeuge, France	Renault Solar Project	2012
6	11.9	Batilly, France	Renault Solar Project	2012
7	11.8	Figueruelas, Spain	GM facility	2008
8	11	Martorell, Spain	Seat al Sol, SEAT facility	2010–2013
9	10.6	Flins, France	Renault Solar Project	2012
10	10.5	Sandouville, France	Renault Solar Project	2012

Source: PVRESOURCES

In order to reduce the cost and increase the PV system efficiency, many countries have developed noise barriers and PV modules integration and use bi-facial PV modules. Some also use AC PV modules to reduce line loss and other advantage

Tab. 10.2: World's top ten PV noise barrier systems.

Rank	Power	Location	Description	On-grid (year)
1	2.65 MW	Aschaffenburg, Germany	Noise barrier A3	2009
2	1 MW	Töging am Inn, Germany	Noise barrier A94	2007
2	1 MW	Bollberg Thuringia, Germany	Noise barrier along motorway	2015
4	833 kW	Oppeano, Italy	Noise barrier SS 434 Transpolesana	2010
5	730 kW	Marano d'Isera, Italy	Noise barrier A22	2009
6	600 kW	Freising (Munich), Germany	Noise barrier A92	2003–2009
7	365 kW	Freiburg, Germany	Noise barrier B31	2006
8	283 kW	Bürstadt, Germany	Noise barrier B57	2010
9	216 kW	Amstelvee, Dutch	Noise barrier A9	1998
10	180 kW	Vaterstetten, Germany	Noise barrier along railway tracks	2004

Tab. 10.3: Largest PV power station in each year.

Year	PV power station	Nation	Capacity（MW）
1982	Lugo	USA	1
1985	Carrisa Plain	USA	5.6
2005	Bavaria Solarpark	Germany	6.3
2006	Erlasee Solar Park	Germany	11.4
2008	Olmedilla PV Park	Spain	60
2010	Sarnia PV Power Plant	Canada	97
2011	Huanghe Hydropower Golmud Solar Park	USA	200
2012	Agua Caliente Solar Project	USA	290
2014	Topaz Solar Farm	USA	550
2015	Solar Star	USA	579
2015	Longyangxia Hydropower PV Station	China	850

Tab. 10.4: Top 10 PV power plants in the world.

Rank	Capacity (MW)	Nation	Name	Power generation (GWh/year)	Commission year
1	850	China	Longyangxia Hydropower PV Station	824	2015
2	648	India	Kamuthi Solar Power Project	648	2016
3	579	USA	Solar Star		2015
4	550	USA	Topaz Solar Farm	1301	2014
5	550	USA	Desert Sunlight Solar Farm	1287	2015
6	500	China	Yellow River Hydropower Geermu PV Power Station		2014
7	458	USA	Copper Mountain Solar Facility	1087	2015
8	345	India	Charanka Solar Park		2012
9	300	France	Cestas Solar Farm	380	2015
10	290	USA	Agua Caliente Solar Project	626	2014

technics. With the continuous advancement of science and technology, the PV noise barrier system, which is a combination of the booming PV power generation and the great potential noise barrier, has a very broad prospect for development.

10.1.4.3 Floating PV system

PV system uses a lot of land resources. The development of PV will be subject to certain restrictions, especially for countries with a limited land area. Similarly, the land in central and eastern China is relatively scarce, but PV power stations can be built using water areas such as lakes, reservoirs, and fish ponds. Therefore, in recent years, floating PV power plants have developed rapidly.

In addition to saving land, floating PV power plants have many advantages. For example:

- It can improve the power generation efficiency. Because the water body has a cooling effect on the PV modules, it can reduce the module surface temperature.
- The amount of reflected radiation from the surface of the water is larger than that from the ground, which can increase power generation.
- Due to the convenience of water intake, the surface of the PV modules can be washed frequently, which reduces the pollution loss of PV power plants.

– The installation plane is very flat and can be installed at the optimum tilt angle. There is basically no mismatch due to inconsistent shielding and orientation.

It is reported that the floating PV power plant will generate 10% to 15% more power than the equivalent PV power station installed on the ground or on the roof.

There are roughly two types of floating PV array installations. One is to float on the floating base with a fixed cable position. The other is to use a pile driver to fix the pipe pile and install a PV array on the surface when the water is not deep.

Kyocera TCL Solar of Japan has been studying and developing floating PV very early. In 2014, two floating PV power plants with capacities of 1.7 MW and 1.2 MW were built on the Yamakura Dam Reservoir (Fig. 10.52). The third 2.3 MW power station was put into production in May 2015, which has higher power generation than the original design, and power was input to Tokyo Electric Power Company. The 13.7 MW large-scale floating PV power station was completed in March 2018. It uses 50,904 pieces of 270 W modules, covers an area of 180,000 square meters, and has an annual production of electricity of 16,170K kWh, which is equivalent to burning 19,000 barrels of oil, and is sufficient for 5,000 households.

Fig. 10.52: Japan Yamakura Dam Reservoir floating PV power station.

South Korea made full use of the characteristics of floating PV power plants, and built a "sunflower floating PV power plant" (Fig. 10.53). It applied tracking and rotation systems to track the movement of the sun. The prototype power station has a capacity of 465 kW and an area of 8,000 m^2. A total of 1,550 modules consisting of 72 polysilicon cells per module are used, and the power generation is reported to increase by 22% compared with the fixed installation on the ground.

Fig. 10.53: Korean sunflower floating PV power station.

China Huainan Nihe floating PV power plant (Fig. 10.54) was designed by Hebei Energy Engineering Design Co Ltd. and built by Panyang PV Power Generation Co Ltd. It has floating arrays with a 12° tilt angle and adopts a modular subarray and centralized-connection design. In order to compare various devices, the poly c-Si double glass PV modules, mono c-Si double glass modules, and poly c-Si single glass anti-PID modules each having 265 W per module were used. This site uses

Fig. 10.54: China Huainan Nihe 20 MW floating PV power plant.

630 kW centralized inverters and 80 kW string inverters. This floating power station consists of six 3 MW systems and four 1.5 MW systems.

China Huzhou Xianghui 100 MW floating PV project was built by Talesun Solar Technologies Co. Ltd. with a total investment of about 1 billion RMB. The project occupies a total of 200 acres of fish ponds, uses more than 50,000 piles, 16 extension towers, a booster substation, boosts to 110,000 volts, and finally connects to the power grid. The first phase of the project has been successfully connected to the grid. After the project is completed, the average annual power generation is about 100 M kWh. It can save 35,000 tons of standard coal, reduce CO_2 emissions by approximately 92,000 tons, and reduce dust emissions by 280 tons.

At present, the world's largest floating PV power station is the Cixi floating PV power station in Zhejiang Province, which was invested with 1.8 billion RMB by Hangzhou Fengling Power Technology Co., Ltd., and was built by Norseman Energy Technology (Beijing) Co., Ltd. with a capacity of 200 MW, covering about 300 acres of water surface. It uses more than 750,000 pieces of modules, 320 inverters, 160 transformers, and more than 130,000 piles. The grid-connected power generation began on December 31, 2016. It is estimated that the annual power generation will be 220 M kWh, which will save 63,000 tons of standard coal and reduce 593,300 tons of CO_2 emissions (Fig. 10.55).

Fig. 10.55: 200 MW floating PV power station in Cixi.

10.1.4.4 Agricultural PV

Agricultural PV is the application of PV in modern agricultural production activities such as greenhouses and livestock and poultry breeding.

The solar PV greenhouse is a new kind of greenhouse that lays PV modules on the sunny side of the greenhouse. It not only has the ability to generate electricity, but also provides a suitable plant growing temperature and light. It can also use translucent solar modules as needed to provide the plants growing environment.

In 2012, Sardiniab of Italy built the world's largest PV greenhouse at the time, with an initial investment of 80 million euros. Its capacity is 20 MW, with 84,000 poly c-Si modules covering an area of 26 hectares. In addition to supplying greenhouse irrigation, heating, and lighting, the electricity generated is fed into the grid. In the greenhouse, 10 hectares of land were planted with some of the finest varieties of rose crops, and the rest were planted with vegetables, which yielded good economic benefits.

In Sicily, 50 kW solar modules installed in six greenhouses with 50 m × 8 m each, totaling 300 kW, and plants are grown in the greenhouse (Fig. 10.56) [12].

Fig. 10.56: Solar greenhouse in Sicily, Italy.

Shanxi Xinfu 50 MW PV Eco-Agriculture Greenhouse Project is located in Shanxi Province of China. It covers an area of 114 acres and was invested and constructed by Shanghai Aerospace Automotive Electromechanical Co., Ltd. The project officially started in July 2015 and was successfully connected to the grid on December 29, 2015. With an annual power generation of 67 M kWh, the project can reduce CO_2 emissions by about 64,000 tons per year, SO_2 emissions by about 66.8 tons, nitrogen oxides by about 66.8 tons, and dust emissions by 6.7 tons, and save 21,000 tons of standard coal.

The PV agriculture greenhouse demonstration project was designed by Hebei Energy Engineering Design Co Ltd (Fig. 10.57) and is located at an economic forest land about 1 km from Yuanlong Village, Ningxia Autonomous Region of China. The

Fig. 10.57: PV Agricultural Greenhouse in Yuanlong Village, Nixia Province.

project covers an area of about 83 acres and consists of 588 agricultural greenhouses. There are six types of greenhouses, such as glass covering, jointed-rows, double membrane double mesh, spring warm half-transparent half-opaque, single row, and herbal sheds. The plants include mushrooms, flowers, vegetables, and herbs. The three specifications of 250, 135, and 100 W PV modules are installed on the roof. The total designed capacity of PV is 200 MW, and the first phase is to build 30 MW. This project is expected to achieve a power generation of 1,013 M kWh within its lifetime.

Tea is a shade-loving plant. PV arrays that are installed above the tea trees can shield the tea from the excessive sun and does not need to occupy the land. The Yunnan Menghai 50 MW Tea trees PV power station (Fig. 10.58), which was invested with 425 million RMB by Xishuangbanna Hengding New Energy Development Co., Ltd. and designed by the Yunnan Electric Power Design Institute. The project was completed at the end of 2015 and achieved a good economic and social benefit.

10.1.4.5 Solar road

The roads are all over the world. Studies have shown that only 10% of the time on the road is spent driving, and most of the time is directly exposed to the sun. Therefore, it has been proposed to install PV modules on roads to generate electricity long time ago.

The Netherlands government cooperated with private companies and academia, and built a solar-powered bicycle lane named "SolaRoad" (Fig. 10.59) in 2014. The total length is nearly 100 m, and the generated energy can provide electricity

Fig. 10.58: 50 MW tea farm PV power station in Menghai County, Yunnan Province.

Fig. 10.59: Solar bike path under construction in Netherlands.

for three households. The Netherlands became the first country in the world to apply PV modules for road pavement.

In October 2015, COLAS, which is engaged in the construction of a transportation infrastructure in France, announced that after 5 years of research, the company and the French National Solar Energy Research Institute have jointly developed and used a polymer adhesive with only 7 mm in thickness to combined small PV modules and bonded to the road surface. Figure. 10.60 shows the ongoing load test of the solar highway. According to calculations by the French Agency for Environment and Energy Management, solar roads with a length of 1 km can support daily public lighting in a small town with 5,000 residents. Measurements show that a 20 m² solar road are sufficient to supply electricity to a French home, and a 100 m² solar road can provide 100,000 km driving by an electric car in 1 year. The French government plans to raise 220–440 million USD for solar road projects by an added gasoline taxes. It plans to build a 1,000 km long solar road within 5 years and use solar energy to provide electricity to cities.

Fig. 10.60: Solar road load test in France.

Mr. Scott Brushaw and Ms. Julie Brushaw from USA have proposed the concept of solar roads as early as 2006, and they believe that solar panels can be used instead of traditional asphalt to make full use of the sun. In the same year, they established a company called Solar Roadways and invented special PV modules after years of research and improvement. The hexagonal shape of the component 30 cm on each side consists of six layers, from top to bottom are, respectively, tempered glass, LED light indicator layer, PV layer, base layer made from recyclable materials, and flood discharge and water storage layer. Each module can generate 7.6 Wh of electricity a day. In addition, various road indication systems can be provided by the LED lighting layer, and a large amount of electric energy can be stored for quickly melting. Finally, there is an important function, that is, the solar road can be used as a huge charging device to wirelessly charge electric vehicles that are driven and parked on it.

According to a report, a one-mile, four-lane solar road can generate enough electricity every day to support 500 homes. If all the roads and parking lots in USA are covered with solar panels, they will generate about 1.34 trillion kWh of electricity, which is about three times the amount of electricity consumed in USA in 2009.

This kind of assembly has been built in the garage with a test prototype with an area of 3.7 m × 11 m (Fig. 10.61) and is ready for further testing on US Route 66.

Fig. 10.61: A prototype solar road in the USA.

Although the prospects of the solar road proposed by the USA are very attractive, there are still many obstacles. The first is that the cost is too high. The initial cost of each solar panel is US\$ 6,900, which means that laying 29,000 square miles of solar roads in the USA will cost about US\$ 56 trillion, which is a tremendous number. At the same time, there are problems such as service life and maintenance costs. In addition, some people have suggested that the LED performance during daytime is not ideal, also will accumulated energy be enough to melt snow and ice on the road, and the question, moreover, wireless charging for electric vehicles has yet to be practiced. Anyway, there is still a long way to go for solar roads to be fully promoted. It is believed that with the development of science and technology, solar road may eventually be realized.

10.1.5 Centralized grid-connected system

This kind of PV system usually has a large scale, and the generated electric energy is all feed into the grid, which is equivalent to a conventional power station. Because it is a centralized operation and management, it can take advantages of scale economy, advanced technology, and unified grid control, to reach a lower cost of power generation.

In recent years, due to advances in technology and reduction in PV module price, coupled with the implementation of support policies in many countries, large-scale PV power plants have begun to be built in large numbers, and the scale has rapidly expanded, forming booming market of the large-scale PV plant construction. At the end of 1982, the world's largest PV power plant had a capacity of 1 MW, with a record capacity of 6.3 MW in 2005. By June 2015, the capacity of the Sun Star project in the Antelope Valley in Los Angeles, California alone was 579 MW. China's Longyangxia Hydropower PV Project completed the first phase of 320 MW in December 2013, and completed the second phase of 530 MW in 2015, setting a new world record with a total capacity of 870 MW. The development of the largest PV power plants in the calendar year is shown in Tab. 10.3.

According to the Wikipedia, as of 2015, there are 50 PV power plants with a global capacity of more than 100 MW, among which 18 are from the USA, eight from China, six from India, three from Germany, and two from Chile, Canada, France, and Japan, respectively. Ukraine, Philippines, Pakistan, Honduras, South Africa, Australia, and Thailand each has one. The top 10 PV power plants in the world are shown in Tab. 10.4.

The Longyangxia Hydropower PV Station built by Huanghe Hydropower Development Co. Ltd., is a hydro-PV hybrid system, which has a total installed PV capacity of 850 MW, on an area of 20.4 km^2. It consists of 3.74 million PV modules, 10,000 combiner boxes, 1,630 centralized inverters, 1,426 string inverters, and more than 880 transformers. It is currently the largest grid-connected PV power station in the world (Fig. 10.62).

The biggest feature of Longyangxia Hydropower PV Station is the co-operation of hydropower and PV power generations. Its installed hydropower capacity is 1,280 MW and the storage capacity is 24.7 billion cubic meters. When the sunlight is strong, the electricity generated by the 850 MW PV station is sent to the grid through the hydropower station's 330 kV transmission line, and the hydropower can be deactivated or decreased. When the weather changes, at night, the hydropower can be automatically adjusted or increased to meet the power grid dispatch requirement and obtain a stable and reliable power supply. The main advantages are as follows:
- Save 70% of hydropower reserve capacity, about 400,000–600,000 kW.
- Enhance the ability to adjust peak and frequency: 18% on sunny days, 9% on cloudy days, and 5% on rainy days.
- Raise the electricity output transmission capacity by 22.4%.
- Improve power quality of the power grid: balance, high quality, and safety.

Fig. 10.62: Longyangxia Hydropower PV Station.

In September 2016, Adani Power of India announced that it took 8,500 installation workers in Tamil Nadu to spend 8 months to build a PV power plant with a total capacity of 648 MW (Fig. 10.63). The total investment of the power station is 45.5 billion rupees (US\$ 679 million). The power station uses a total of 2.5 million

Fig. 10.63: 648 MW PV power station in India.

PV modules, 576 inverters, 154 transformers, and 6,000 km of electrical cables. It is connected to a 400 kV substation, and the electricity generated can meet the demand of 150,000 households.

The third-largest Solar Star PV power station (Fig. 10.64), with a capacity of 579 MW, is located close to Rosamond, California, USA. The construction began in 2013 and was completed in June 2015. It covers an area of 13 km^2 and uses 1.7 million mono c-Si PV modules produced by Sunpower Corp. The system uses single-axis trackers, and it can increase power generation by up to 25%, generate enough electricity to satisfy 255,000 households. The generated power is sold to Southern California Edison utility company. It has an annual reduction of CO_2 emissions by 570,000 tons, which is equivalent to the emissions of 108,000 vehicles on the road.

Fig. 10.64: Solar Star PV power station in the USA.

It can be seen that the PV power plants built all over the world are rapidly developing toward very large scale (VLS), which reflects the fact that the cost of PV power generation is continuously declining and is already to approach the level of competition with conventional power. According to a prediction of Energy From The Desert (IEA-PVPS Task8: 2015), by the end of the twenty-first century, PV power will account for 1/3 of the world's primary energy supply. By 2100, the cumulative installed capacity of PV will reach 133TW, of which half of them will be the VLS-PV power plants (as shown in Fig. 10.65). The VLS-PV plants will play an increasingly important role in the energy supply structure.

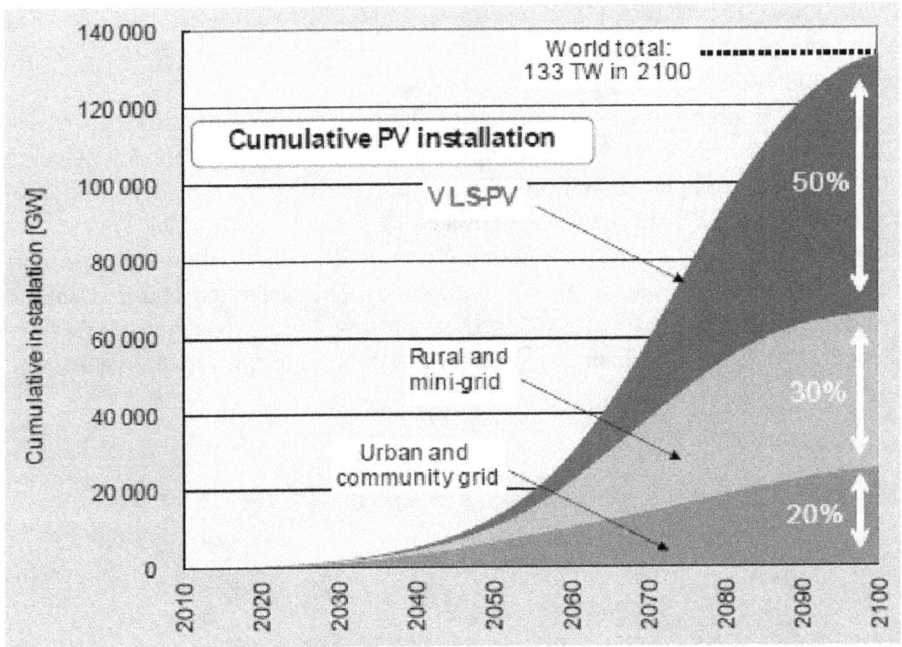

Fig. 10.65: VLS-PV development roadmap.
Source: Roadmap of Energy from the desert (IEA-PVPS Task8: 2015)

10.1.6 Hybrid PV system

In the case of off-grid, a system with PV and other kinds of power generation technologies working collectively to supply electric power to the loads is called a hybrid PV system.

10.1.6.1 PV-diesel hybrid power system

Due to the current high price of PV systems, if PV power generation is used to meet the demand for larger loads, off-grid PV systems must support load operation in the worst winter weather conditions. This requires a large capacity of PV array and storage battery. However, when the solar irradiation is large in summer, excess power could only be wasted. In order to solve this contradiction problem, a diesel generator can be configured as a standby power source. The system is usually powered by a PV system, and when the solar irradiation is insufficient in winter, the diesel generator is started to supply power, which can often save initial investment. Of course, this requires certain conditions, such as the large difference between the local winter and summer solar irradiation, the ability to equip diesel engines and other equipment, and to ensure the reliable supply of diesel fuel, as well as technicians who operate and maintain diesel generators, and so on.

10.1.6.2 PV–wind hybrid power system

Both wind energy and solar energy have weaknesses such as low energy density and poor stability and are affected by geographical locations, seasonal changes, and day-night alternations. However, solar energy and wind energy have a certain degree of complementarity in time and area. In the daytime, when the sunlight is the strongest, the wind is usually smaller. After dusk, the illumination is very weak, and the wind power increases because the ground surface temperature changes greatly. In summer, the solar irradiation is high and the wind is small; in winter, the solar irradiation is weak and the wind is large. Solar power generation is stable and reliable, but the current cost is high. Wind power generation costs are lower, but the wind speed fluctuation is large and the reliability of power supply is poor. If we combine these two together, we can benefit from each other and achieve both day-night power supply and cost reduction. Of course, since PV and wind power have their own characteristics, in the design of the system, we must fully grasp the meteorological data and carefully optimize the design to achieve good results. This type of system is also commonly used to power street lights (Fig. 10.66).

Fig. 10.66: Wind–PV hybrid system.

The PV-wind hybrid power generation system is usually composed of PV array, wind turbine, controller, battery, inverter, and accessories. The schematic diagram is shown in Fig. 10.67.

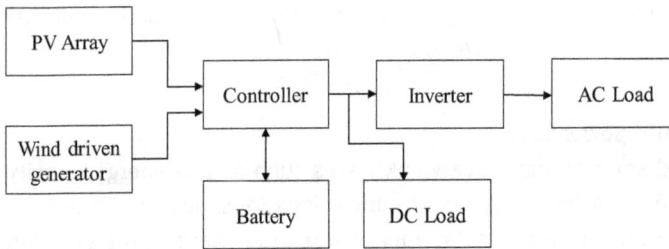

Fig. 10.67: Schematic diagram of wind–PV hybrid system.

10.2 Space PV power station

When sunlight hits the ground through the atmosphere, it loses about 1/3 of its energy. In addition, if a PV system is placed in the geostationary orbit of space, the night time will occur once a day for 45 days before and after the vernal equinox and the autumnal equinox, and the maximum duration will not exceed 72 minutes, so the accumulated night time will be less than 4 days a year, which indicates that 99% of the year is sunshine. In comparison, PV modules that are installed on the Earth ground surface spend half time during night and do not work, and the sunshine on the module is tilted, except at noon on the ground during the daytime. The daily solar energy received in space is about 32 kWh/m^2, and the average daily solar energy received on earth ground is only 2–12 kWh/m^2. Therefore, if we can build a PV power station in space, the result will be much better than on the ground.

As early as in 1968, an American engineer Glaser proposed a creative idea of building a Solar Power Satellite (SPS) on the geostationary orbit of 36,000 km above the ground. Imagine using billions of PV modules placed on huge slabs to generate electricity under sunlight and convert it into radio microwaves and send it to receiving stations on earth surface. After receiving on the ground, microwaves can be returned to direct current or alternating current and transmitted to users.

At the end of the 1970s, there was a global oil crisis. The US government organized experts to conduct feasibility studies and a "1979 Reference SPS System" was proposed. It is assumed that the system consists of 60 space PV stations with 5 GW each, each of them is 10 km long, 5 km wide, and the total output power is 300 GW. The system has 3 million tons' mass. One end of the PV system is connected to a microwave transmitting antenna which has a diameter of 1 km. The status control system keeps the PV array always facing the sun. The pointing mechanism keeps the transmitting antenna always aligned with the earth station. The power generated by the PV station is converted into microwaves by a microwave generator and then sent to the ground via an antenna. The operating frequency of the microwave is 2.45 GHz or 5.8 GHz. The ground receiving antenna is a 13 km × 10 km elliptical area covering about 10,000 hectares and consists of many half-wave dipole antennas.

The microwave energy received by the antenna is converted into alternating current. Since the area of the ground antenna is very large, the power density of the microwave beam reaching the ground is very small, the center of the microwave beam is approximately 20 mW/cm^2 and the edge is only 0.1 mW/cm^2. Therefore, the microwave beam will not cause harm to people, animals, and crops. The ideal receiving antenna is shaped like a grid, and it is raised high with a column. The mesh can let air, water, and sunlight pass through. In this way, the land below the antenna can grow crops, grazing cattle, and sheep or other uses as usual. The US 1979 reference SPS proposed technical parameters are shown in Tab. 10.5.

Tab. 10.5: US "1979 Reference SPS System" technical parameters.

System composition	Number of satellites	60
	Power (GW)	60×5
	Lifetime (year)	30
Space power station	Single satellite mass (107 kg)	3 ~ 5
	Size	10 km × 5 km × 0.5 km
	Material	Carbon fiber composite material
	Orbit	36,000 km high stationary orbit
PV system	PV material	Si or GaAs
Power transmission system	Transmitting antenna diameter (km)	1
	Frequency (GHz)	2.45
	Ground-receiving antenna size	13 km×10 km (ellipse)

As the entire system is too large, it will require about US$ 250 billion of investment and 18,000 person-years of in-orbit workload (equivalent to 600 astronaut assembly workers working in space for 30 years). The study was discontinued in 1981.

From 1995 to 1997, the USA NASA also organized experts to conduct a new round of research and demonstration. A total of 29 different scenarios were analyzed and compared, of which the SunTower and "SolarDisk" were mostly favored. The problem was still that the investment is huge and some technical problems have yet to be resolved.

In the summer of 1995, celebrities in the aerospace community from 16 countries conducted a study entitled "Development Vision 2020" at the International Space University. The research report put forward four basic plans, one of which is to develop the SPS system, and it is expected that solar-powered satellites will enter the practical stage in 2010–2020.

Japan established the SPS research group in 1987 and completed the model design of the "SPS 2000" satellite in 1993. A variety of SPS solutions were later introduced. One of the distributed tethered satellite proposal is tethered SPS weighted 26,562 MT totally, size of 2.5 km x 2.375 km x 10 km with total of 250 PV modules, and total of 1.36 GW output power; each PV module weighted 42.5 MT, tether length 10 km, panel size of 100 m x 95 m; each power generation/transmission module weighted 4.25 kg, and power output 230 W. This whole system is easier to assemble and maintain, but its weight is still too high for launching.

Later, it was proposed to use local materials and build solar power stations on the moon. This would greatly reduce launch cost.

Some European countries and China are also conducting extensive research on space PV stations. In addition to the large amount of funds required for the development of space PV power stations, there are still many technical issues that need to be solved. However, with the development of society and the advancement of science and technology, these problems will be gradually solved, and PV space power stations will eventually be launched. It can be imagining that one day, if we can really use a large number of space PV power stations, it will be possible to solve the human power supply problem forever.

References

[1] Erik H. Lysen. Pico solar PV systems for remote homes. Report IEA-PVPS T9-12:2012 ISBN: 978-3-906042-09-1 January 2013.

[2] Janet L Sawin. et al., Renewables 2016·global status report. ISBN: 978-3-9818107-0-7. REN21 Secretariat. Paris 2016. www.ren21.net.

[3] Michael Taylor. Eun young So. Solar PV costs and markets in Africa. ISBN: 978-92-95111-47-9. IRENA Bonn, 2016.

[4] J. Schlangen and J. C. Aguillon. 1992, PV solar home systems in Lebek-west Java – Indonesia [J]. Proceedings of the 11th E.C. Photovoltaic Solar Energy Conference, Montreux: Switzerland, 1539–1541.

[5] Crowley L. A. Utility owned remote PV systems, [J]. Solar Today, March/April, 1993, 29–30.

[6] Teresa D. Morales. Design of small PV (PV) solar-powered water pump systems. USA department of agriculture technical note No. 28, October 2010.

[7] R. William. Young Jr. History of applying photovoltaics to disaster relief. Florida Solar Energy Center. FSEC-CR-934-96 http://www.fsec.ucf.edu/en/publications/pdf/FSEC-CR-934-96.pdf.

[8] Caifu Wu, Jianxuan Zhang, Yukai Chen, Photovoltaic energy supply and lighting system overview (Second Edition). Taibei: Chuan Hwa Publishing Ltd, 2007.

[9] Silke Krawietz, Jef Poortmans. BIPV position paper. Secretariat of the European technology and innovation platform PVs, December 2016. http://www.etip-pv.eu.

[10] Patrina Eiffert, Gregory J Kiss. Building-integrated PV designs for commercial and institutional structures. A sourcebook for architects. NREL/BK-520-25272.February 2000.

[11]   P. Bellucci. et al., Assessment of the PV on noise barrier along national roads in Italy. 3rd World Conference on PV Energy Conversion. May 2003, Osaka Japan.
[12]   Filippo Sgro. Efficacy and efficiency of Italian energy policy: The case of PV systems in greenhouse farms[J]. Energies 2014, 7, 3985–4001. doi:10.3390/en7063985.

Exercises

10.1   Briefly describe the five types of applications of ground-based PV systems and illustrate them separately.

10.2   What are the main components and functions of the off-grid PV system?

10.3   What are the two forms of PV with buildings? Please explain separately.

10.4   What are the advantages of BIPV? What are the difficulties in its implementation?

10.5   When is a PV hybrid system suitable for use? What kinds of hybrid PV systems are currently used?

10.6   If PV power is a driving force entirely, what kind of transportation means is easier to achieve? What are the reasons?

10.7   Explain the advantages and disadvantages of solar street lights. At present, the use of solar street lights is not satisfactory, and what are the main reasons?

10.8   What do you think of the prospects to building a space PV power plant?

Chapter 11
Benefit analysis of PV power generation

In modern society, without electricity supply, it is impossible to promote scientific and technological progress, and it is difficult to improve people's living standards. In areas where conventional power grids cannot be reached, the scope and scale of application of PV systems are increasingly expanding. They have played an important role in solving power generation and electricity supply problems in unpowered areas and have achieved significant social and economic benefits. In some areas, PV power generation has reached a level that can compete with conventional power price, which indicates that solar power has entered a new era and will occupy a growing share in the energy field. However, there are still some problems which need to be further explored for how to evaluate the actual benefits of PV power generation.

11.1 Economic benefits of PV power generation

11.1.1 The cost of PV power generation

There are many kinds of power generation methods now. In order to objectively compare the economic benefits of various power generation methods, one of the most important indicators is the levelized cost of electricity (LCOE). It refers to the cost of the production unit of electricity (usually 1 kWh) and can be expressed by the following equation:

$$\text{LCOE} = \frac{C_{\text{total}}}{E_{\text{total}}} \tag{11.1}$$

where C_{total} is the sum of investment cost; E_{total} is the total amount of electricity actually produced.

11.1.1.1 Calculation of LCOE

11.1.1.1.1 Methods of calculating LCOE
(1) A general method is to use the equation published in "A manual for the economic evaluation of energy efficiency and renewable energy technologies" by the National Renewable Energy Laboratory (U.S) in 1995 [1]:

https://doi.org/10.1515/9783110524833-011

$$LCOE = \frac{TLCC}{\sum_{n=1}^{N} \left[\frac{E_n}{(1+d)^n} \right]} \tag{11.2}$$

where E_n is energy output in year n; d is discount rate; N is analysis period; TLCC is total life cycle cost.

TLCC can be calculated with the following equation:

$$TLCC = \sum_{n=0}^{N} \frac{C_n}{(1+d)^n}$$

where C_n is investment cost in the nth year of the lifetime, including financial cost, expected residual value, nonfuel operating and maintenance cost, replacement cost, and energy consumption cost;

d is the discount rate.

(2) In order to meet the needs of different energy technologies, RETScreen's financial analysis model brings in standard financial terms that are cited from many financial textbooks, such as Brealey and Myers (1991), or Garrison et al. (1990) and makes the following assumptions:

The start of investment year is the 0th year.

Calculated cost and loan start from the 0th year, and the inflation rate is calculated from the first year.

Cash flow occurs at the end of each year.

The cost of energy production is determined based on the idea of avoiding a net present value of zero, and an equation can be obtained in this condition:

$$NPV = \sum_{n=0}^{N} \frac{\tilde{C}_n}{(1+r)^n} = 0$$

where NPV is the net present value; r is the discount rate; n is year; $\tilde{C}_n$ is the after-tax cash flow, which is determined by the following formula:

$$\tilde{C}_n = C_n - T_n$$

where T_n is the nth year's tax; C_n is the nth year pre-tax cash flow, determined by

$$C_n = C_{in,n} - C_{out,n}$$

$C_{in,n}$ is the cash income:

$$C_{in,n} = C_{ener}(1+r_e)^n + C_{capa}(1+r_i)^n + C_{RE}(1+r_{RE})^n + C_{GHG}(1+r_{GHG})^n$$

C_{ener} is the annual sales income of energy (electricity); C_{capa} is the annual income generated from the increase of capacity; C_{RE} is the annual income of renewable energy products; r_{RE} is the growth rate of renewable energy; C_{GHG} is the income of greenhouse gas (GHG) emission reduction from clean development mechanism (CDM) indicator sales; r_{GHG} is the growth rate of GHG emission reductions.

In the final year of project completion, the increase in expenditure due to inflation should be added to the right side of the equation.

Cash expenditure $C_{out,n}$ is determined by

$$C_{out,n} = C_{O\&M}(1+r_i)^n + C_{fuel}(1+r_e)^n + D + C_{per}(1+r_i)^n$$

where $C_{O\&M}$ is the operating and maintenance cost of the year; r_i is the inflation rate; C_{fuel} is the cost of fuel or electricity for the year; r_e is the energy growth rate; C_{per} is the periodic cost of the system payments; D is the number of debt service payments for the year, determined by

$$D = Cf_d \frac{i_d}{1 - \frac{1}{(1+i_d)^{N'}}}$$

where C is the total project initial investment cost; f_d is the debt ratio; i_d is the annual interest rate of the effective debt; N' is the debt year.

The above method for calculating the cost of power generation can be applied to various power generation technologies. However, different power generation methods have different influence factors. Therefore, the specific content involved is also not the same.

11.1.1.1.2 Calculating PV LCOE

There are mainly several methods below about calculating PV LCOE:

(1) A white paper published by SunPower in 2008, proposed a simplified calculation formula for grid-connected PV LCOE [2]:

$$LCOE = \frac{C_{pro} + \sum\limits_{n=1}^{N} \frac{AO}{(1+DR)^n} - \frac{RV}{(1+DR)^n}}{\sum\limits_{n=1}^{N} \frac{E_{ini}(1-SDR)^n}{(1+DR)^n}} \tag{11.3}$$

where C_{pro} is the initial investment of the system; AO is the annual operating cost; DR is the discount rate; SDR is the system decay rate; RV is the residual value; N is the number of years of system operation. However, in the formula, it did not fully reflect the impact of taxes, subsidies, and other related factors. In 2010, it also published a modified formula:

$$\text{LCOE} = \frac{\text{PCI} - \sum_{n=1}^{N} \frac{\text{DEP}+\text{INT}}{(1+\text{DR})^n}\text{TR} + \sum_{n=1}^{N} \frac{\text{LP}}{(1+\text{DR})^n} + \sum_{n=1}^{N} \frac{\text{AO}}{(1+\text{DR})^n}(1-\text{TR}) - \frac{\text{RV}}{(1+\text{DR})^n}}{\sum_{n=1}^{N} \frac{E_{\text{ini}}(1-\text{SDR})^n}{(1+\text{DR})^n}} \quad (11.4)$$

where PCI is the project cost minus any investment tax credits or subsidies; DEP is the depreciation value; INT is the interest rate; LP is the loan payment; TR is the tax rate.

(2) Branker et al. studied and summarized the previous research on PV LCOE. In "A review of solar photovoltaic levelized cost of electricity" published in 2011, it was proposed that the sum of LCOE present value multiplied by the sum of generated energy should be equal to the net present value of cost, i.e. [3]:

$$\sum_{t=0}^{T} \left[\frac{\text{LCOE}_t}{(1+r)^t} \times E_t \right] = \sum_{t=0}^{T} \frac{C_t}{(1+r)^t}$$

Therefore,

$$\text{LCOE} = \frac{\sum_{t=0}^{T} \frac{C_t}{(1+r)^t}}{\sum_{t=0}^{T} \frac{E_t}{(1+r)^t}}$$

Net cost should include cash flow, such as initial investment (through equity or debt financing). The debt financing requires payment of interest, operating, and maintenance cost (PV system has no fuel cost). The government incentives should be included in the cash flow. Therefore, the calculation of net cost should also consider financing, taxation, and incentives, and expand the initial definition. If the LCOE is to be compared to the grid price, it must include all cost (including transmission and connection cost, etc.), so future projects must carry out dynamic sensitivity analysis. In the absence of consideration of incentives, the PV LCOE can be expressed by following equation:

$$\text{LCOE} = \frac{\sum_{t-0}^{T} \frac{I_t+O_t+M_t+F_t}{(1+r)^t}}{\sum_{t-0}^{T} \frac{E_t}{(1+r)^t}} = \frac{\sum_{t-0}^{T} \frac{I_t+O_t+M_t+F_t}{(1+r)^t}}{\sum_{t-0}^{T} \frac{S_t(1-d)^t}{(1+r)^t}} \quad (11.5)$$

where T is the project lifetime (year); t is the number of the year; E_t is the annual power generation amount; I_t is the investment cost of the system in year t; O_t is the operation cost in year t; M_t is the maintenance cost in year t; F_t Is the interest expense in year t; r is the discount rate; S_t is the power generation amount in year t; d is the decay rate.

(3) In the "PV LCOE in Europe 2014–2030" report published on June 23, 2015 by the Photovoltaic LCOE Working Group of the European Photovoltaic Technology Platform, the following equation is used to calculate the cost of PV LCOE [4]:

$$LCOE = \frac{CAPEX + \sum_{t=1}^{n}\left[OPEX(t)/(1+WACC_{NOM})^{t}\right]}{\sum_{t=1}^{n}\left[Utilization_{0} \cdot (1-Degradation)^{t}/(1+WACC_{Real})^{t}\right]} \qquad (11.6)$$

where t is time (in years); n is economic lifetime of the system (in years); CAPEX = total investment expenditure of the system, made at $t = 0$ (in €/kWp); OPEX(t) is operation and maintenance expenditure in year t (in €/kWp); $WACC_{Nom}$ is nominal weighted average cost of capital (per annum); $WACC_{Real}$ is real weighted average cost of capital (per annum); $Utilisation_{0}$ is initial annual utilization in year 0 without degradation (in kWh/kWp), and

$$WACC_{Real} \text{ is} (1+WACC_{Nom})/(1+Inflation) - 1$$

Inflation is the annual inflation rate.

11.1.1.2 Specific influencing factors
Although there are many forms of methods to calculate PV LCOE, the general principle is actually the ratio of the total input cost to the total power generated, as shown in eq. (11.1). When actually calculating the cost of power generation, due to the financial analysis involving PV system output and investment income, the annual cash flow during the lifetime should be analyzed in accordance with finance requirements and by financial professionals, and the influencing factors are quite complex. For a rough estimate, it is feasible to make some simplification. If inflation factor is not taken into account, the two major components can be roughly divided into the following two parts.

11.1.1.2.1 Input section
The input mainly includes the following aspects:

$$\sum C_{total} = \sum C_{ini} + \sum C_{O\&M} + \sum C_{rep} + \sum C_{int} - \sum C_{CDM} - \sum C_{sub} \qquad (11.7)$$

where $\sum C_{total}$ is the total investment cost of the project during the lifetime. According to the current technology of c-Si PV modules, it assumes that the service life can reach 25 years, so the working lifetime of PV plants can be set as 25 years. As the technology advances, it may be gradually extended in the future.

$\sum C_{ini}$ is the initial investment cost, including cost of purchasing or renting for all equipment, supporting components and land; construction of facilities, civil works (foundation, power distribution room, central control room, dormitory, roads), transportation, construction and installation during the construction of PV stations, as well as grid-connection, design, management, and other related costs.

$\sum C_{O\&M}$ is the operation and maintenance cost, including raw material consumption, operation and maintenance cost, repair cost, management staff salary, and welfare and other expenses.

$\sum C_{rep}$ is the cost of replacing equipment and devices. Some equipment and components in the system have a working lifetime less than 25 years, so these equipment and devices need to be replaced before the end of the system working lifetime. For example, a typical inverter has a working lifetime of 10–15 years, so it needs to be replaced once in the middle. However, the working life of some brands of inverters can now reach 25 years. The warranty of inverter can also be extended by increasing price. In this way, it does need to consider this part of the cost.

In addition, the cost of dismantling, cleaning up, and other aftermath work should be considered after the end of the lifetime of PV power station.

$\sum C_{int}$ is paid interest cost. The construction of a large-scale PV station requires a lot of investment. Generally, it needs to make a loan from a bank. This means that it is necessary to pay interest to the bank year by year. Strictly speaking, changes in loan interest rates and inflation must be taken into consideration.

The last two terms in eq. (11.7) are actually revenues. Since PV power is a clean energy source and reduces carbon dioxide emissions, $\sum C_{CDM}$ is an income derived from trading CDM credits; $\sum C_{sub}$ is obtained from government subsidies and tax credits or deductions.

11.1.1.2.2 Output section

The output of grid-connected PV station is mainly the revenue generated from the sale of electrical power based on net price. This depends to a large extent on the amount of power generated by the PV plant. Obviously, this is related to the local solar irradiation conditions and the energy efficiency ratio of the system. The annual power generation of the PV power station can be calculated according to eq. (8.4):

$$E = H_t P_0 \cdot PR$$

If the degradation of solar cell is considered, the total power generation during the lifetime is calculated according to eq. (8.5):

$$E_{total} = H_t P_0 \cdot PR \sum_{t=0}^{T} (1-d)^t$$

where d is module degradation rate.

LCOE can be obtained by dividing the input to the output. However, this is only the balance that assumes the investment just be recovered during the working life. LCOE is not a feed-in price to the grid company. PV stations are required to pay taxes by government regulation, if there is a profit. Therefore, when determining feed-on price, profit and tax must be added in addition to the cost of power generation.

Example 11.1 It is planned to invest in the construction of a 20 MW PV station in the Lanzhou area of Gansu Province, China. The source of funding will be a 20% equity fund and 80% financing from a bank loan. The debt period is 20 years and the annual interest rate is 4.9%. Assume that no funding from government or other sources has been obtained. In order to simplify the calculation, it is assumed that factors such as inflation and interest rate changes are not considered. The system performance ratio is taken as 0.80, and the annual decay rate of the module is 0.8%. What is the LCOE?
Solution 1. Not consider the discount rate first and make a rough estimate.

(1) Basic data

(a) Expected initial investment cost $\sum C_{ini}$ is listed in Tab. 11.1.

Tab. 11.1: Initial investment for a 20 MW PV station.

Item	Unit	Amount	Unit price (yuan)		Total price ($\times 10^4$ yuan)		Others
			Equipment	Installation	Equipment	Installation	
PV module	W	22,119,020	3.20	0.12	7,078	265	
Bracket	t	1,083.63	6,500	2,053.32	704	223	
Inverters and other equipment	set	20	250,000	11,487.52	500	23	
DC combine box	unit	266	4,500	521.48	120	14	
Transformers and accessories					500	102	
PV cable engineering						581	
Booster equipment and installation engineering					373	51	
Control protection equipment and installation					553	74	
Other equipment and installation engineering					68	160	
Power plant infrastructure engineering and construction					1,834		

Tab. 11.1 (continued)

Item	Unit	Amount	Unit price (yuan)		Total price (x10⁴ yuan)		Others
			Equipment	Installation	Equipment	Installation	
Project construction land fee							198
Project construction management fee							424
Production preparation fee							100
Survey and design fee							96
Other taxes							95
Total above					9,896	3,406	913
Sum					14,215		

It can be seen that the initial investment is about 143 million yuan, of which PV module is close to 50%. The discussion above is only a rough estimation. Actual expenditure may be different.

(b) The lifetime of the PV station is 25 years, and the operation and maintenance cost is 1 million yuan per year.

(c) Cost for replacement of equipment and components (such as the inverter needs to be replaced once) is 5 million yuan.

(d) Loan interest: The project self-raised funds accounted for 20% of the initial investment, totaling 28.6 million yuan, and the remaining 114.4 million yuan is bank loan. The repayment period is 20 years, and interest rate is 4.9%.

(e) Calculate power generation

The longitude and latitude of Lanzhou are 103.73° and 36.03°, respectively, and the average annual exposure irradiance is 1401.6 kWh/m² on the horizontal surface. If the installation is at the optimum tilt angle of 25°, the average daily irradiation dose on the tilted plane is 4.077 kWh/m²/day (as shown in Table 8.1), then the annual average peak sunshine duration on the local slope are $H_t = 4.077$ (h/day) × 365 (days) = 1488.1 (h).

(2) Specific calculations

Based on eq. (11.7):

The initial cash cost of the system: $\sum C_{ini} = 28.6$ million yuan.

Maintenance cost: $\sum C_{O\&M} = 25 \times 100 = 25$ million.

Replace equipment cost: $\sum C_{rep} = 5$ million.

Payment of interest: loan 114.4 million yuan, annual interest rate 4.9%, repayment period 20 years. Using Excel's PMT formula, it can be calculated that the annual installment of the loan is 9.102 million yuan, and the total interest was

$$\sum C_{int} = 20 \times 9.102 \text{ million yuan} = 182.04 \text{ million yuan}$$

Therefore, the total investment cost of the PV power plant during its lifetime is

$$\sum C_{total} = (28.6 + 25 + 5 + 182.04) \text{ million} = 240.64 \text{ million yuan}$$

Calculate the total power generation based on eq. (8.5):

$$E_{total} = H_t P_0 \cdot PR \sum_{n=0}^{N} (1-r)^n = 1488.1 \times 20 \times 0.8 \times \sum_{n=0}^{24} (1-0.008)^n = 541406.5 \text{ MWh}$$

Therefore:

$$LCOE = \frac{C_{total}}{E_{total}} = \frac{140.64 \text{ million yuan}}{541406.5 \text{MW} \times \text{h}} = 0.44 \text{ yuan/kWh}$$

Solution 2. Although the above calculation is relatively simple, it does not consider the time value of money. Funds that are deposited in banks have interest, which means that the value of 10 yuan last year is a little higher than the value of 10 yuan this year. The difference between the two is reflected in the discount rate. Assuming a discount rate of 10%, then use $10/(1+10\%) = 9$, which means that this year's 10 yuan is only equivalent to 9 yuan last year. Similarly, $10/(1+10\%)2 = 8.3$, which means that this year's 10 yuan is only equivalent to 8.3 yuan of the previous year. Therefore, the financial net cash flow of each year in the project lifetime should be discounted according to a given standard. The conversion of the rate to the sum of present value at the beginning of construction should be calculated according to formula (11–7), as follows:
The initial cash cost of the system:

$$\sum I_t = 28.60 \text{ million yuan}$$

Assume that the annual discount rate is 3%. Since the 25-year project lifetime starts from the year of completion, the integral upper and lower limits are calculated from 0 to 24.
Maintenance cost present value:

$$\sum_{0}^{T} \frac{O_t}{(1+r)^t} = \sum_{t=0}^{24} \frac{100}{(1+0.03)^{24}} = 17.416 \text{ million yuan}$$

In the 12th year, the current value of 5 million inverters is replaced once, then the present value is

$$\sum_{0}^{T} \frac{M_t}{(1+r)^t} = \frac{M}{(1+r)^{11}} = \frac{500}{(1+0.03)^{11}} = \frac{500}{1.384} = 3.614 \text{ million yuan}$$

Calculate loan interest: loan amount 114.4 million yuan, repayment period 20 years, interest rate 4.9%. The PMT formula of Excel can be used to find the annual amortization of the loan at the fixed interest rate of $F_t = 9.102$ million yuan, and then the value of the formula is calculated according to Excel's PV formula:

$$\sum_{0}^{T} \frac{F_t}{(1+r)^t} = \sum_{t=0}^{19} \frac{9.102}{(1+0.03)^t} = 139.47 \text{ million yuan}$$

It can be obtained that the present value of the total investment of PV plants in the lifetime as:

$$\sum C_{total} = 28.6 + 17.416 + 3.614 + 139.47 = 189.1 \text{ million yuan}$$

The present value of the total power generation of PV plants during the lifetime is:

$$E_{total} = H_t P_0 \cdot PR \sum_{t=0}^{T} \frac{(1-d)^t}{(1+r)^t} = 1,488.1 \times 20 \times 0.8 \times \sum_{t=0}^{24} \frac{(1-0.008)^t}{(1+0.03)^t} = 393120 \text{ MWh}$$

Therefore:

$$LCOE = \frac{C_{total}}{E_{total}} = \frac{189.1 \text{ million yuan}}{393120 \text{ MWgh}} = 0.48 \ yuan/kWh$$

However, it must be emphasized here again that none of the factors such as the residual value of the fixed assets are taken into account. The calculated cost is only the PV power generation cost, not the feed-in price.

11.1.2 History and prospects of PV power generation cost

For a long time, many people think that the cost of PV generation is very high and cannot compete with conventional power generation. Therefore, it is hard to promote the large-scale application of PV generation. In fact, this method of simply comparing the price of electricity is unscientific, because the current price of conventional electricity does not reflect the actual production cost. To reduce the burden on consumers, all governments have subsidized the energy industry, according to the International Energy Agency (according to the "World Energy Outlooks" published in the 2011–2016 edition of IEA), the energy subsidies provided by various countries in recent years are shown in Tab. 11.2. It can be seen that even if subsidies for renewable energy are increased in recent years, it is still less than half of fossil energy.

Tab. 11.2: Energy subsidies by countries (billion US$).

Year	2007	2008	2009	2010	2011	2012	2013	2014	2015
Fossil energy	342	554	300	409	523	544	550	493	325
Renewable energy	39	44	60	66	88	101	120	135	150

It is true that governments have offered strong financial and policy support for conventional power generation for more than a decade, renewable energy sources such as wind and PV have reason to expect the government to provide continued financial and policy support until they can compete with conventional power generation. According to the Energy Watch Group's estimate, in 2008, fuel and electricity consumption in the world ranged from 5,500 to 7,500 billion US dollars, of which 8.5%

to 11.8%, or about 650 billion US dollars from subsidies. It includes the subsidies for fossil fuel production as 550 billion US dollars and 100 billion for fossil fuel producers. If one year's subsidies for conventional energy are used for renewable energy, about 200 GW of PV generation systems can be installed.

At the end of the twentieth century, in order to promote the development of the PV industry, some countries (such as Germany, etc.) were subsidized by the government and adopted a policy of purchasing PV power at a higher price. The maximum purchase price was 0.57 EUR/kWh for a residential rooftop PV system with a capacity of 3–5 kW. It made a investment for users to install PV system very attractive. This prompted the rapid expansion of PV installations in Germany and made Germany a leading country in the world on PV applications.

In addition, there are many indirect costs that are not included in the cost of conventional power generation. For example, the impact on the local environment and the cost of global climate change caused by GHG emissions have so far been difficult to quantify and include in the cost of power generation. Even if carbon trading is carried out according to the market-based CDM prescribed by the Kyoto Protocol, the market price of CO_2 is still low. In 2014, it was only about 9 US$/ton in Europe. In order to address climate change issue, the "Paris Agreement" requires that the global average temperature be controlled within 2 °C of the pre-industrial level and that requires the concentration of GHG in the atmosphere be kept below 450 ppm. This will require efforts to reduce GHG emissions. This goal is called a 2 °C scenario or a 450 scenario. There is also a hiRen scenario that is based on the 2 °C scenario, slower development of nuclear energy and CO_2 capture and storage (CCS) technologies, giving priority to PV and wind power generation. The 2014 version of the IEA Photovoltaic Technology Roadmap predicts that the global CO_2 market price by 2050 under the 2 °C scenario and the hiRen scenario are shown in Tab. 11.3.

Tab. 11.3: CO_2 Market Price Forecast (US$/t$CO_2$).

Year	2020	2030	2040	2050
2 °C scenario	46	90	142	160
hiRen scenario	46	115	152	160

According to WEO 2016 forecast, the market price of CO_2 will be 20 US$/ton in 2020, 100 US$/ton in 2030, and 140 US$/ton in 2040. Even if the market price of CO_2 is in the range of 10 to 20 €/ton, and the global CO_2 emission rate for power generation is 0.6 kg/kWh on average, the cost of PV generation should be reduced by an additional 0.006 to 0.012 euro/kWh.

In November 2016, a report published by the Fraunhofer Institute in Germany stated that in 1990, the price of 10–100 kW rooftop PV systems in Germany was approximately 14,000 €/kW, and by the end of 2015, the price of such systems was approximately 1,270 €/kW. It has dropped by 90% in 25 years. The experience curve for price changes over 35 years (also known as the learning curve) shows that when the cumulative module production increases by a factor of two, the price will be reduced by 23% (as shown in Fig. 11.1).

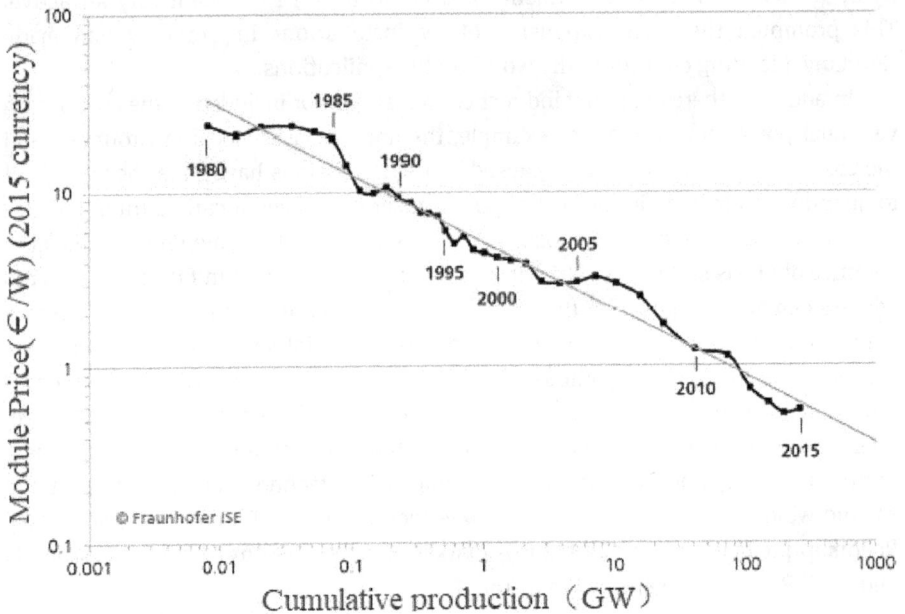

Fig. 11.1: Learning curve of PV module price vs production.

The IEA proposed in the Photovoltaic Technology Roadmap "Technology Roadmap Solar Photovoltaic Energy 2014 Edition": In order to achieve the goal of controlling global average temperature within 2 °C above pre-industrial levels in 2050, PV will account for 16% of the electricity supply [5]. It is necessary to accumulatively install PV capacity of 4,600 GW, annual power generation of 6,300 TWh, and reduction of CO_2 emissions by 4 Gt per year. PV power cost will be reduced by 25% by 2020, 45% by 2030, and 65% by 2050, which is approximately 40 to 160 US$/MWh (as shown in Tab. 11.4).

Due to the limited reserves of fossil fuels, the price of conventional power generation is bound to increase gradually. However, with the rapid expansion of the PV generation market and large-scale commercial production, the price of PV module and BOS will gradually decrease. With the advancement and development

Tab. 11.4: PV generation cost forecast to 2050 under hiRen scenario ($/MWh).

Year		2013	2020	2025	2030	2035	2040	2045	2050
Newly built large-scale system	Minimum	119	96	71	56	48	45	42	40
	Average	177	133	96	81	72	68	59	56
	Maximum	318	250	180	139	119	109	104	97
Rooftop PV system	Minimum	135	108	80	63	55	51	48	45
	Average	201	157	121	102	96	91	82	78
	Maximum	539	422	301	231	197	180	171	159

of science and technology, the performance and quality of related products will also continue to increase. The cost of PV will be further reduced. PV price decreased very quickly. In 2016, in the "2017–2018 Photovoltaic Power Plant Construction Tendering Request for Production Bid," the lowest bid for the UAE project is 0.058 US$/kWh, the Peruvian project is 0.048 US$/kWh, and the Mexico project median price is 0.045 US$/kWh. In May 2016, the lowest quote for an 800 MW PV station project in Dubai was 0.0299 US$/kWh. In a bid of 120 MW PV plant in Chile, a Spanish company won the bid at the price of 0.0291 US$/kWh. The recent quotation for the Sweihan project in Abu Dhabi in the United Arab Emirates was as low as 0.0242 US$/kWh. The average electricity price in the USA in 2015 was approximately 0.10 US$/kWh (AEO 2016). It can be seen that in many regions, PV power generation has been fully capable to compete with conventional power generation, and PV price tends to fall faster than forecasted.

11.2 PV energy payback time

11.2.1 Energy payback time

Solar energy is a clean, nonpolluting energy. However, in the process of manufacturing PV systems, it consumes certain energy. Some people doubt whether the energy generated by PV systems can compensate for the energy consumed in the manufacturing process. Some people even think that PV generation is not worth at all.

One of the indicators to measure the effectiveness of an energy system is the Energy Payback Time (EPBT), which is defined as the ratio of the total energy input during the lifetime of the energy system to the annual energy generated during the operation of the system. Both terms use the same unit and are expressed in equivalent primary energy or electrical energy:

$$\text{EPBT} = \frac{E_{\text{in}}}{E_g} = (E_{\text{mat}} + E_{\text{manu}} + E_{\text{trans}} + E_{\text{inst}} + E_{\text{EOL}}) / \left(\frac{E_{\text{agen}}}{\eta_G} - E_{\text{aoper}} \right) \qquad (11.8)$$

where E_{in} is the total energy input during the lifetime of the energy system, including the total energy externally input needed for manufacturing, installation, operation, dismantling, and waste treatment at the end of the lifetime; E_g is the annual output energy of the energy system during its operation; E_{mat} is the primary energy consumed for producing energy system materials; E_{manu} is the primary energy consumed for manufacturing energy systems; E_{trans} is the primary energy consumed by transportation materials during the energy system's lifetime; E_{inst} is the primary energy consumed for installing energy systems; E_{EOL} is the primary energy consumed after the end of the energy system's end of life cycle; E_{agen} is the annual energy production of the energy system; E_{aoper} is the annual energy consumption and the amount of primary energy consumed by the energy system; and η_G is the efficiency of the average primary energy conversion into electricity at the consumer end.

Due to different conditions of fuels and technologies used by various countries, the average conversion efficiency of primary energy into equivalent annual energy (E_{agen}) is not the same, taking 0.29 in the USA and 0.31 in Western Europe.

Primary energy is defined as the energy (such as coal, crude oil, natural gas, and uranium) that is present in natural resources without any artificial conversion and needs to be converted and transported to become available energy. The unit of EPBT is years. Obviously, the shorter the EPBT, the better.

The energy that device can generate during its lifetime can be calculated by multiplying the energy E_g produced by this device each year with its lifetime. If this energy is less than the energy consumed when manufacturing the device, the device cannot be used as an energy source, such as a commonly used battery. Although under certain conditions, the battery can provide power to the outside, but even if the energy consumed in the manufacturing process is not considered, only the energy input when charging is greater than the energy output when discharging, so the battery is not an energy source.

The EPBT of PV system depends on a series of complicated factors [6]. The input energy depends on many factors, such as the type of PV cell (such as mono c-Si, poly c-Si, a-Si, or other type of cells), process, packaging materials and methods, the frame and bracket, and BOS materials and processes, and sometimes batteries. In addition, the energy required to dismantle the system and dispose of the waste at the end of installation, operation and final life cycle also need to be added, especially considering the energy input by human labor.

The energy output from a PV system is also related to many factors, such as the service life and performance and efficiency of PV cells and associated components, the type of PV system (off-grid or grid-connected), local geography, and weather conditions, whether the design of the system is reasonable, whether the tilt angle of the array is appropriate, whether the installation process is improper, and the maintenance management. In addition, there are some indirect factors that are not directly related to the power generation system itself.

Although the influencing factors are complex, it is still possible to grasp the main factors based on theoretical research and practical investigations and conduct a comprehensive analysis.

11.2.2 Overview of worldwide situations

In early 1990s, a large number of countries began to conduct analysis and research on the EPBT of PV systems. Here are some representative studies.

11.2.2.1 Alsema E A

Prof. Alsema E A, from Utrecht University in the Netherlands, has published many research reports on the EPBT of PV systems. He is one of the most famous researchers in benefits of PV systems. Later many research papers cited his conclusions. In his *"Energy Pay-Back Time of Photovoltaic Energy Systems: Present Status Prospects"* published in July 1998, he analyzed and compared more than a dozen of the commonly accepted references published in various countries, pointing out that in some literatures, the energy required to manufacture the modules was different: $2,400–7,600$ MJ/m^2 for poly c-Si modules, and $5,300–16,500$ MJ/m^2 for mono c-Si modules [7]. This is partly due to the different parameters of the production process, such as the thickness of the silicon wafer and the slice cutting loss.

Alsema established the "best estimate" method for the energy required for poly c-Si, mono c-Si, thin-film modules, and associated components. The main idea is as follows:

- The energy consumed in the manufacturing of PV system component should use the general "equivalent primary energy" unit, which is the amount of primary energy (or fuel) needed to produce these components.
- Assume that the efficiency of converting primary energy into electricity is 35%, so that 1 MJ of primary energy can be converted into 0.097 kWh of electricity.
- The efficiency of poly c-Si, mono c-Si, thin-film modules is assumed to be 13%, 14%, and 7%, respectively.
- Assuming that the annual solar radiation received by the system is 1,700 kWh/m^2, the system performance ratio is 0.75.

On this basis, he examined and analyzed the production process of PV modules and proposed energy consumption estimates for various processes of crystalline silicon modules, as shown in Tab. 11.5.

Table 11.5 assumes that around 2.0–2.4 kg of poly c-Si material is required for each square meter of module. The low and high estimates are mainly due to the

Tab. 11.5: Energy consumption estimates at various processes for c-Si modules.

Process	Poly c-Si PV module		Mono c-Si PV module		Unit
	Low	High	Low	High	
Silicon raw material production	450	500	500	500	MJ/m^2 module
Silicon purification	1,800	3,800	1,900	4,100	MJ/m^2 module
Crystallization and molding 1		5,350		5,700	MJ/m^2 module
Crystallization and molding 2	750	750	2,400	2,400	MJ/m^2 module
Wafer	250	250	250	250	MJ/m^2 module
Battery process	600	600	600	600	MJ/m^2 module
Module assemble	350	350	350	350	MJ/m^2 module
Module summary without frame	4,200	11,600	6,000	13,900	MJ/m^2 module
Module summary without frame	35	96	47	109	MJ/W

difference in the production process of silicon materials. The energy consumption during the purification of silicon is 900–1,700 MJ/kg, and the energy consumption during Czochralski technique is 500–2,400 MJ/kg, so two estimates are given. The low estimate is based on the lower purity of the purified silicon and does not consider the initial crystallization process; the high estimate assumes that the silicon purity is relatively high and includes the energy consumption of the initial crystallization process at 2,400 MJ/kg. Assuming that in the process of forming a silicon ingot, the yield of polycrystalline silicon is 64%, the yield of single crystal silicon is 60%, and in the process of cutting into a thickness of 350 µm, the yield of silicon wafer is 60%. In the second-stage single crystal molding, the consumed energy of 1,100 MJ/kg is considered to be lower than that in the first-stage single crystal molding because the final silicon rod size is relatively small (6 in), and the PV material has lower requirements than the electronic-grade material. With standard screen-printing technology and glass/Tedlar packaging, the battery production process requires 600 MJ/m^2 of energy, and the module package consumes 350 MJ/m^2. The main uncertainty is 400 MJ/m^2 of additional energy for lighting and the environmental control of the module packaging workshop. It is also considered that the yields for the production of batteries and modules are 95% and 97%, respectively, so that the energy consumption of crystalline silicon battery modules can be in the range of 4,200 to 13,900 MJ/m^2.

For frameless a-Si modules, the required energy range is between 710 and 1,980 MJ/m^2. The difference lies in the different substrate/package materials and whether the manufacturing equipment is considered. There is also an indefinite factor in the environmental control of the lighting and module packaging plant's extra energy (estimated range is 80–800 MJ/m^2). Table 11.6 shows the estimated energy consumed for each process of the a-Si module with an assumption that the module efficiency of a-Si is 6%. If it is a stainless steel thin strip substrate, the estimated energy consumption should increase 150 MJ/m^2.

Tab. 11.6: Energy consumption estimates at various processes for a-Si module.

	Energy needed (MJ/m^2)	Ratio (%)
Battery material	50	4
Substrate material + encapsulates	350	29
Battery/ module produce	400	33
Extra energy	250	21
Manufacture	150	13
Module summary (without frame)	1,200	100
Module summary (without frame)	20 MJ/W	

Regarding the Balance of System (BOS), a 3.3 kW solar power plant in Serre, Italy, was analyzed in the literature. The results showed that the primary energy is about 1,900 MJ/m^2. However, most of the products combined with buildings consume only about 600 MJ/m^2 of primary energy. The reason for the high value of the former is that it is installed in open areas and uses more concrete and steel. It is not expected that there will be a large decline in the future. On the contrary, if a PV system is installed on an existing building structure, the analysis shows that the primary energy required for the tilted roof may be reduced to 400 MJ/m^2, and is only 200 MJ/m^2 for the facade wall. This can be achieved by reducing the amount of material used or using a large amount of recycled material, particularly aluminum. The BOS EPBT for the roof system is currently about 700 MJ/m^2 on average and can be reduced to 500 MJ/m^2 in the future.

The cable was not considered in this analysis, but the impact on most systems is negligible.

In the current system, the module frame has a significant influence on energy consumption. Since the amount of used aluminum frame is different, the energy consumption range is 300–770 MJ/m^2. In many cases, frameless modules can be used in future PV applications.

The storage battery is an important component of the off-grid PV system. Some documents describe the required energy range of 25–50 MJ/kg. The low estimate includes only the input energy required by the material and does not include the energy consumed during battery manufacturing. The energy is estimated to be 9–16 MJ/kg. Lead-acid battery has an energy density of approximately 40 Wh/kg, which can provide 0.6–1.2 MJ of energy per Wh of capacity. The intermediate value of 0.9 MJ/Wh is used in the analysis. The energy required for each part of the PV system is shown in Tab. 11.7.

Tab. 11.7: The energy required for each part of PV systems.

Component	Unit	Year of 1997	Year of 2007
Module frame (aluminum)	MJ/m^2	300–770	0
Rack (ground)	MJ/m^2	1,900	1,800
Rack (rooftop)	MJ/m^2	500–1,000	350–700
Rack (façade)	MJ/m^2	600–700	200–550
Inverter (3.3 kW)	MJ/kW	0.5	0.5
Battery (lead-acid)	MJ/kW	0.6–1.2	0.6–1.2

To calculate the EPBT, in addition to the above-mentioned energy to manufacture various components, it is also necessary to consider the entire external input required for the installation, operation, dismantle the system, and process the waste at the end of the lifetime. However, these energies are relatively small and basically negligible. At the same time, when calculating the energy output from the PV system, the operating conditions of the PV system must also be determined. Table 11.8 shows the operating conditions for calculating the PV system EPBT.

Tab. 11.8: Operating condition for PV system EPBT calculation.

	Grid-connected PV system	Off-grid PV system
Annual solar irradiation (kWh/m^2)	1,700	1,900
Annual system energy generation (kWh/W)	1.28	1.3
Storage battery size (Ah@12 V)	0	70
Number of changed storage battery during lifetime	Null	5
Other auxiliary energy efficiency (%)	35	25 (diesel engine)

Through the above analysis and calculation, the final conclusion is shown in Fig. 11.2. The figure shows EPBT at present (in 1997) and in the future (in 2007) for three types of PV systems: rooftop grid-connected installation, ground-based grid-connected installation, and off-grid household use PV system, with poly c-Si and a-Si modules. For the poly c-Si module system, the EPBT is different from the high estimate of 8 years to the low estimate of 3 to 4 years. The future roof-mounted PV system can expect the EPBT of the poly c-Si module and the a-Si module to be reduced to 1.7 years and 1.2 years.

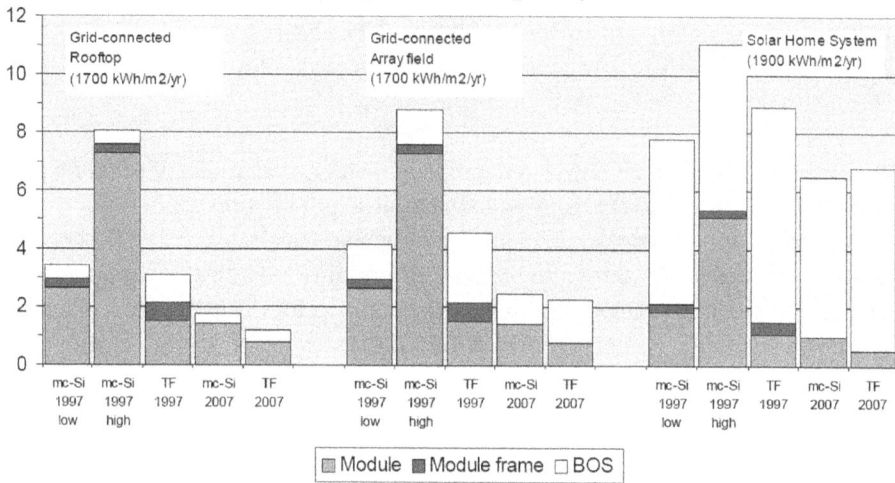

Fig. 11.2: EPBT for three types of PV systems.
Source: Alsema 1998.

11.2.2.2 IEA-PVPS joint report

The International Energy Agency (IEA-PVPS Task 10), the European Photovoltaic Technology Platform (EPTP) and the European Photovoltaic Industry Association (EPIA) conducted a survey about existing research on energy input of PV systems worldwide and issued a joint report in May 2006: "*Compared Assessment of Selected Environmental Indicators of Photovoltaic Electricity in OECD Cities*" [8]. This report provided clear and detailed profiles, calculated, and analyzed the PV power generation of total of 41 major cities in 26 OECD countries. The annual power generation, EPBT, and PV power equivalent to the reduction of CO_2 emissions of these cities were listed in details. For EPBT calculation, the energy consumption included energy needed for all PV system component (not only PV modules, but also supporting components, connecting cables, and electronic devices, etc.) It illustrated the EPBT depended on local solar irradiation conditions. The obtained conclusion is that the EPBT for rooftop grid-connected PV systems is 1.6–3.3 years and 2.7–4.7 years if installed vertically toward the equator. The best is Perth, Australia, and the worst is Edinburgh, UK.

Gaiddon B et al. subsequently published an article *"Environmental Benefits of PV Systems in OECD Cities,"* expounded the basis and method of analysis, and pointed out that the above conclusions were mainly directed to the use of standard poly c-Si modules and grid-connected PV systems in cities [9].

Since the orientation and tilt angle of the PV array have a significant impact on the power generation of the grid-connected PV system, the following two common situations are discussed in consideration of the specific application of BIPV in the city.

(1) Facing the equator, a rooftop grid-connected PV system with a tilt angle of 30°.

(2) Installed vertically towards the equator, that is, with a tilt angle of 90°, as used for curtain walls.

Then, calculate the annual power generation of the poly c-Si grid-connected PV system with unit power (1 kW) according to the local solar radiation data.

During the entire processing, manufacturing, and installation of the PV system, energy is consumed. According to the statistics of nine modern PV manufacturing plants in Europe and America, the power consumption for the grid-connected polycrystalline silicon PV system is shown in Tab. 11.9.

Tab. 11.9: Energy consumption of grid-connected poly c-Si PV system.

Component	Energy consumption (kWh/kW)
Module	2,205
Rack	91
Accessory parts	229
Total	2,525

Finally, calculate the EPBT of the grid-connected PV system in 41 cities. The EPBT range of the OECD country's PV system was shown in Tab. 11.10.

11.2.2.3 IEA-PVPS Task 12

In January 2015, the International Energy Agency IEA-PVPS Task 12 published a study report *"Life Cycle Inventories and Life Cycle Assessments of Photovoltaic Systems,"* which analyzed the EPBT of various types of PV systems [10]. It is pointed out that the early researchers estimated a wide range of primary energy consumption for silicon

Tab. 11.10: PV system EPBT of OECD nations.

	Minimum (year)	Maximum (year)
Roof-mounted PV system	1.6	3.3
Façade-mounted PV system	2.7	4.7

PV modules: 2,400–7,600 and 5,300–16,500 MJ/m^2 for poly c-Si and mono c-Si modules. Alsema estimated 4,200 and 5,700 MJ/m^2 of primary energy for poly c-Si and mono c-Si modules, respectively, and the corresponding EPBTs of 2.5 years and 3.1 years. Meijer et al. reported that a poly c-Si module with an efficiency of 14.5% made of electronic grade silicon consumed more primary energy (4,900 MJ/m^2), and therefore the EPBT is 3.5 years under the solar irradiation in the Netherlands (1,000 kWh/m^2/year). Jungbluth reports that the efficiency of poly c-Si module made with 50% of solar grade silicon and 50% of electronic grade silicon and with a thickness of 300 μm is 13.2%, while that of monocrystalline silicon modules is 14.8%. Under the environmental conditions in Switzerland of 1,100 kWh/m^2/year, the EPBT is 3–6 years, depending on the configuration of the PV system (tilt angle, horizontal or vertical, placement, etc.).

Later, Alsema and de Wild-Scholten used the raw material data of the manufacturing industry to analyze the roof PV system. It was found that the primary energy consumptions of poly c-Si and mono c-Si module were only 3,700 and 4,200 MJ/m^2, respectively, which was far lower than the previous studies. Fthenakis and Alsema also reported that for products produced from 2004 to 2005, under the conditions of 1,700 kWh/m^2/year irradiation in Southern Europe and a performance ratio of 0.75, the EPBT of the rooftop application is 2.2–2.7 years, where the balance of system (BOS) EPBT is 0.3 years. De Wild-Scholten recently estimated the EPBT of modules with relatively thin cells and a more efficient process to be approximately 1.8 years.

For thin-film modules, Kato et al. studied the life cycle of CdTe modules in 2001 and estimated that for the module production facilities with annual capacity of 10, 30, and 100 MW, respectively, the primary energy consumed was 1523, 1234, and 992 MJ/m^2, respectively. However, these numbers dropped significantly when many large-scale manufacturing plants started production. Fthenakis and Kim (2006) estimated that the primary energy consumed was only 1,200 MJ/m^2 based on the actual 2005 production of the First Solar 25 MW prototype plant. In the USA, where the average annual irradiation is 1,800 kWh/m^2/year, the ground-mounted CdTe PV system has an EPBT of approximately 1.1 years, which includes BOS EPBT of 0.3 years. Raugei et al. estimated that the amount of primary energy consumed was about 1,100 MJ/m^2 based on data of the year 2002 from Antec Solar's 10 MW plant in Germany. Fthenakis recently estimated that the EPBT was only about 0.87 years based on data from First Solar's manufacturing plant in Frankfurt, Germany.

Figure 11.3 shows currently available EPBT data in the public domain. It is calculated based on the condition that the module is installed on the roof in Southern Europe, the solar irradiation is assumed as 1,700 kWh/m²/year, and performance ratio of 0.75. However, these data are basically from 2006 and do not represent the current situation. For example, the thicknesses of mono c-Si and poly c-Si wafers were 270 µm and 240 µm, respectively, while the thickness is about 200 µm at present. The efficiency is also higher than that in before, so the EPBT should now be shorter.

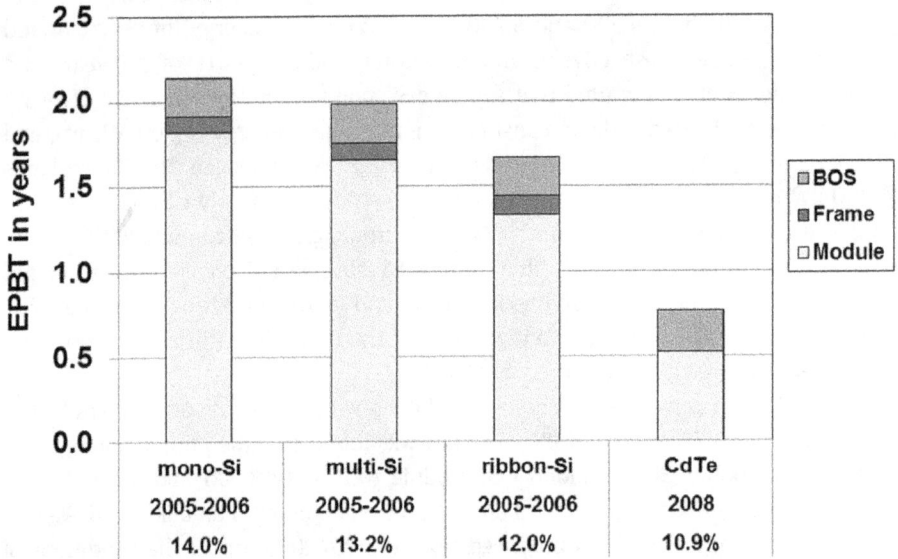

Fig. 11.3: EPBT for rooftop PV systems.

Figure 11.4 is a new estimate on the reports of Fthenakis et al. (2009) and de Wild Scholten (2009), and was based on REC Solar's new data with the thicknesses of the mono c-Si and poly c-Si wafers as 180 µm and 200 µm, respectively. This estimate may not reflect the average industrial production situation. The CdTe modules in Figs. 11.3 and 11.4 were produced by First Solar's old factory in Frankfurt, which was the largest manufacturer of CdTe modules. At present, the efficiency has reached 11.7%, which is higher than that shown in these Figures. Therefore, the EPBT is also shortened accordingly now. These modules are also installed on the roof condition where the solar irradiation of Southern Europe is 1,700 kWh/m² per year, and performance ratio is assumed as 0.75. For mono c-Si, poly c-Si, and CdTe modules, the EPBT is 1.7, 1.7, and 0.8 years, respectively.

According to a report published by the Fraunhofer Institute in November 2016, the EPBT for PV system in north Europe is about 2.5 years, and 1.5 years or less in

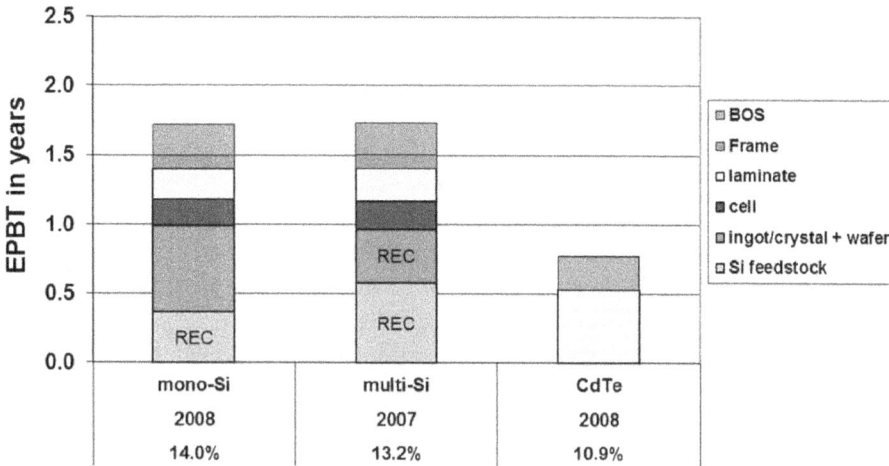

Fig. 11.4: EPBT of three relatively new modules.

south Europe. For instance, the EPBT of the poly c-Si PV system in Sicily, Italy is only one year. For a concentrator PV system in south Europe, the EPBT can be less than 1 year.

11.2.3 Calculation methods for relevant parameters

11.2.3.1 Annual output energy from a PV system

The effective power generated by an off-grid PV system not only depends on the capacity of the PV array, the local meteorological, and geographical conditions, and the installation and operation conditions, but also limited by the capacity and the days of autonomy of the storage battery. The situation is complicated. Therefore, only the grid-connected PV system is discussed here.

The energy output per unit power of the grid-connected PV system can be calculated with eq. (8.4):

$$E_g = H_t \cdot P_0 \cdot PR$$

where E_g is the annual power output per unit power of PV system with unit of kWh/kWy; H_t is the annual peak sunshine hours, which is calculated as total amount of solar radiation (kWh/m^2) received by the tilted PV array during the whole year divided by 1 kW/m^2; P_0 is the rated power of PV system, and is set as 1 kW; and PR is the system performance ratio.

In order to simplify the calculation, the degradation of module efficiency is generally not considered.

Since the orientation and tilt angle of the PV array have a significant impact on the power generation of the grid-connected PV system, considering the specific application of BIPV in cities, the IEA-PVPS joint report analyzed two cases in which the array tilt angle is 30° and the vertical installation, that is, the tilt angle is 90°. However, the latitudes are quite different in different cities. It is unreasonable to set the tilt angle as 30° for all situations. Therefore, the cases need to be modified and the following two common situations are discussed.

(1) For grid-connected PV systems installed facing the equator with the best tilt angle, the best tilt angle refers to the tilt angle corresponding to the maximum solar irradiation received by the local area throughout the year.

(2) PV arrays are mounted vertically toward the equator, with an inclination of 90°.

Refer to the data in the above IEA-PVPS joint report and calculate with PR = 75%.

11.2.3.2 The total energy input E_{in} during the lifetime of the PV system

During the production process, different type of PV modules consumes different amounts of electricity for each rated power. Different processes and production scales also have an impact on energy consumption. Comparing several module types commonly used at present, it is found that with the same rated power, mono a-Si module consumes the most energy, followed by the poly c-Si module, and a-Si module consumes the least energy. The poly c-Si module is discussed below.

For the poly c-Si grid-connected PV system, the average energy consumed is 2,525 kWh/kW, based on result listed in Tab. 11.9.

11.2.3.3 Calculate EPBT

The EPBT can be calculated as follows:

$$\text{EPBT} = \frac{E_{in}}{E_g}$$

In order to assess the environmental effect on grid-connected PV systems, 28 major cities in China were analyzed and calculated according to the above technical indicators. The amount of solar irradiation on the local horizontal plane is an average of measurement data according to the "Annual Data of China Meteorological Radiation" published by China Meteorological Center from 1981 to 2000. Then according to Klein (1981) method, calculating the monthly mean solar irradiation on different tilted plane, and comparing them to obtain the maximum solar irradiation H_t that can be received locally throughout the year. The corresponding tilt angle is the optimum tilt angle of grid-connected PV arrays. It is also possible to determine the amount of solar irradiation received on an array throughout the year when mounted vertically towards the equator. The rest parameters were determined as described above. The EPBT for grid-connected PV systems in China major cities is shown in Tab. 11.11.

Tab. 11.11: EPBT for grid-connected PV systems in China major cities.

City	Latitude (degree)	Optimum tilt angle (degree)	Average daily irradiation (kWh/m²/day)		EPBT (year)	
			Mounted with optimum tilt angle	Mounted vertically	Mounted with optimum tilt angle	Mounted vertically
Haikou	20.02	10	3.8915	2.0771	2.37	4.46
Guangzhou	23.10	18	3.1061	1.8398	2.97	5.02
Kunming	25.01	25	4.4239	2.6973	2.09	3.42
Fuzhou	26.05	16	3.3771	1.8991	2.73	4.86
Guiyang	26.35	12	2.6526	1.4715	3.48	6.28
Changsha	28.13	15	3.0682	1.7156	3.01	5.38
Nanchang	28.36	18	3.2762	1.8775	2.82	4.91
Chongqing	29.35	10	2.4519	1.3345	3.76	6.92
Lhasa	29.40	30	5.8634	3.6935	1.57	2.50
Hangzhou	30.14	20	3.183	1.8853	2.90	4.90
Wuhan	30.37	19	3.1454	1.8536	2.94	4.98
Chengdu	30.40	11	2.4536	1.3863	3.76	6.66
Shanghai	31.17	22	3.5999	2.1761	2.56	4.24
Hefei	31.52	22	3.3439	2.0351	2.76	4.53
Nanjing	32.00	23	3.3768	2.0804	2.73	4.44
Xi'an	34.18	21	3.3184	2.0009	2.78	4.61
Zhengzhou	34.43	25	3.8807	2.4450	2.38	3.78
Lanzhou	36.03	25	4.0771	2.5495	2.26	3.62
Jinan	36.36	28	3.8241	2.4754	2.41	3.73
Xining	36.43	31	4.558	3.0242	2.03	3.05
Taiyuan	37.47	30	4.1961	2.7699	2.20	3.33
Yinchuan	38.29	33	5.0982	3.4324	1.81	2.69
Tianjin	39.06	31	4.0736	2.7473	2.27	3.36
Beijing	39.56	33	4.2277	2.9121	2.18	3.17
Shenyang	41.44	35	4.0826	2.8643	2.26	3.22

Tab. 11.11 (continued)

City	Latitude (degree)	Optimum tilt angle (degree)	Average daily irradiation (kWh/m²/day)		EPBT (year)	
			Mounted with optimum tilt angle	Mounted vertically	Mounted with optimum tilt angle	Mounted vertically
Urumqi	43.47	31	4.2081	2.7818	2.19	3.32
Changchun	43.54	38	4.4700	3.2617	2.07	2.83
Harbin	45.45	38	4.2309	3.0740	2.18	3.00

It can be seen that among major cities in China, the grid-connected PV modules that face the equator and are installed vertically or installed with the optimum tilt angle in Lhasa have the shortest EPBT, with only 1.57 and 2.50 years, respectively, while modules installed in Chongqing have the longest EPBT of 3.76 years and 6.92 years, respectively. In the calculations, the externally input energy needed for transportation, installation, operation, and dismantling the system and dispose of the waste at the end of the lifetime is not taken into account. However, according to the analysis, the allocated energy required for rated power of PV module is not large and only has little effect on EPBT. Obviously, for mono c-Si modules, the EPBT will increase slightly, while for the thin film modules, it will decrease.

It is also necessary to point out that the energy consumption of 2,525 kWh/kW for producing poly c-Si module grid-connected PV system in the calculation was based on production level before 2006. Due to improvement in science and technology, the current data are much lower. The EPBT is also much smaller.

Example 11.2 To build a poly c-Si module grid-connected PV system in Dunhuang area, Gansu Province, assuming that the system performance ratio is 0.75, what is the EPBT of this PV system?
Solution According to Table 8.1, the optimal tilt angle of the grid-connected PV system in Dunhuang area is 35°, and the solar irradiation on the PV array is 5.566 kWh/m² day.
Assume that the energy consumed by PV system is 2,525 kWh/kW and substitute that into the eq. (11.8):

$$EPBT = \frac{E_{in}}{E_g} = \frac{2,525}{5.566} \times 1 \times 0.75 = 1.66 \text{ year}$$

In summary, during the whole lifetime of the PV system (currently 25 years, it is expected to increase to 35 years later), the generated energy is much larger than the energy input for its manufacturing, transportation, installation, and operation. Moreover, with the development of technology, the energy consumed by PV systems will continue to decline, and the EPBT will be further reduced. PV is indeed an effective clean energy that needs to be vigorously promoted.

11.3 PV power generation to reduce CO_2 emission

The "United Nations Framework Convention on Climate Change" proposed to comprehensively control the emission of GHG such as carbon dioxide, in order to deal with the negative effects of global warming on the economy and society. The "*Paris Agreement*" has come into effect on November 4, 2016. This agreement calls for the joint efforts of all countries in the world to reduce GHG emissions, to control the global average temperature increase within 2 °C compared to the period of pre-industrialization, and to continue to strive to limit the temperature increase to 1.5 °C.

Climate scientists have observed a dramatic increase in the concentration of CO_2 in the atmosphere over the past century. Over the past decade, the annual increase of concentration of CO_2 is 2 ppm. Methane (CH_4) and nitrous oxide (N_2O) have also increased dramatically. The total artificial GHG emissions in 2010 was (49 ± 4.5) Gt CO_2 equivalents, of which CO_2 accounted for 76% at (38 ± 3.8) Gt; CH_4 accounted for 16% at (7.8 ± 1.6) Gt CO_2 equivalents; N_2O accounted for 6.2% at (3.1 ± 1.9) Gt CO_2 equivalents; various fluorine-containing gases accounted for 2.0% at (1.0 ± 0.2) Gt CO_2 equivalents. The total global CO_2 emission in 2013 was 32.3 Gt with an increase of 2.2% over 2012. There are 6.5 million people die each year worldwide as a result of air pollution.

11.3.1 Greenhouse gas emitted by electricity generation

Among various human activities, the use of energy sources emits the most GHG. According to the 2010 statistical data, the emission of CO_2 from energy sources accounted for 68% of the total. Among the GHG components, CO_2 accounts for 90%, CH_4 accounts for 9%, and N_2O accounts for 1% (as shown in Fig. 11.5). The composition of GHG is mainly CO_2, so other GHG are often converted to CO_2 equivalents, and the effect of reducing GHG is measured by how much CO_2 equivalent is reduced. In order to discuss the effect of application of PV on reduction of GHG, only the GHG emissions from power generation will be studied.

At present, most of the electric power plants in the world use fossil fuel, which generates large amounts of GHG when it burns and causes environmental pollution. PV power is clean energy without any waste, so it can avoid the GHG emitted by the local power plants that generate the same amount of electricity. This is one of the important social benefit of PV systems.

Generally, there are two indicators used to evaluate PV systems effect on reduction of CO_2 emissions.

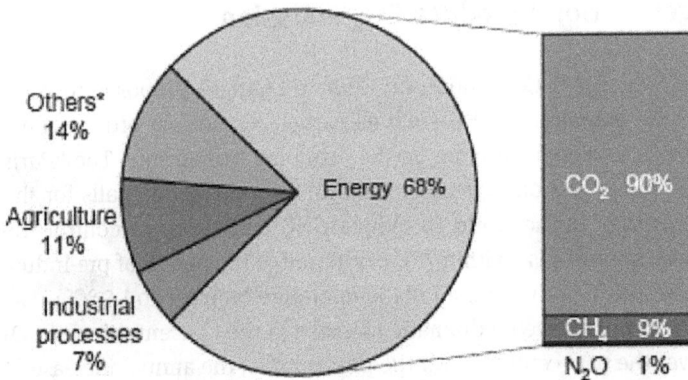

Fig. 11.5: Estimated shares of global anthropogenic GHG, 2010.
* Others include large-scale biomass burning, post-burn decay, peat decay, indirect N_2O emissions from nonagricultural emissions of NO_x and NH_3, Waste, and Solvent Use.
Source: CO_2 emissions from fuel combustion highlights (2016 Edition).

11.3.2 CO_2 emission factors

The CO_2 emission factors (EF) is a factor that quantifies the amount of gas emissions or gas removals per unit of activity. The following two EFs are often used to analyze the social benefit of PV systems.

11.3.2.1 Fuel CO_2 emission factor

The fuel CO_2 EF is defined as the amount of CO_2 emitted from generating 1 kWh of electric energy is called the CO_2 EF of this fuel, and the unit is g $CO_2/(kWh)$.

The amount of CO_2 emitted by different fuels during combustion is different. For clean energy such as hydropower, solar energy, wind energy, and geothermal energy, CO_2 emissions can be considered zero when generating electricity. Nuclear power generates very little emissions and can also be treated as zero emissions.

The average GHG EFs for different types of fuels in OECD countries during 2010–2014 that reported in the "CO_2 *Emissions from Fuel Combustion Database Documentation (2016 edition)*" published by the International Energy Agency (IEA) in October 2016 is shown in Tab. 11.12 [11].

It can be seen that the CO_2 EFs of different types of fuels vary greatly, and the combustion of blast furnace gas emits the largest amount of GHG.

11.3.2.2 CO_2 emission factor for electricity generation

The concept of CO_2 EF can also be used to evaluate the severity of a country's emissions of GHG. It is defined as the number of CO_2 emissions emitted by all power stations with multiple fuels in the whole country for each 1 kWh of electricity

Tab. 11.12: Implied carbon emission factors from electricity generation for selected products.

Fuel	gCO$_2$/kWh	Fuel	gCO$_2$/kWh
Anthracite*	875	Peat*	765
Coking coal*	820	Natural gas	405
Other bituminous coal	870	Crude gas*	590
Sub-bituminous coal	940	Refinery gas*	450
Lignite	1,030	Liquefied petroleum gases*	525
Gas works gas*	335	Kerosene*	625
Coke oven gas*	390	Gas/diesel oil*	715
Blast furnace gas*	2,425	Fuel oil	670
Other recovered gases*	1,590	Petroleum coke*	930
Oil Shale*	1,155	Municipal waste (nonrenew.)*	1,200

*The electricity output from these products represents less than 1% of electricity output in the average of OECD member countries for the years 2010–2014. Values will be less reliable and should be used with caution.

generated. This is also the reduction of CO$_2$ emissions per 1 kWh of electricity produced by the PV system.

The CO$_2$ EF for electricity generation can be calculated based on the energy generated by different types of fuels. The amount of CO$_2$ emitted per fuel can be calculated by multiplying the power generated by each fuel used in power generation with the corresponding fuel EF. The sum of CO$_2$ emitted by various fuel is the total CO$_2$ emissions for electricity generation. The CO$_2$ EF for electricity generation is obtained by dividing the total CO$_2$ emissions for electricity generation by the total electricity generation of various fuels (including hydropower, nuclear power, and renewable energy with a CO$_2$ EF of zero) for that year.

There are pure power plants and combined heat and power (CHP) plants coexisting in many countries. The CO$_2$ generated by heat should not account for electricity generation CO$_2$ emission, but the difference is not significant. In 2013, the CO$_2$ EF between pure power plants and CHP plants in OECD countries was only 3%.

The EF for electricity generation cannot be determined on the basis of the fuel type of an individual power plant. Because a country may use multiple fuels to generate electricity, the average CO$_2$ EF for mixed electricity generation should be used. It is similar that evaluating the effect of PV on reducing emission cannot be judged based on the EF of one region. If there is no thermal power plant at a certain place and only clean energy such as hydroelectric power generation etc., the CO$_2$ EF of local power generation is zero. It is obviously unreasonable to judge the CO_2 emission reduction as zero, so the scope of evaluation should be entire country.

The CO_2 EFs in each country vary widely due to different fuel compositions used. The IEA "CO_2 *Emissions from Fuel Combustion Highlights (2016 edition)*" lists the CO_2 EF in various regions over the years, as shown in Tab. 11.13.

Tab. 11.13: Key indicator of CO_2 emission factors in specific regions [gCO_2/kWh].

Region	1990	1995	2000	2005	2010	2013	2014	% change 90–14
World	533	533	533	546	530	526	519	−3%
OECD countries	509	492	488	478	442	430	421	−17%
Africa	681	699	663	645	625	602	615	−10%
Asia (excluding China)	634	672	685	671	687	661	685	8%
Middle East	742	814	708	688	678	685	678	−9%
China (incl. Hong Kong)	911	918	893	878	759	710	681	−25%

It can be observed that the CO_2 EF of China is relatively high due to the high percentage of burning coal in the electricity generation. According to the "*International Energy Outlook 2016*" Index H12 to H17, the fuel power generation of various countries in 2012 is summarized and shown in Tab. 11.14 [12]. It can be seen that among the total power generation in China, the proportion of coal-fired power generation exceeds three-fourth, which are also relatively large in countries such as India and Australia. Therefore, the CO_2 EF is relatively high, and the impact on the environment is also greater. However, the proportion of hydropower and renewable energy in Brazil, Canada, and other countries is relatively large, and the CO_2 EF is relatively low. With the advancement of technology, changes in the types of fuels, the increase in power generation efficiency of power plants, and the promotion and application of clean energy, global CO_2 EFs will gradually decrease.

The world average CO_2 EF has been slightly less than 0.6 kg/kWh for many years, so the average CO_2 EF is taken as 0.6 kg/kWh in a lot of documents.

11.3.3 PV potential mitigation

The PV CO_2 potential mitigation (PM) is another important indicator to evaluate the reduction of CO_2 emissions by PV systems. It is defined as the amount of CO_2 mitigation by output power of a given unit rated power PV system. That is the equivalent

Tab. 11.14: Various fuel electricity generation amount and its proportion by region and country in 2012 (billion kW).

	Petroleum (liquid-fired electricity)		Natural gas-fired electricity		Coal (coal-fired electricity)		Nuclear electricity		Hydroelectric and other renewable electricity		Total power generation
	Power generation	Proportion (%)	Power generation	Proportion (%)	Power generation	Proportion (%)	Power generation	Proportion (%)	Power generation	Proportion (%)	
USA	23	0.6	1,228	30.3	1,514	37.3	769	19.0	520	12.8	4,055
Canada	7	1.1	63	10.2	60	9.7	89	14.4	397	64.4	616
Japan	170	17.6	373	38.5	285	29.4	17	1.8	122	12.6	968
South Korea	20	4.0	105	21.0	225	44.9	144	28.7	7	1.4	501
Australia and New Zealand	4	1.4	55	19.7	164	58.8	0	0	55	19.7	279
Russia	26	2.6	494	48.8	159	15.7	166	16.4	168	16.6	1013
China	6	0.1	81	1.7	3,587	75.2	93	1.9	1,004	21.0	4,771
India	21	2.0	88	8.4	753	71.6	30	2.9	160	15.2	1,052
Brazil	17	3.2	41	7.6	12	2.2	15	2.8	451	83.8	538

Source: *EIA: International Energy Outlook 2016.*

amount of CO_2 emissions to the power output of a PV system with a unit rated power (usually 1 kW) during its lifetime. The unit is gCO_2/kW.

Obviously, in addition to the local CO_2 EF, the PV CO_2 PM depends on the amount of electricity generated by the local PV system. In order to simplify the calculation, the PV module efficiency degradation is usually not taken into consideration. The calculation method is using output electrical energy of the PV system with unit rated power of 1 kW during its lifetime multiply the CO_2 EF (gCO_2/kWh).

$$PM = H_t \cdot P_0 \cdot PR \cdot N \cdot EF \tag{11.9a}$$

where N is the number of years of the lifetime; EF is the CO_2 EF; others can be found in eq. (8.4).

The power output of a unit-rated PV system during its lifetime is not only related to the local meteorological and geographical conditions, but also the type of PV system (grid connection or off-grid system), the array tilt angle, the performance ratio of the system, and other related factors. For comparison, the IEA-PVPS Task 10, EPTP, and EPIA issued a joint report in 2006 that evaluated 41 major cities in 26 OECD countries. The technical condition is grid-connected roof PV system facing the equator; title angle of 30° and 90°, performance ratio of 0.75, and the PV system lifetime of 30 years. As a result, the PV CO_2 PM is best in Perth, Australia. During the entire lifetime, PM = 40 tCO_2/kW at a 30° tilt angle, and 23.5 tCO_2/kW at a vertical installation. This is due to the large amount of solar irradiation in the area and the large amount of CO_2 emitted by local power plants when generating electricity. The lowest is Oslo, Norway, which is almost 0. This is because there is almost no GHG emitted by local power plant and the solar irradiation is low.

For the 28 major cities in China, two corrections were made with reference to the technical conditions of the IEA-PVPS Joint Report (2006).

(1) Considering that the PV system consumes energy in the manufacturing process, which also generates GHG. Therefore, the amount of CO_2 emissions during the energy payback time should be deducted. According to Tab. 11.9, from the statistical result of nine PV manufacturers in Europe and America, for a 1 kW grid-connected poly c-Si PV system, 2,525 kWh of electricity is consumed in the manufacturing process, so the formula for calculating the potential of CO_2 emission reduction from PV should be changed to:

$$PM = (H_t \cdot PR \cdot N - 2,525) \cdot EF \tag{11.9b}$$

(2) Based on the IEA-PVPS joint report, it is unreasonable to set the tilt angle of each locale as 30°, because geographical and meteorological conditions vary from place to place. Therefore, two kinds of situations that grid-connected PV systems are installed toward the equator with optimum tilt angle and tile angle of 90° are analyzed and calculated, respectively. The method for calculating and determining the optimal tilt angle and corresponding solar irradiation is

the same as that in 11.2.3, and the solar irradiation received all year round on the PV array that being installed toward the equator at tile angle of 90° can also be obtained. The system performance ratio is chosen as 75%, and the lifetime of the PV system is 30 years. According to Tab. 11.13, in 2014, China's CO_2 EF = 681 gCO_2/kWh. The results are shown in Tab. 11.15.

Tab. 11.15: CO_2 potential mitigation of grid-connected PV system in China (tCO_2/kW).

City	Optimum tilt angle	Vertical installation	Region	Optimum tilt angle	Vertical installation
Haikou	20.05	9.91	Nanjing	17.17	9.93
Guang-zhou	15.65	8.58	Xi'an	16.84	9.48
Kunming	23.02	13.37	Zhengzhou	19.99	13.67
Fuzhou	17.18	8.91	Lanzhou	21.09	12.54
Guiyang	13.12	6.52	Jinan	19.42	12.14
Changsha	15.44	8.78	Xining	23.78	15.21
Nanchang	15.44	8.78	Taiyuan	21.75	13.78
Chongqing	12.00	5.75	Yinchuan	26.80	17.49
Lhasa	31.08	18.95	Tianjin	21.06	13.66
Hangzhou	16.09	8.83	Beijing	21.93	14.59
Wuhan	15.88	8.65	Shenyang	21.12	14.31
Chengdu	12.01	6.04	Urumqi	21.82	13.85
Shanghai	18.42	10.46	Changchun	23.28	16.53
Hefei	16.98	9.66	Harbin	21.94	15.48

It can be seen that the PV system in Lhasa has the largest CO_2 PM. For grid-connected PV system installed with optimum tilt angle and installed vertically, during its lifetime, every newly installed 1 kW PV system can reduce CO_2 emissions by 31.08 and 18.95 t, respectively. The CO_2 PM in Chongqing is the least, with 12.00 and 5.75 t for system with optimum tilt angle and 90°, respectively. At the same time, it can also be seen that the effect of tilt angle on PV power generation system installed in Chongqing is larger than that in Lhasa. The ratio of the electrical energy generated by grid-connected PV system installed at the optimum tilt angle to that with vertically installed modules is 61.0% in Lhasa and 47.9% in Chongqing, so the corresponding CO_2 PM also has to change proportionately. This is due to the large proportion of direct radiation in total solar radiation in the Lhasa area.

Example 11.3 A poly c-Si grid-connected PV system built in Dunhuang is installed at optimum tilt angle. Assume that the performance ratio of the system is 0.75, the lifetime is 30 years, and the CO_2 EF is 764 g/kWh. How many tons of CO_2 emissions is reduced by 1 kW of this PV system?
Solution According to Table 8.1, the optimum tilt angle of the grid-connected PV system in Dunhuang is 35°, and the solar irradiation is 5.566 kWh/m^2/d. Assume the electricity consumed for the PV system is 2,525 kWh/kW.

Substituting these parameters into eq. (11.9a):

$$PM = (H_t \cdot PR \cdot N - 2,525) \cdot EF = (5.566 \times 365 \times 1 \times 0.75 \times 30 - 2,525) \times 0.764 = 33 \text{ t}$$

In summary, because PV is a clean energy source with zero CO_2 emissions, it can avoid environmental pollution caused by fossil fuels used in conventional power plants. With the massive promotion and application of PV, its effect of reducing CO_2 emissions will gradually appear, and it will surely exert significant economic and social benefits. In the Solar Generation 6 research reports jointly published by Greenpeace and the European Photovoltaic Industry Association (EPIA), the forecast of CO_2 emission reduction by 2050 was predicted with Reference, Accelerated and Paradigm shift scenario. The results are shown in Tab. 11.16.

Tab. 11.16: Prediction of PV CO_2 reduction by 2050 in three scenarios.

Year	2008	2009	2010	2015	2020	2030	2040	2050
				Reference scenario				
CO_2 reduction (with 600 g CO_2/ kWh) [annual Mio tCO_2]	10	15	19	33	57	123	226	337
Avoided CO_2 since 2003 [cumulative Mio tCO_2]	35	50	69	208	438	1,300	3,031	5,911
				Accelerated scenario				
CO_2 reduction (with 600 g CO_2/ kWh) [annual Mio tCO_2]	10	15	20	73	254	853	1,693	2,670
Avoided CO_2 since 2003 [cumulative Mio tCO_2]	61	75	95	327	1,160	6,580	19,153	41,460
				Paradigm scenario				
CO_2 reduction (with 600 g CO_2/ kWh) [annual Mio tCO_2]	5	15	20	113	540	1,358	2,603	4,047
Avoided CO_2 since 2003 [cumulative Mio tCO_2]	56	70	90	404	2,014	11,085	30,559	64,890

Note: The emission factor is 600 gCO_2/kWh.

11.4 Other benefits of PV power generation

11.4.1 Increase job opportunities

The development of the PV industry can also provide jobs and increase employment. A large number of employees are required in the design, manufacturing, transportation, installation, and maintenance of PV systems. In the research, production, sales, and installation of PV systems, the proportion of employees required varies from country to country. In general, 30 full-time workers are required to complete the construction of a PV system with an average capacity of 1 MW on a worldwide basis. Based on this estimate, there were about 228,000 people worked in PV industry in 2009. The jobs directly engaged in PV research, development, manufacturing, and installation in IEA-PVPS countries in 2015 are shown in Tab. 11.17 according to the IEA-PVPS published reports "*Trends 2016 IN Photovoltaic Applications Survey Report of Selected IEA Countries Between 1992 and 2015*" [13].

Tab. 11.17: Jobs engaged in PV Industry in IEA-PVPS Countries in 2015.

Country	Job	Compared with 2014
USA	208,859	22%
Japan	128,900	2%
Malaysia	21,717	89%
Australia	14,620	0%
France	8,300	−12%
Canada	8,100	0%
Switzerland	5,700	−2%
Spain	5,000	−33%
Austria	2,936	−9%
Norway	966	25%
Sweden	830	15%

In recent years, due to the rise of China's PV industry, the number of people engaged in the PV industry in Europe has decreased. According to "*Solar PV Jobs & Value Added in Europe*" report published in November 2015 by Solar Power Europe, the proportions and the number of jobs in the PV industry chain at different periods in the EU 28 countries are shown in Tab. 11.18.

Tab. 11.18: Ratio and number of jobs in PV industry in different periods in 28 EU countries.

Year	2008	2014	2020*
Modules	8%	2%	2%
cell	4%	1%	1%
Polysilicon	4%	1%	1%
Wafer	4%	1%	1%
Inverter	6%	2%	2%
BOS components	7%	7%	6%
Engineers, researchers, administrators	50%	36%	33%
Installer	16%	29%	25%
Operation management	1%	21%	31%
Employed population	178,879	109,650	136,096

*Editor's note: Percentage total is not equal to 100%.

According to the *"Renewable Energy and Jobs – Annual Review 2016"* issued by the International Renewable Energy Agency (IRENA), the total employment of the global PV industry in 2015 reached 2.727 million with an increase of 11% over the previous year [14]. Among them, 1.7 million people are from China; 377,000 people are from Japan, which increases 28% over 2014; 194,000 people are from the USA; 103,000 people are from India; 38,000 people are from Germany; 21,000 people are from France, and 8.4 million people are from other EU countries. In off-grid PV systems, microsystems are more difficult to count, mainly off-grid household systems are taken into consideration. In Bangladesh, there are 700,000 sets of household systems that were added in 2015, with a cumulative installed capacity of 4.5 million sets. It is estimated that the number of employees engaged in PV industry reached 127,000, of which one-fourth are manufacturing, the rest are sales, installation, and after-sales services.

According to China's *"Thirteenth Five-Year Plan for Solar Energy,"* it is estimated that by 2020, China's solar energy industry (including heat utilization) can provide about 7 million jobs.

With the rapid development of the PV industry, it will continue to increase employment and will play an active role in promoting regional economic development and improving people's living standards.

11.4.2 Save fuel sources

Conventional power generation requires the burning of fossil fuels, while PV does not consume any fuel and can save natural resources. Due to the differences in the energy generated by various fuels, when comparing the quantity and quality of energy, in order to facilitate the calculation, the coal, petroleum, natural gas, etc. are all converted to a standard coal in a certain proportion based on international conventions. It is stipulated that the lower heating value of 1 kg standard coal is 29.31 MJ, so that energy of different types and different contents can be converted into standard coal according to their heating values.

Generally, there are two kinds of data: the standard coal consumption for electricity generation and the standard coal consumption for electricity supply from power statistical reports of each country. Since the power plant itself consumed some electricity it generates, therefore, the amount of electricity generation is greater than the actual supply. Obviously, it should use the standard coal consumption for electricity supply to evaluate the benefit of PV systems on coal reduction.

According to the "Basic Statistics of Power Statistics" published by the China Electricity Council, the standard coal consumption for electricity supply by power plants with capacity of 6000 kW and above is shown in Tab. 11.19.

Tab. 11.19: Standard coal consumption for power plants of 6,000 kW and above from 2010 to 2015 in China.

Year	2010	2011	2012	2013	2014	2015
Standard coal consumption (g/kWh)	333	329	325	321	319	315

It can be seen that for each 1 kWh of electricity generated in China in 2015, it is equivalent to saving 315 g of standard coal. Moreover, with the advancement of technology, the standard coal consumption for supply electricity will continuously decrease.

11.4.3 Reduce power transmission loss

PV systems can produce electricity as long as the sun is available. They are distributed power sources and do not require long distance power transmission and distribution equipment, which reduces line losses. According to the "SET for 2020" report, the value of PV can be increased implicitly 0.5 Euro cents/kWh in Europe based on this reduction.

11.4.4 Ensure reliable energy supply

Once the PV system is installed, it can provide stable and reliable power at a fixed price for at least 25 years. There is no shortage of fuel resources or transportation issues, nor will it be affected by the price fluctuations of fuels (such as oil, natural gas, and coal) in the international market like conventional power plants. However, as fossil fuels gradually reduce their reserves, their price will rise steadily. In Europe, due to this point, the value of PV can be increased from 0.015 to 0.031 Euro/kWh implicitly, depending on the price fluctuations of oil, natural gas, and coal.

In summary, PV technology is booming and flourishing. Of course, PV power generation still has a long way to go before it can replace conventional power generation. There are still a lot of technical and nontechnical obstacles that need to be conquered. But with the social development and technological advancement, the scale of PV will continue to expand, the cost will also gradually decrease, and more and more significant economic and social benefits will be achieved, which will surely play an important role in the future energy consumption structure. It can be expected that by the end of the twenty-first century, solar power will become the main source of electricity supply, and a bright new era of solar energy will finally come.

References

[1]   Walter Short, J Daniel and Thomas Holt Packey. A manual for the economic evaluation of energy efficiency and renewable energy technologies. March 1995 NREL/TP-462-5173. http://www.nrel.gov/csp/troughnet/pdfs/5173.pdf.

[2]   Seth B. Darling. et al., Assumptions and the levelized cost of energy for photovoltaics. Energy & Environmental Science, 2011, 4(9), 3077–3704.

[3]   K Branker, M. J. M. Pathak and J M Pearce. A review of solar photovoltaic levelized cost of electricity. Renewable & Sustainable Energy Reviews, 2011, 15, 4470–4482. http://dx.doi. org/10.1016/j.rser.2011.07.104.

[4]   Gaëtan Masson Vartiainen and Christian Breyer. PV LCOE in Europe 2014-2030·Final Report. 23 July 2015, European photovoltaic technology platform. DOI: 10.13140/RG.2.1.4669.5520. http://wwweupvplatform.org.

[5]   Cédric Philibert. Technology roadmap solar photovoltaic energy 2014 edition. International Energy Agency, 2014. http://www.iea.org/books.

[6]   E. A. Alsema, P. Frankl and K. Kato. 1998, Energy pay-back time of photovoltaic energy systems: Present status and prospects, 2nd World Conference on Photovoltaic Solar Energy Conversion, Vienna, 6–10 July.

[7]   E A Alsema, P Frankl and K Kato. Energy pay-back time of photovoltaic energy systems: Present status and prospects. 2nd World Conference on Photovoltaic Solar Energy Conversion, Vienna, July 1998. http://www.projects.science.uu.nl/nws/publica/ Publicaties1998/98053.pdf.

[8]   B Gaiddon, M Jedliczka and H. Villeurbanne. Compared assessment of selected environmental indicators of photovoltaic electricity in OECD Cities. IEA PVPS Task 10, Activity 4.4 Report IEA-PVPS T10-01:2006, May, 2006.

[9]   B. Gaiddon and M. Jedliczka. Environmental Benefits of PV Systems in OECD Cities, (IEA-PVPS), May, 2006.

[10] Rolf Frischknecht, et al. Life cycle inventories and life cycle assessments of photovoltaic systems. Report IEA-PVPS 12-04:2015, ISBN 978-3-906042-28-2.

[10] Eero Vartiainen. et al., PV LCOE in Europe 2014-2030·final report. 23 June 2015. European PV technology platform steering committee PV LCOE working group http://www. eupvplatform.org.

[11] Fatih Birol. et al., CO_2 emissions from fuel combustion Highlights (2016 edition). Statistics. October 2016. http://data.iea.org/payment/products/115-CO2-emissions-from-fuel-combus tion-2016-preliminary-edition.aspx.

[12] John Conti. International energy outlook 2016. DOE/EIA-0484(2016) May 2016. http://www. eia.gov/forecasts/ieo.

[13] I Stefan Nowak. Trends 2016 in photovoltaic applications survey report of selected IEA countries between 1992 and 2015. ISBN 978-30906042-45-9 Report IEA PVPS T1-30:2016. 31/10/2016. http://www.iea-pvps.org/index.php?id=256.

[14] Rabia Ferroukhi. et al., Renewable Energy and Jobs – Annual Review 2016, IRENA, ISBN 978-92-95111-89-9 http://www.irena.org/publications/2016/May/Renewable-Energy-and-Jobs–Annual-Review-2016

Exercises

11.1   How to calculate the cost of photovoltaic system? What is the difference with the on-grid tariff?

11.2   What are the methods to reduce the cost of PV systems?

11.3   There is a 2 MW PV power plant with a total investment of 18 million yuan. The performance ratio is 0.75, and the annual solar irradiation of the PV array is 5,000 MJ/m^2. Calculate the annual power generation and the cost for each kWh of the generated electricity (accurate to cent)? (assuming no consideration of module degradation and discount rate and other factors).

11.4   The unit cost of a 10 MW PV power plant is 10,000 yuan/kW, the solar radiation energy on the PV array is 1340 $kWh/m^2 \cdot year$, and the performance ratio of the system is 75% (without considering the degradation of the module). What is the cost of static electricity generation based on a 25-year service life? If 80% of the funding source is bank loans, the loan period is 25 years, and the annual interest rate is 6%, what is the cost of power generation?

11.5   What is the energy payback time of a PV system? How to calculate it?

11.6   What is the CO_2 emissions factor? How to determine it?

11.7   What is the PV CO_2 potential mitigation (PM)? How to calculate it?

11.8   The amount of solar radiation on the PV array of a polysilicon grid-connected PV system in one place is 3.5 $kWh/m^2 \cdot day$, the performance ratio of the system is 0.8, and the energy consumed for construction and operation is 2,400 kWh/kW. If the degradation of solar cell conversion efficiency is ignored, what is the system's energy payback time?

11.9 The polysilicon grid-connected PV system in a certain place has a lifetime of 25 years and a CO_2 EF of 743 g/kWh. Other conditions are the same as those in Exercise 11.8. What is the PV CO_2 potential mitigation for this PV system?

Practice answers

Chapter 2

2.2	−37.5°; 60°
2.4	45.6°, 45.9°, 44.4°
2.5	−69°, 69°, −58.9°, 58.9°, 9.2 h
2.6	34.1°
2.10	1.5, 60°
2.12	1.11

Chapter 3

3.5	1.91 eV
3.6	1.03 eV, cannot
3.9	5.2 A
3.10	30.3 V
3.11	16.7%
3.12	95 W, 11.8%

Chapter 5

5.9	42.5%

Chapter 8

8.5	12,000 W·h
8.6	222.2 A·h
8.7	1,168 MW·h
8.8	Annual power generation (MW·h)

Years	1	2	3	4	5
	13,202.9	13,097.2	12,992.5	12,888.5	12,785.4
Years	6	7	8	9	10
	12,683.2	12,581.7	12,481	12,381.2	12,282.1
Years	11	12	13	14	15
	12,183.9	12,086.4	11,989.7	11,893.8	11,798.6

https://doi.org/10.1515/9783110524833-012

(continued)

Years	16	17	18	19	20
	11,704.3	11,610.6	11,517.7	11,425.6	11,334.2
Years	**21**	**22**	**23**	**24**	**25**
	11,243.5	11,153.6	11,064.3	10,975.8	10,888
Total			300,245.7 (MW·h)		

8.9 5.38×10^{10} m^2
8.10 8.55×10^7 kW
8.11 10 in series and 2 series in parallel
8.12 (1) Three modules are needed in parallel, each of which is made up of 36 solar cells in series
8.13 14.1 m
8.14 3.09 m

Chapter 11

11.3 2.08×10^6 kWh; 0.43 yuan/kWh
11.4 0.4 yuan/kWh; 0.88 yuan/kWh
11.8 2.35 years
11.9 17.2 t/kW

Index

https://doi.org/10.1515/9783110524833-013